Named Organic Reactions

Named Organic Reactions

Thomas Laue and **Andreas Plagens**

Technical University, Braunschweig, Germany

Translated into English by Dr. Claus Vogel

Universität Magdeburg, Germany

JOHN WILEY & SONS

Chichester • New York • Weinheim • Brisbane • Singapore • Toronto

Originally published by B. G. Teubner in
1995 : *Namen- und Schlagwort-Reaktionen der
Organischen Chemie*, 2nd Ed.

Other Wiley Editorial Offices

John Wiley & Sons, Inc., 605 Third Avenue,
New York, NY 10158-0012, USA

WILEY-VCH Verlag GmbH, Pappelallee 3,
D-69469 Weinheim, Germany

Jacaranda Wiley Ltd, 33 Park Road, Milton,
Queensland 4064, Australia

John Wiley & Sons (Asia) Pte Ltd, Clementi Loop #02-01,
Jin Xing Distripark, Singapore 129809

John Wiley & Sons (Canada) Ltd, 22 Worcester Road,
Rexdale, Ontario M9W 1L1, Canada

British Library Cataloguing in Publication Data

A catalogue record for this book is available from the British Library

ISBN 0 471 97142 1

Typeset in 10/12pt Times by Laser Words, Madras, India
Printed and bound in Great Britain by Biddles, Guildford & King's Lynn
This book is printed on acid-free paper responsibly manufactured from sustainable forestation,
for which at least two trees are planted for each one used for paper production.

Contents

Introduction	*ix*
Acyloin Ester Condensation	1
Aldol Reaction	4
Alkene Metathesis	10
Arbuzov Reaction	12
Arndt–Eistert Synthesis	13
Baeyer–Villiger Oxidation	16
Bamford–Stevens Reaction	19
Beckmann Rearrangement	22
Benzidine Rearrangement	24
Benzilic Acid Rearrangement	26
Benzoin Condensation	27
Bergman Cyclization	29
Birch Reduction	33
Blanc Reaction	36
Bucherer Reaction	37
Cannizzaro Reaction	40
Chugaev Reaction	42
Claisen Ester Condensation	45
Claisen Rearrangement	48
Clemmensen Reduction	52
Cope Elimination Reaction	54
Cope Rearrangement	56
Corey–Winter Fragmentation	59
Curtius Reaction	61
1,3-Dipolar Cycloaddition	64
[2 + 2] Cycloaddition	67
Darzens Glycidic Ester Condensation	71
Delépine Reaction	73
Diazo Coupling	74
Diazotization	77
Diels–Alder Reaction	78

Di-π-Methane Rearrangement   86
Dötz Reaction   88

Elbs Reaction   92
Ene Reaction   93
Ester Pyrolysis   97

Favorskii Rearrangement   100
Finkelstein Reaction   102
Fischer Indole Synthesis   103
Friedel–Crafts Acylation   106
Friedel–Crafts Alkylation   110
Friedländer Quinoline Synthesis   114
Fries Rearrangement   116

Gabriel Synthesis   120
Gattermann Synthesis   123
Glaser Coupling Reaction   125
Glycol Cleavage   127
Gomberg–Bachmann Reaction   129
Grignard Reaction   132

Haloform Reaction   139
Hantzsch Pyridine Synthesis   141
Heck Reaction   144
Hell–Volhard–Zelinskii Reaction   147
Hofmann Elimination Reaction   149
Hofmann Rearrangement   153
Hunsdiecker Reaction   155
Hydroboration   157

Japp-Klingemann Reaction   161

Knoevenagel Reaction   164
Knorr Pyrrole Synthesis   168
Kolbe Electrolytic Synthesis   170
Kolbe Synthesis of Nitriles   172
Kolbe–Schmitt Reaction   173

Leuckart–Wallach Reaction   175
Lossen Reaction   176

Malonic Ester Synthesis   178
Mannich Reaction   182
McMurry Reaction   184

Meerwein–Ponndorf–Verley Reduction   187
Michael Reaction   189
Mitsunobu Reaction   192

Nazarov Cyclization   195
Neber Rearrangement   197
Nef Reaction   198
Norrish Type I Reaction   200
Norrish Type II Reaction   203

Ozonolysis   206

Paterno–Büchi Reaction   209
Pauson–Khand Reaction   210
Perkin Reaction   213
Peterson Olefination   215
Pinacol Rearrangement   217
Prilezhaev Reaction   218
Prins Reaction   220

Ramberg–Bäcklund Reaction   223
Reformatsky Reaction   224
Reimer–Tiemann Reaction   226
Robinson Annulation   228
Rosenmund Reduction   232

Sakurai Reaction   234
Sandmeyer Reaction   236
Schiemann Reaction   237
Schmidt Reaction   239
Sharpless Epoxidation   242
Simmons–Smith Reaction   244
Skraup Quinoline Synthesis   246
Stevens Rearrangement   248
Stork Enamine Reaction   250
Strecker Synthesis   253

Tiffeneau-Demjanov Reaction   255

Vilsmeier Reaction   258
Vinylcyclopropane Rearrangement   260

Wagner–Meerwein Rearrangement   263
Weiss Reaction   265
Willgerodt Reaction   267

Williamson Ether Synthesis   269
Wittig Reaction   271
Wittig Rearrangement   275
Wohl–Ziegler Bromination   277
Wolff Rearrangement   279
Wolff–Kishner Reduction   281
Wurtz Reaction   282

Index   285

Introduction

Name reactions are still an important element of organic chemistry. A thorough knowledge of such reactions is essential for the chemist. The scientific content behind the name is of great importance, and the names are used as short expressions in order to ease spoken as well as written communication in organic chemistry. Furthermore named reactions are a perfect aid for learning the principles of organic chemistry. This is not only true for the study of chemistry as a major subject, but also when studying chemistry as a minor subject, e.g. for the student of biology or pharmaceuticals.

This book—*Named Organic Reactions*—is not meant to completely replace an organic chemistry textbook. It is rather a reference work on name reactions, that is also suitable for easy reading and learning, as well as for revision for an exam in organic chemistry. The book deals with 134 of the most important reactions in organic chemistry; the selection is based on their importance for modern preparative organic chemistry, as well as for a modern organic chemistry course.

In particular the reactions are arranged in alphabetical order, and treated in a consistent manner. The name of the reaction serves as a heading, and is followed by a subtitle giving a one sentence description of the reaction. Then follows a formula scheme depicting the overall reaction, and an initial paragraph with an introductory description of the reaction.

The major part of each section deals with mechanistic aspects; however, for didactic reasons in most cases not with too many details. Side-reactions, variants and modified procedures with respect to product distribution and yields are described. Recent as well as older examples of the application of a particular reaction or method are given, together with references to the original literature. Those examples are not aimed at a complete treatment of every aspect of a particular reaction, but are rather selected from a didactic point of view.

At the end of each section, a list of references is given. In addition to the very first publication, and to review articles, references to recent and very recent publications are often given. This is meant to encourage work with, and give access to, the original literature, review articles and reference works. The reference to the very first publication of a reaction is aimed at showing the origin of the particular name, and how the reaction has been explored or developed. With the outlining of modern examples and listing of references, this book is directed at the advanced student as well as doctoral candidates.

Special thanks go to Prof. Dr. H. Hopf (University of Braunschweig, Germany) for his encouragement and his critical reading of the manuscript. In addition we are indebted to Mr. Daniel Geuenich, Dr. Helmut Lipka, Dipl.-Chem. Jörg Michalski and Dr. Claus Vogel, as well as to the editorial staff of John Wiley & Sons, Chichester.

Acyloin Ester Condensation

α-Hydroxyketones from carboxylic esters

$$2\ \text{RCOR'} \xrightarrow{\ \text{Na}\ } \underset{\textbf{1}}{} \quad \underset{}{\text{R}-\text{C}=\text{C}-\text{R}}\ \overset{\text{NaO}\ \ \text{ONa}}{} \xrightarrow{\ \text{H}_2\text{O}\ } \underset{\textbf{2}}{\text{R}-\underset{\text{H}}{\overset{\text{HO}}{\text{C}}}-\overset{\text{O}}{\text{C}}-\text{R}}$$

Upon heating of a carboxylic ester **1** with sodium in an inert solvent, a conden-sation reaction can take place to yield a α-hydroxy ketone **2** after hydrolytic workup.[1–3] This reaction is called *Acyloin condensation*, named after the prod-ucts thus obtained. It works well with alkanoic acid esters. For the synthesis of the corresponding products with aryl substituents (R = aryl), the *Benzoin condensa-tion* of aromatic aldehydes is usually applied.

For the mechanistic course of the reaction the diketone **5** is assumed to be an intermediate, since small amounts of **5** can sometimes be isolated as a minor product. It is likely that the sodium initially reacts with the ester **1** to give the radical anion species **3**, which can dimerize to the dianion **4**. By release of two alkoxides R'O⁻ the diketone **5** is formed. Further reaction with sodium leads to the dianion **6**, which yields the α-hydroxy ketone **2** upon aqueous workup:

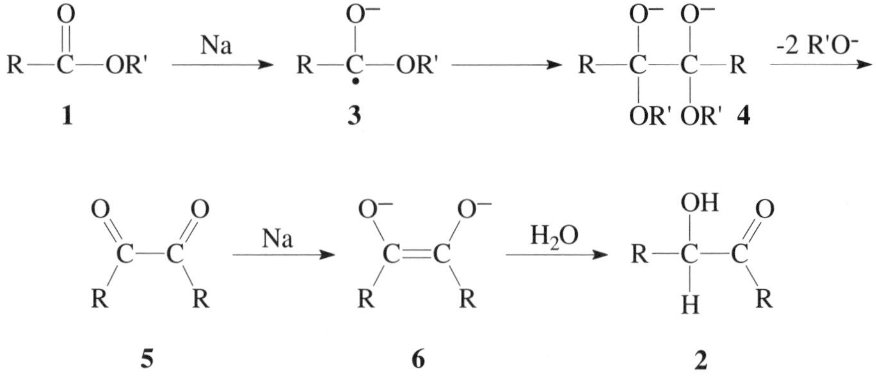

An intramolecular reaction is possible with appropriate substrates containing two ester groups, leading to the formation of a carbocyclic ring. This reaction is especially useful for the formation of rings with ten to twenty carbon atoms, the yield depending on ring size.[4] The presence of carbon–carbon double or triple bonds does not affect the reaction. The strong tendency for ring formation with appropriate diesters is assumed to arise from attachment of the chain ends to the sodium surface and thereby favoring ring closure.

A modified procedure, which uses trimethylsilyl chloride as an additional reagent, gives higher yields of acyloins and is named after Rühlmann.[5] In the presence of trimethylsilyl chloride, the *bis*-O-silylated endiol 7 is formed and can be isolated. Treatment of 7 with aqueous acid leads to the corresponding acyloin 2:

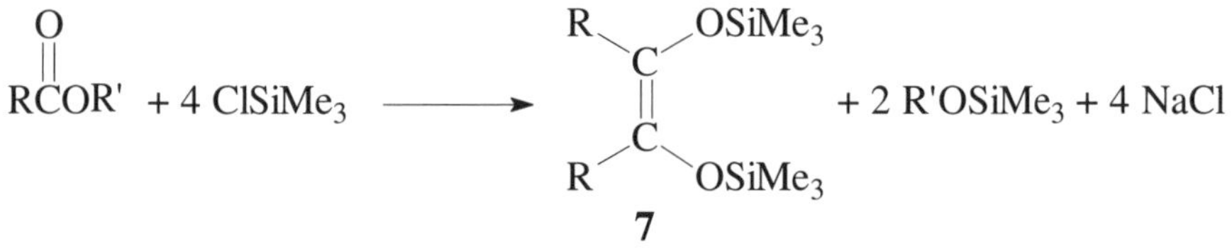

This modification has become the standard procedure for the acyloin ester condensation. By doing so, the formation of products from the otherwise competitive *Dieckmann condensation* (*Claisen ester condensation*) can be avoided. A product formed by ring closure through a Dieckmann condensation consists of a ring that is smaller by one carbon atom than the corresponding cyclic acyloin.

As an example of ring systems which are accessible through this reaction, the formation of [n]paracyclophanes[6] like 8 with $n \geq 9$ shall be outlined:

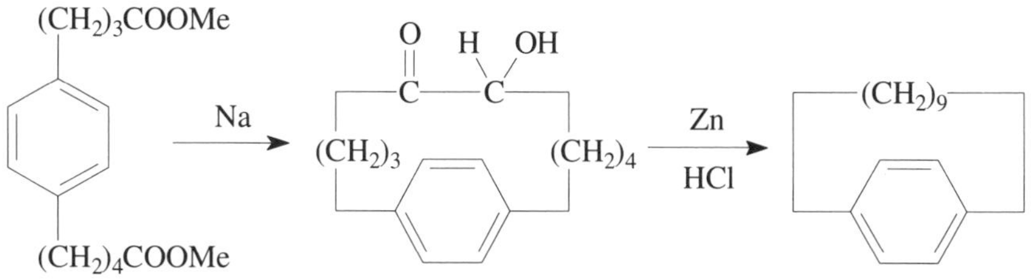

A spectacular application of the acyloin ester condensation was the preparation of catenanes like **11**.[7] These were prepared by a statistical synthesis; which means that an acyloin reaction of the diester **10** has been carried out in the presence of an excess of a large ring compound such as **9**, with the hope that some diester molecules would be threaded through a ring, and would then undergo ring closure to give the catena compound:

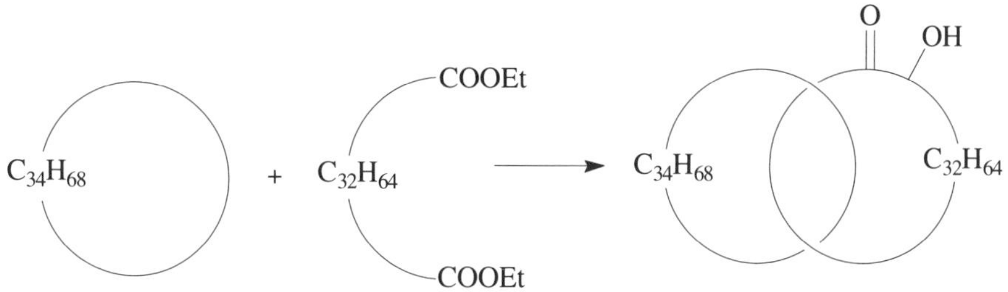

As expected, the yields of catenanes by this approach are low, which is why improved methods for the preparation of such compounds have been developed.[8] The acyloins are often only intermediate products in a multistep synthesis. For example they can be further transformed into olefins by application of the *Corey–Winter fragmentation*.

1. A. Freund, *Justus Liebigs Ann. Chem.* **1861**, *118*, 33–43.
2. S. M. McElvain, *Org. React.* **1948**, *4*, 256–268.
3. J. J. Bloomfield, D. C. Owsley, J. M. Nelke, *Org. React.* **1976**, *23*, 259–403.
4. K. T. Finley, *Chem. Rev.* **1964**, *64*, 573–589.
5. K. Rühlmann, *Synthesis* **1971**, 236–253.
6. D. J. Cram, M. F. Antar, *J. Am. Chem. Soc.* **1958**, *80*, 3109–3114.
7. E. Wasserman, *J. Am. Chem. Soc.* **1960**, *82*, 4433–4434.
8. J.-P. Sauvage, *Acc. Chem. Res.* **1990**, *23*, 319–327.

Aldol Reaction

Reaction of aldehydes or ketones to give β-hydroxy carbonyl compounds

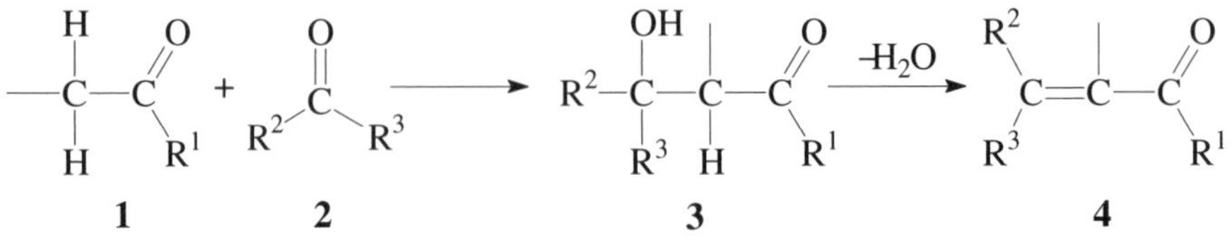

The addition of the α-carbon of an enolizable aldehyde or ketone **1** to the carbonyl group of a second aldehyde or ketone **2** is called the *aldol reaction*.[1,2] It is a versatile method for the formation of carbon–carbon bonds, and is frequently used in organic chemistry. The initial reaction product is a β-hydroxy aldehyde (aldol) or β-hydroxy ketone (ketol) **3**. A subsequent dehydration step can follow, to yield an α,β-unsaturated carbonyl compound **4**. In that case the entire process is also called *aldol condensation*.

The aldol reaction as well as the dehydration are reversible. In order to obtain the desired product, the equilibrium might have to be shifted by appropriate reaction conditions (see below).

The reaction can be performed with base catalysis as well as acid catalysis. The former is more common; here the enolizable carbonyl compound **1** is deprotonated at the α-carbon by base (e.g. alkali hydroxide) to give the enolate anion **5**, which is stabilized by resonance:

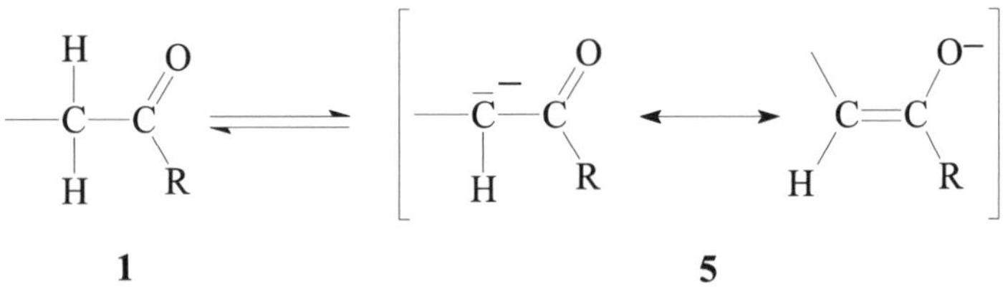

The next step is the nucleophilic addition of the enolate anion **5** to the carbonyl group of another, non-enolized, aldehyde molecule **2**. The product which is obtained after workup is a β-hydroxy aldehyde or ketone **3**:

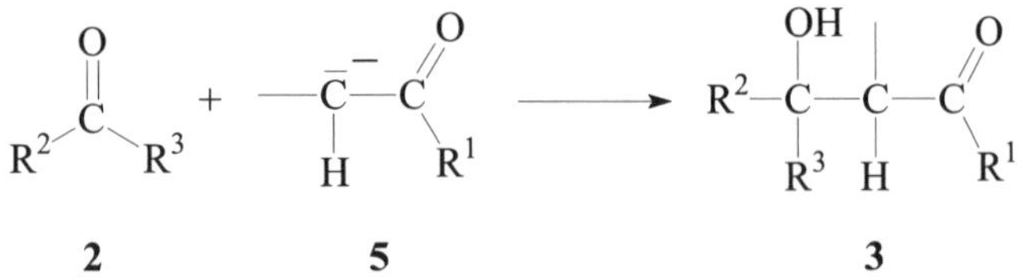

In the acid-catalyzed process, the enol **6** reacts with the protonated carbonyl group of another aldehyde molecule **2**:

If the initially formed β-hydroxy carbonyl compound **3** still has an α-hydrogen, a subsequent elimination of water can take place, leading to an α,β-unsaturated aldehyde or ketone **4**. In some cases the dehydration occurs already under the aldol reaction conditions; in general it can be carried out by heating in the presence of acid:

Several pairs of reactants are possible. The aldol reaction between two molecules of the same aldehyde is generally quite successful, since the equilibrium lies far to the right. For the analogous reaction of ketones, the equilibrium lies to the left, and the reaction conditions have to be adjusted properly in order to achieve satisfactory yields (e.g. by using a Soxhlet extractor).

With unsymmetrical ketones, having hydrogens at both α-carbons, a mixture of products can be formed. In general such ketones react preferentially at the less substituted side, to give the less sterically hindered product.

A different situation is found in the case of *crossed aldol reactions*, which are also called *Claisen–Schmidt reactions*. Here the problem arises, that generally a mixture of products might be obtained.

From a mixture of two different aldehydes, each with α-hydrogens, four different aldols can be formed—two aldols from reaction of molecules of the same aldehyde + two crossed aldol products; not even considering possible stereoisomers (see below). By taking into account the unsaturated carbonyl compounds which could be formed by dehydration from the aldols, eight different reaction products might be obtained, thus indicating that the aldol reaction may have preparative limitations.

If only one of the two aldehydes has an α-hydrogen, only two aldols can be formed; and numerous examples have been reported, where the crossed aldol reaction is the major pathway.[2] For two different ketones, similar considerations do apply in addition to the unfavorable equilibrium mentioned above, which is why such reactions are seldom attempted.

In general the reaction of an aldehyde with a ketone is synthetically useful. Even if both reactants can form an enol, the α-carbon of the ketone usually adds to the carbonyl group of the aldehyde. The opposite case—the addition of the α-carbon of an aldehyde to the carbonyl group of a ketone—can be achieved by the *directed aldol reaction*.[3,4] The general procedure is to convert one reactant into a preformed enol derivative or a related species, prior to the intended aldol reaction. For instance, an aldehyde may be converted into an aldimine **7**, which can be deprotonated by lithium diisopropylamide (LDA) and then add to the carbonyl group of a ketone:

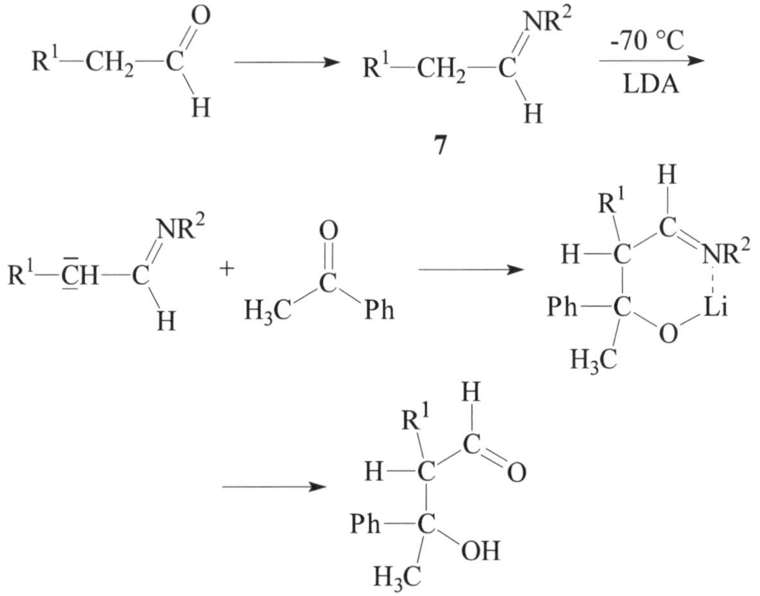

By using the directed aldol reaction, unsymmetrical ketones can be made to react regioselectively. After conversion into an appropriate enol derivative (e.g. trimethylsilyl enol ether **8**) the ketone reacts at the desired α-carbon.

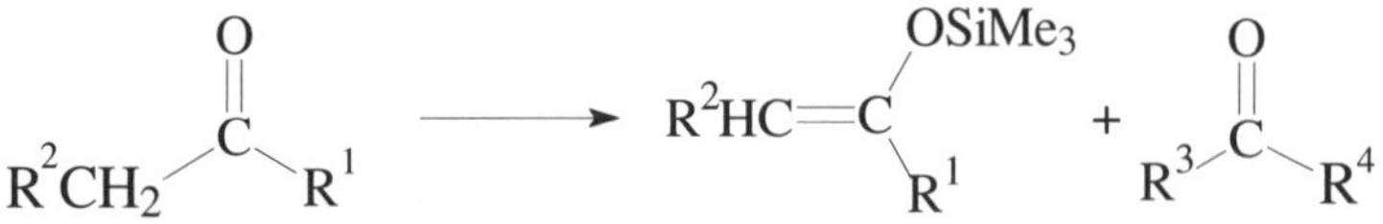

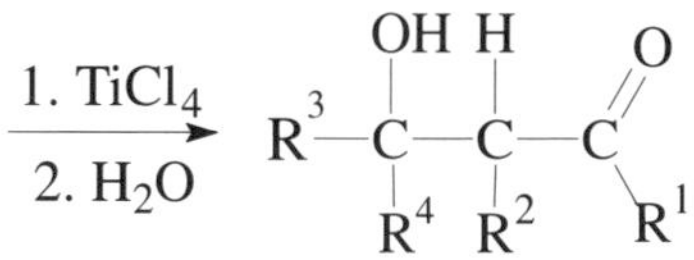

$$\xrightarrow[\text{2. H}_2\text{O}]{\text{1. TiCl}_4} \quad \underset{\underset{R^4}{|} \underset{R^2}{|}}{R^3 - \overset{\overset{OH}{|}}{C} - \overset{\overset{H}{|}}{C} - \overset{\overset{O}{\|}}{C} \diagdown R^1}$$

An important aspect is the control of the stereochemical outcome.[5-7] During the course of the reaction two new chiral centers can be created and thus two diastereomeric pairs of enantiomers (*syn/anti* resp. *erythro/threo* pairs) may be obtained.

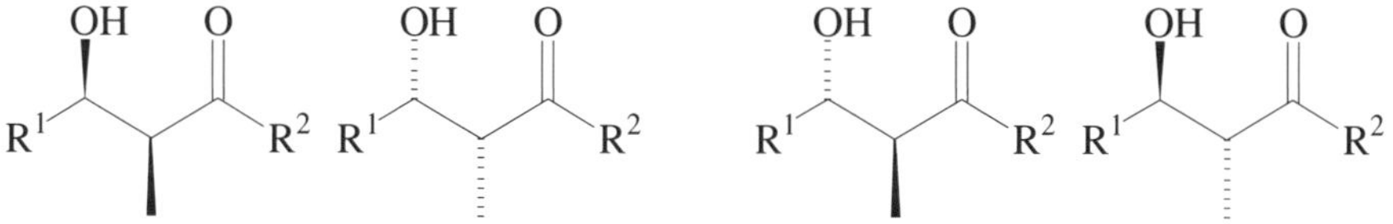

syn / erythro   *anti / threo*

The enantiomers are obtained as a racemic mixture if no asymmetric induction becomes effective. The ratio of diastereomers depends on structural features of the reactants as well as the reaction conditions as outlined in the following. By using properly substituted preformed enolates, the diastereoselectivity of the aldol reaction can be controlled.[7] Such enolates can show *E*-or *Z*-configuration at the carbon–carbon double bond. With *Z*-enolates **9**, the *syn* products are formed preferentially, while *E*-enolates **12** lead mainly to *anti* products. This stereochemical outcome can be rationalized to arise from the more favored transition state **10** and **13** respectively:

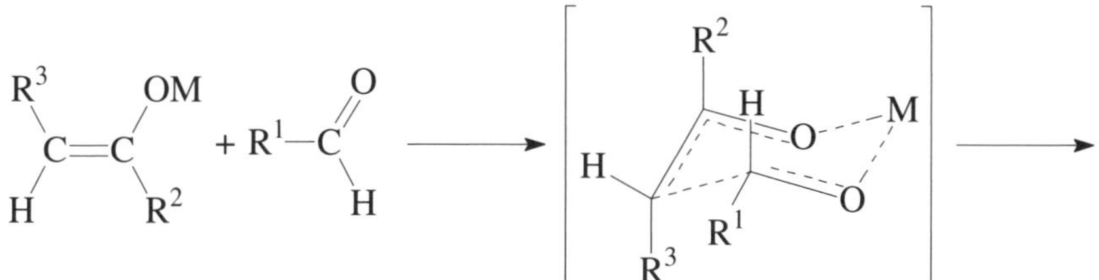

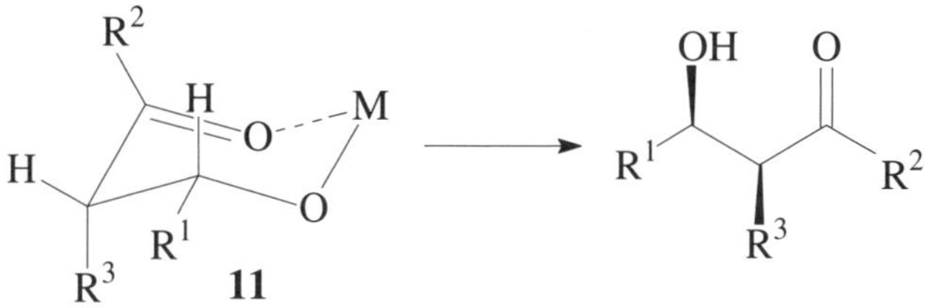

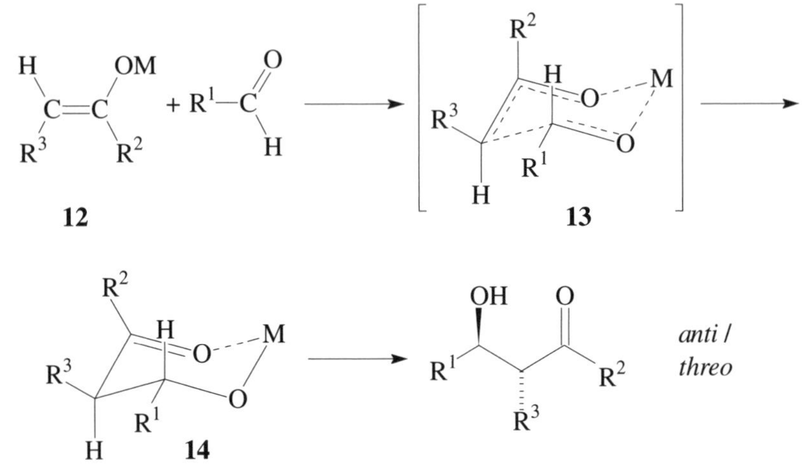

12 **13**

14 *anti / threo*

Under conditions which allow for equilibration (thermodynamic control) however, the *anti*-product is obtained, since the metal-chelate **14** is the more stable. As compared to **11** it has more substituents in the favorable equatorial position:

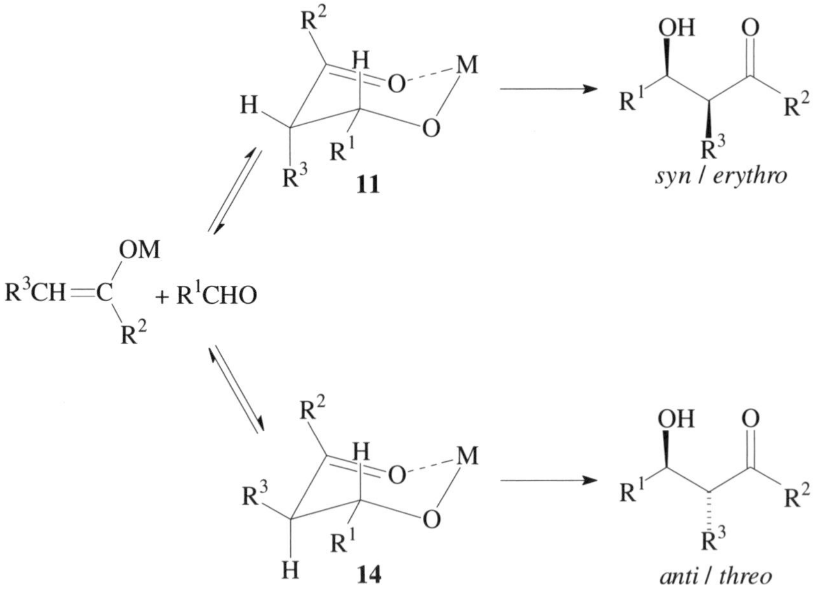

11 *syn / erythro*

$R^3CH=C$ ...OM ...R^2 + R^1CHO

14 *anti / threo*

With an appropriate chiral reactant, high enantioselectivity can be achieved, as a result of *asymmetric induction*.[8] If both reactants are chiral, this procedure is called the *double asymmetric reaction*,[6] and the observed enantioselectivity can be even higher.

Especially with the ordinary aldol reaction a number of side reactions can be observed, as a result of the high reactivity of starting materials and products. For instance, the α,β-unsaturated carbonyl compounds **4** can undergo further aldol reactions by reacting as vinylogous components. In addition compounds **4** are potential substrates for the *Michael reaction*.

Aldehydes can react through a hydride transfer as in the *Cannizzaro reaction*.

Moreover aldoxanes **15** may be formed; although these decompose upon heating to give an aldol **3** and aldehyde **1**:

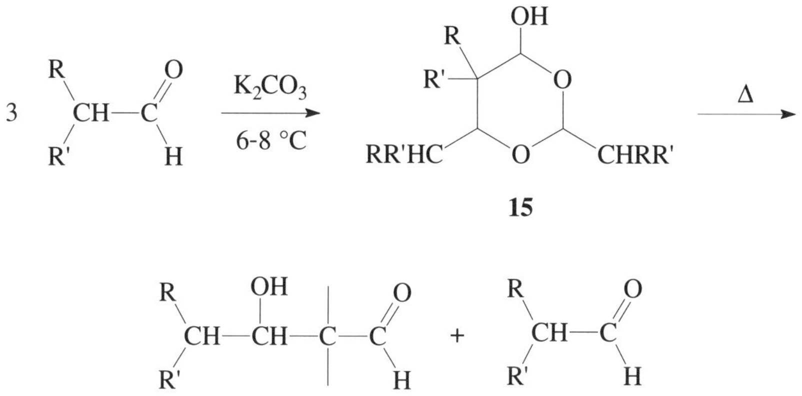

Aldols can form dimers; e.g. acetaldol **16** dimerizes to give paraldol **17**:

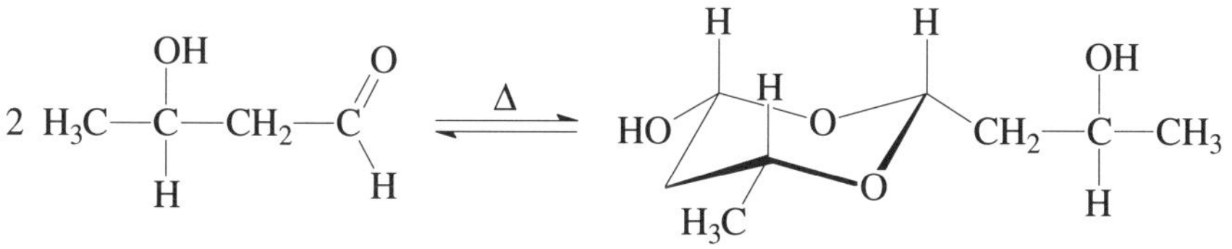

Because of the many possible reactions of aldols, it is generally recommended to use a freshly distilled product for further synthetic steps.

Besides the aldol reaction in the true sense, there are several other analogous reactions, where some enolate species adds to a carbonyl compound. Such reactions are often called *aldol-type reactions*; the term aldol reaction is reserved for the reaction of aldehydes and ketones.

1. M. A. Wurtz, *Bull. Soc. Chim. Fr.* **1872**, *17*, 436–442.
2. A. T. Nielsen, W. J. Houlihan, *Org. React.* **1968**, *16*, 1–438.
3. G. Wittig, H. Reiff, *Angew. Chem.* **1968**, *80*, 8–15; *Angew. Chem. Int. Ed. Engl.* **1968**, *7*, 7.
4. T. Mukaiyama, *Org. React.* **1982**, *28*, 203–331;
    T. Mukaiyama, S. Kobayashi, *Org. React.* **1994**, *46*, 1–103.

5. C. H. Heathcock, *Science* **1981**, *214*, 395–400.
6. S. Masamune, W. Choy, J. S. Petersen, L. S. Sita, *Angew. Chem.* **1985**, *97*, 1–31; *Angew. Chem. Int. Ed. Engl.* **1985**, *24*, 1.
7. C. H. Heathcock in *Modern Synthetic Methods 1992* (Ed.: R. Scheffold), VHCA, Basel, **1992**, p. 1–102.
8. D. Enders, R. W. Hoffmann, *Chem. Unserer Zeit* **1985**, *19*, 177–190.

Alkene Metathesis

Interchange of alkylidene groups of alkenes—metathesis of olefins

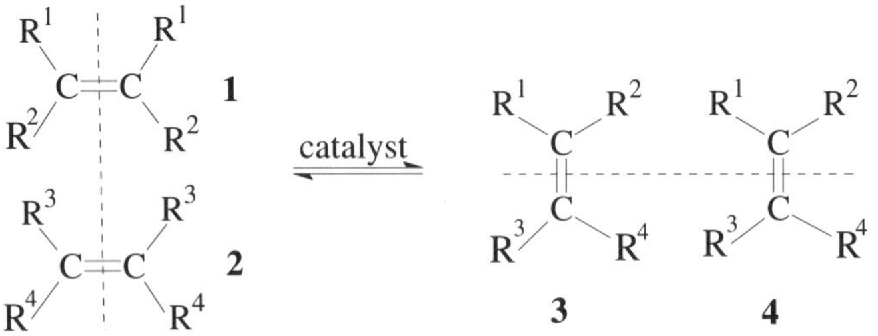

When mixtures of alkenes **1** and **2** or unsymmetrically substituted alkenes are treated with certain catalysts, a mixture of possible metathesis products (including *E/Z* isomers) is obtained by a process called *alkene metathesis*.[1,2] Special applications are for example the *ring-closing metathesis* and the *ring-opening metathesis polymerization* (see below).

This reaction proceeds by a catalytic chain mechanism.[2–4] A metal catalyst and an alkene **4** react to give a metal carbene complex **5**. This complex can add to the alkene **6** to give a metallacyclobutane **7**. The four-membered ring intermediate decomposes, yielding the new alkene **8** and the new metal carbene complex **9**:

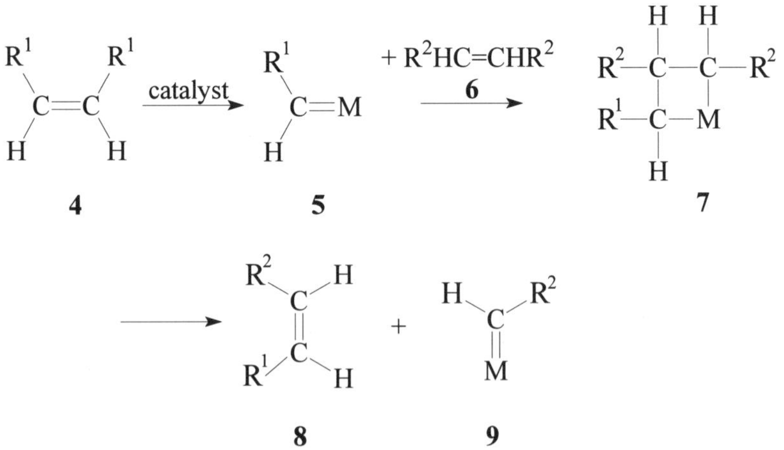

The synthetic utility may be limited because of the formation of a mixture of products.[5,6] For example pent-2-ene **10** reacts to give a statistical mixture of alkenes (here shown without taking into account E/Z-isomers):

10

50 %   25 %   25 %

Yields of desired products can often be improved by choosing the appropriate catalyst; e.g. a catalyst which selectively activates terminal alkenes. Furthermore an equilibrium can be shifted by removing one reaction product from the reaction mixture. An example for the catalytic *ring-closing metathesis*[9-12] is the formation of cyclohexene and ethylene by an intramolecular metathesis reaction of 1,7-octadiene[2] **11**. The gaseous product ethylene can be allowed to escape from the reaction mixture, thus shifting the equilibrium and leading to a higher yield of the cyclization product cyclohexene **12**.

11   **12**

The alkene metathesis reaction has gained industrial importance.[7,8] A major field is the production of key chemicals for polymer and petrochemistry. The alkene metathesis allows for using starting materials from renewable resources, e.g. unsaturated fatty acid esters from fats and oils, isolated from agricultural raw materials, can be used for the production of tensides.

Another application is the use of alkene metathesis catalysts for the preparation of special polymers from cycloalkenes by the *ring-opening metathesis polymerization* (ROMP).[11]

As metathesis catalysts various transition metal compounds are used; especially tungsten, molydenum, rhenium and ruthenium compounds, e.g. $WCl_6/SnMe_4$, MoO_3, Re_2O_7, carbene complexes of tungsten, molybdenum and ruthenium[9-12], $MeReO_3$.[13]

1. R. L. Blanks, C. G. Bailey, *Ind. Eng. Chem. Prod. Res. Dev.* **1964**, *3*, 170–173.
2. K. J. Ivin, *Olefin Metathesis*, Academic Press, London, **1983**.
3. J. M. Basset, M. Leconte, *Chemtech* **1980**, *10*, 762–767.
4. N. Calderon, E. A. Ofstead, W. A. Judy, *Angew. Chem.* **1976**, *88*, 433–442; *Angew. Chem. Int. Ed. Engl.* **1976**, *15*, 401.
5. W. B. Hughes, *J. Am. Chem. Soc.* **1970**, *92*, 532–537.
6. N. Calderon, E. A. Ofstead, J. P. Ward, W. A. Judy, K. W. Scott, *J. Am. Chem. Soc.* **1968**, *90*, 4133–4140.

7.   S. Warwel, *Nachr. Chem. Tech. Lab.* **1992**, *40*, 314–320.
8.   R. Streck, *Chemtech* **1989**, *19*, 498–503.
9.   U. Koert, *Nachr. Chem. Tech. Lab.* **1995**, *43*, 809–814.
10. H.-G. Schmalz, *Angew. Chem.* **1995**, *107*, 1981–1984; *ibid. Int. Ed. Engl.* **1995**, *34*, 1833.
11. R. H. Grubbs, *Acc. Chem. Res.* **1995**, *28*, 446–452;
     D. M. Lynn, S. Kanaoka, R. H. Grubbs, *J. Am. Chem. Soc.* **1996**, *118*, 784–790.
12. W. A. Nugent, J. Feldman, J. C. Calabrese, *J. Am. Chem. Soc.* **1995**, *117*, 8992–8998.
13. W. A. Herrmann, W. Wagner, U. N. Flessner, U. Volkhardt, H. Komber, *Angew. Chem.* **1991**, *103*, 1704–1706; *Angew. Chem. Int. Ed. Engl.* **1991**, *30*, 1636–38.

Arbuzov Reaction

Alkyl phosphonates from phosphites

The *Arbuzov reaction*,[1–3] also called the *Michaelis–Arbuzov reaction*, allows for the synthesis of pentavalent alkyl phosphoric acid esters **4** from trivalent phosphoric acid esters **1** (Z,Z′ = R,OR) by treatment with alkyl halides **2**.

Most common is the preparation of alkyl phosphonic acid esters (phosphonates) **4** (Z,Z′ = OR) from phosphorous acid esters (phosphites) **1** (Z,Z = OR). The preparation of phosphinic acid esters (Z = R, Z′ = OR) from phosphonous acid esters, as well as phosphine oxides (Z,Z′ = R) from phosphinous acid esters is also possible.

The reaction mechanism outlined below for phosphorous acid esters analogously applies for the other two cases. The first step is the addition of the alkyl halide **2** to the phosphite **1** to give a phosphonium salt[2] **3**:

This intermediate product is unstable under the reaction conditions, and reacts by cleavage of an O-alkyl bond to yield the alkyl halide **5** and the alkyl phosphonate **4**:

$$\left[\begin{array}{c} OR \\ | \\ RO-P-R' \\ | \\ OR \end{array} \right]^{+} X^{-} \longrightarrow \begin{array}{c} O \\ || \\ RO-P-R' \\ | \\ OR \end{array} + RX$$

$$\textbf{3} \qquad\qquad\qquad\qquad \textbf{4} \qquad\quad \textbf{5}$$

It is a reaction of wide scope; both the phosphite **1** and the alkyl halide **2** can be varied.[3] Most often used are primary alkyl halides; iodides react better than chlorides or bromides. With secondary alkyl halides side reactions such as elimination of HX can be observed. Aryl halides are unreactive.

With acyl halides, the corresponding acyl phosphonates are obtained. Furthermore allylic and acetylenic halides, as well as α-halogenated carboxylic esters and dihalides, can be used as starting materials. If substituents R and R' are different, a mixture of products may be obtained, because the reaction product RX **5** can further react with phosphite **1** that is still present:

$$P(OR)_3 + RX \longrightarrow \begin{array}{c} O \\ || \\ (RO)_2P-R \end{array}$$

$$\textbf{1} \qquad \textbf{5}$$

However with appropriate reaction control, the desired product can be obtained in high yield.[3]

The phosphonates obtained by the Arbuzov reaction are starting materials for the *Wittig–Horner reaction* (*Wittig reaction*); for example, appropriate phosphonates have been used for the synthesis of vitamin A and its derivatives.[4]

Moreover organophosphoric acid esters have found application as insecticides (e.g. Parathion). Some derivatives are highly toxic to man (e.g. Sarin, Soman). The organophosphonates act as inhibitors of the enzyme cholinesterase by phosphorylating it. This enzyme is involved in the proper function of the parasympathetic nervous system. A concentration of 5×10^{-7} g/L in the air can already cause strong toxic effects to man.

1. A. Michaelis, R. Kaehne, *Ber. Dtsch. Chem. Ges.* **1898**, *31*, 1048–1055.
2. B. A. Arbuzov, *Pure Appl. Chem.* **1964**, *9*, 307–335.
3. G. M. Kosolapoff, *Org. React.* **1951**, *6*, 273–338.
4. H. Pommer, *Angew. Chem.* **1960**, *72*, 811–819 and 911–915.

Arndt–Eistert Synthesis

Chain elongation of carboxylic acids by one methylene group

$$R-\overset{\displaystyle O}{\underset{\displaystyle OH}{C}} \quad\longrightarrow\quad R-\overset{\displaystyle O}{\underset{\displaystyle Cl}{C}} \quad\xrightarrow{CH_2N_2}\quad R-\overset{\displaystyle O}{C}-CHN_2$$

$$\mathbf{1} \qquad\qquad\qquad \mathbf{2} \qquad\qquad\qquad \mathbf{3}$$

$$\xrightarrow[\text{Ag}_2\text{O}]{\text{H}_2\text{O}} \quad R-CH_2-\overset{\displaystyle O}{\underset{\displaystyle OH}{C}}$$

$$\mathbf{4}$$

The Arndt–Eistert synthesis allows for the conversion of carboxylic acids **1** into the next higher homolog[1,2] **4**. This reaction sequence is considered to be the best method for the extension of a carbon chain by one carbon atom in cases where a carboxylic acid is available.

In a first step, the carboxylic acid **1** is converted into the corresponding acyl chloride **2** by treatment with thionyl chloride or phosphorous trichloride. The acyl chloride is then treated with diazomethane to give the diazo ketone **3**, which is stabilized by resonance, and hydrogen chloride:

$$R-\overset{\displaystyle |\overline{O}|}{C}=\overset{\displaystyle H}{C}-\overset{+}{N}{\equiv}N|$$

$$R-\overset{\displaystyle O}{\underset{\displaystyle Cl}{C}} + CH_2N_2 \xrightarrow{-HCl} R-\overset{\displaystyle O}{C}-\overset{\displaystyle H}{\underset{\displaystyle}{C}}{}^{-}-\overset{+}{N}{\equiv}N| \qquad \mathbf{3}$$

$$\mathbf{2}$$

$$R-\overset{\displaystyle O}{C}-\overset{\displaystyle H}{C}=\overset{+}{N}=\overset{-}{N}$$

The hydrogen chloride thus produced can in turn react with the diazoketone to yield a α-chloro ketone. In order to avoid this side reaction, two equivalents of diazomethane are used. The second equivalent reacts with HCl to give methyl chloride.[2]

The diazo ketone **3**, when treated with silver oxide as catalyst, decomposes into ketocarbene **5** and dinitrogen N_2. This decomposition reaction can also be achieved by heating or by irradiation with uv-light. The ketocarbene undergoes a *Wolff rearrangement* to give a ketene **6**:

$$R-\overset{\overset{\displaystyle O}{\|}}{C}-\overset{\overset{\displaystyle H}{|}}{\underset{\curvearrowleft}{C}}-N\overset{+}{\equiv}N| \quad \xrightarrow[{-N_2}]{Ag_2O} \quad R-\overset{\overset{\displaystyle O}{\|}}{C}-\overset{\overset{\displaystyle H}{}}{C} \quad \longrightarrow \quad R-CH=C=O$$

$$\textbf{3} \qquad\qquad\qquad\qquad \textbf{5} \qquad\qquad\qquad\qquad \textbf{6}$$

The final step is the reaction of the ketene with the solvent; e.g. with water to yield the carboxylic acid **4**:

$$R-CH=C=O \quad \xrightarrow{H_2O} \quad R-CH_2-\overset{\overset{\displaystyle O}{\|}}{C}{\diagdown}_{OH}$$

$$\textbf{6} \qquad\qquad\qquad\qquad\qquad \textbf{4}$$

If an alcohol R'OH is used as solvent instead of water, the corresponding ester **7** can be obtained directly. In analogous reactions with ammonia or amines (R'NH$_2$) the amides **8** and **9** respectively are accessible.

$$R-CH=C=O$$

$$\textbf{6}$$

$$\xrightarrow{R'OH} \quad R-CH_2-\overset{\overset{\displaystyle O}{\|}}{C}-OR' \quad \textbf{7}$$

$$\xrightarrow{NH_3} \quad R-CH_2-\overset{\overset{\displaystyle O}{\|}}{C}-NH_2 \quad \textbf{8}$$

$$\xrightarrow{R'NH_2} \quad R-CH_2-\overset{\overset{\displaystyle O}{\|}}{C}-NHR' \quad \textbf{9}$$

The reaction is of wide scope (R = alkyl, aryl); however the substrate molecule should not contain other functional groups that can react with diazomethane. With unsaturated acyl halides the yield can be poor, but may be improved by modified reaction conditions.[3]

1. F. Arndt, B. Eistert, *Ber. Dtsch. Chem. Ges.* **1935**, *68*, 200–208.
2. W. E. Bachmann, W. S. Struve, *Org. React.* **1942**, *1*, 38–62.
3. T. Hudlicky, J. P. Sheth, *Tetrahedron Lett.* **1979**, *20*, 2667–2670.

B

Baeyer–Villiger Oxidation

Oxidation of ketones to carboxylic esters

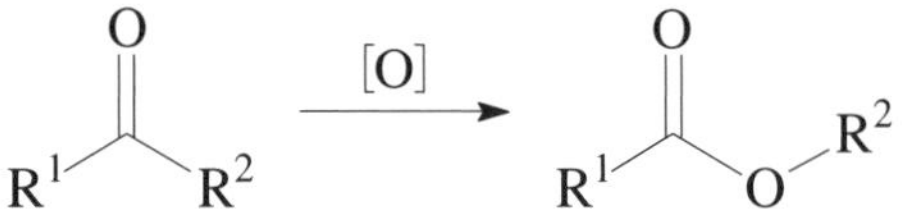

1   **2**

When a ketone **1** is treated with hydrogen peroxyde or a peracid, a formal insertion of oxygen can take place to yield a carboxylic ester **2**. This process is called the *Baeyer–Villiger oxidation*.[1-3]

In a first step the reactivity of the carbonyl group is increased by protonation at the carbonyl oxygen. The peracid then adds to the cationic species **3** leading to the so-called *Criegee intermediate* **4**:

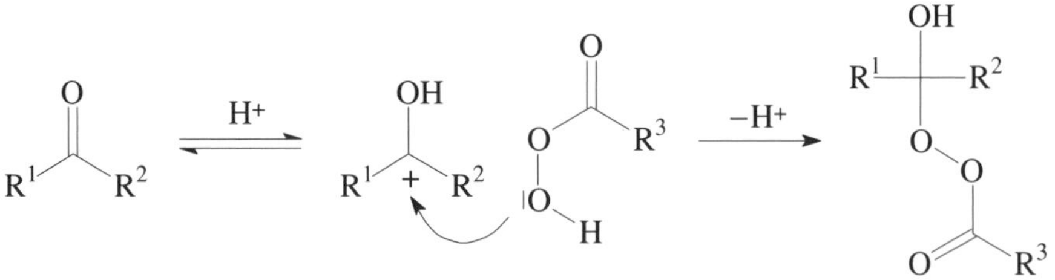

1   **3**   **4**

Cleavage of the carboxylic acid R^3COOH from that intermediate leads to an electron-deficient oxygen substituent with an electron sextet configuration. This deficiency can be compensated through migration of the substituent R^1 or R^2; experimental findings suggest that cleavage and migration are a concerted process. The cationic species **5** which can be thus formed (e.g. by migration of R^2), loses a proton to yield the stable carboxylic ester **2**:

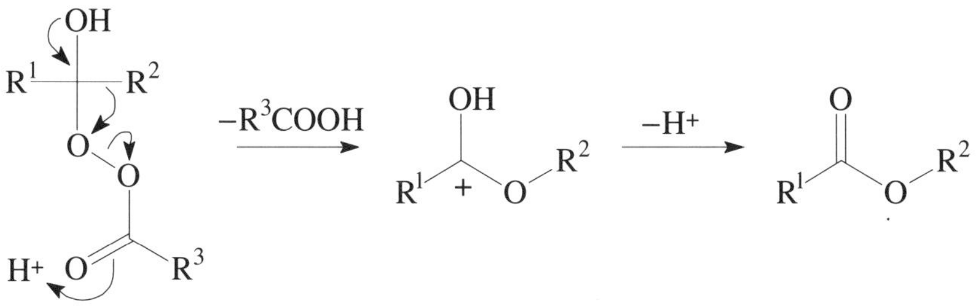

The ease of migration of substituents R^1, R^2 depends on their ability to stabilize a positive charge in the transition state. An approximate order of migration[2] has been drawn: $R_3C > R_2CH > Ar > RCH_2 > CH_3$. Thus the Baeyer–Villiger oxidation of unsymmetrical ketones is regioselective. On the other hand aldehydes usually react with migration of the hydrogen to yield the carboxylic acid.

The reaction mechanism is supported by findings from experiments with ^{18}O-labeled benzophenone **6**; after rearrangement, the labeled oxygen is found in the carbonyl group only:

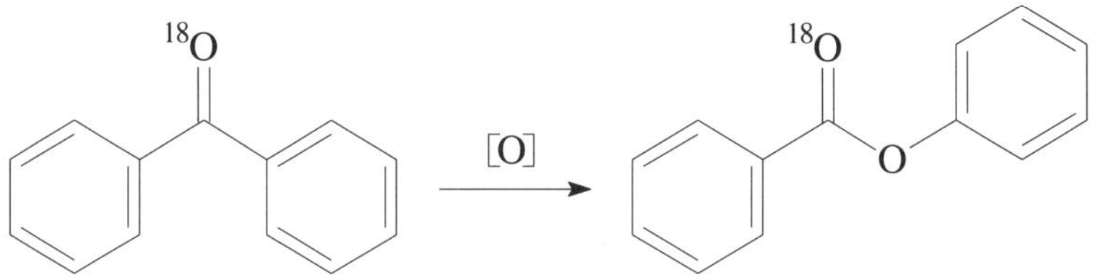

Cyclic ketones react through ring expansion to yield lactones (cyclic esters). For example cyclopentanone **7** can be converted to δ-valerolactone **8**:

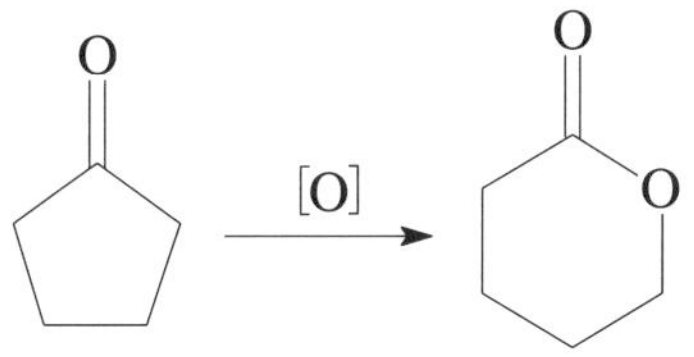

The Baeyer–Villiger oxidation is a synthetically very useful reaction; it is for example often used in the synthesis of natural products. The *Corey lactone* **11** is a key intermediate in the total synthesis of the physiologically active prostaglandins. It can be prepared from the lactone **10**, which in turn is obtained from the bicyclic ketone **9** by reaction with *m*-chloroperbenzoic acid (MCPBA):[4]

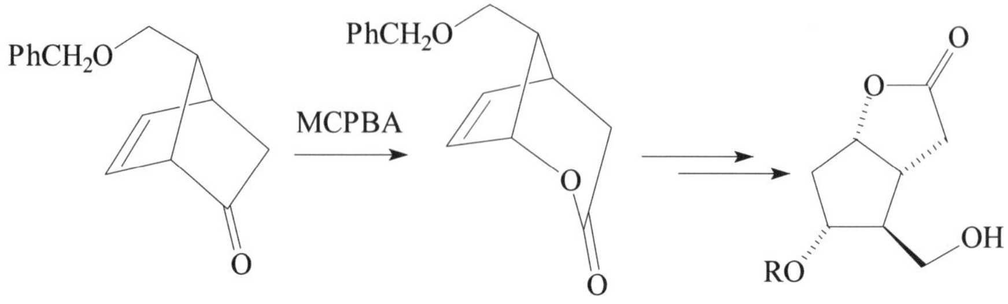

As peracids are used peracetic acid, peroxytrifluoroacetic acid, *m*-chloroperben-zoic acid and others. Hydrogen peroxide or a peracid in combination with triflu-oroacetic acid[5] or certain organoselenium compounds[6] have been successfully employed.

A modern variant is the *enzyme-catalyzed* Baeyer–Villiger oxidation.[7,8] It allows for a reaction under mild conditions in good yields, with one stereoisomer being formed predominantly:

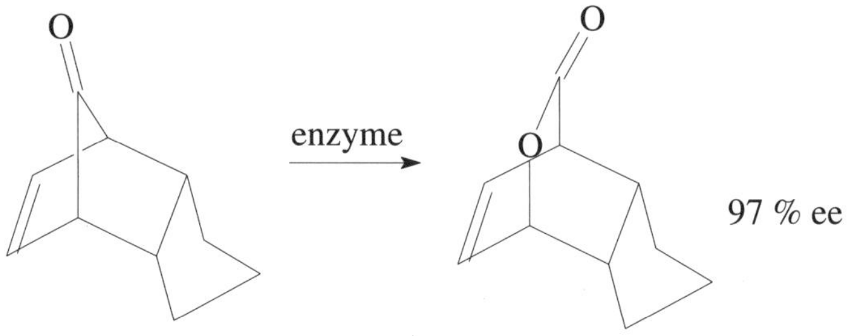

The *Dakin reaction*[2,9] proceeds by a mechanism analogous to that of the Baeyer–Villiger reaction. An aromatic aldehyde or ketone that is activated by a hydroxy group in the *ortho* or *para* position, e.g. salicylic aldehyde **12** (2-hydroxybenzaldehyde), reacts with hydroperoxides or alkaline hydrogen peroxide. Upon hydrolysis of the rearrangement product **13** a dihydroxybenzene, e.g. catechol **14**, is obtained:

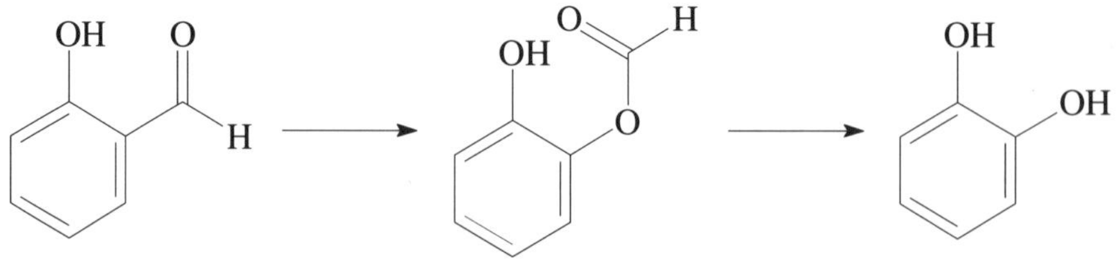

The electron-donating hydroxy substituent is necessary in order to facilitate the migration of the aryl group; otherwise a substituted benzoic acid would be obtained as reaction product.

1. A. v. Baeyer, V. Villiger, *Ber. Dtsch. Chem. Ges.* **1899**, *32*, 3625–3633.
2. C. H. Hassall, *Org. React.* **1957**, *9*, 73–106;
   G. R. Krow, *Org. React.* **1993**, *43*, 251–798.
3. L. M. Harwood, *Polar Rearrangements*, Oxford University Press, Oxford, **1992**, p. 53–59.
4. E. J. Corey, N. M. Weinshenker, T. K. Schaaf, W. Huber, *J. Am. Chem. Soc.* **1969**, *91*, 5675–5677.
5. A. R. Chamberlin, S. S. C. Koch, *Synth. Commun.* **1989**, *19*, 829–833.
6. L. Syper, *Synthesis* **1989**, 167–172.
7. C. T. Walsh, Y.-C. J. Chen, *Angew. Chem.* **1988**, *100*, 342–352; *Angew. Chem. Int. Ed. Engl.* **1988**, *27*, 333.
8. M. J. Taschner, L. Peddada, *J. Chem. Soc., Chem. Commun.* **1992**, 1384–1385.
9. W. M. Schubert, R. R. Kintner in *The Chemistry of the Carbonyl Group* (Ed.: S. Patai), Wiley, New York, **1966**, *Vol. 1*, p. 749–752.

Bamford–Stevens Reaction

Alkenes from tosylhydrazones

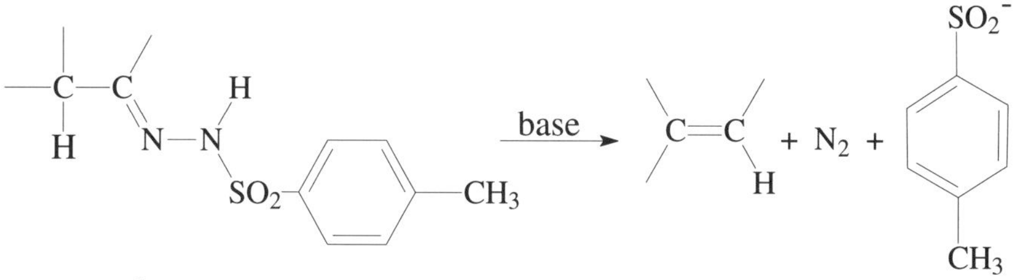

1   **2**

p-Toluenesulfonyl hydrazones **1** (in short tosyl hydrazones) of aliphatic aldehydes or ketones furnish alkenes **2** when treated with a strong base. This reaction is called the *Bamford–Stevens reaction*.[1–3]

Reaction of tosyl hydrazone **1** with a strong base initially leads to a diazo compound **3**, which in some cases can be isolated:

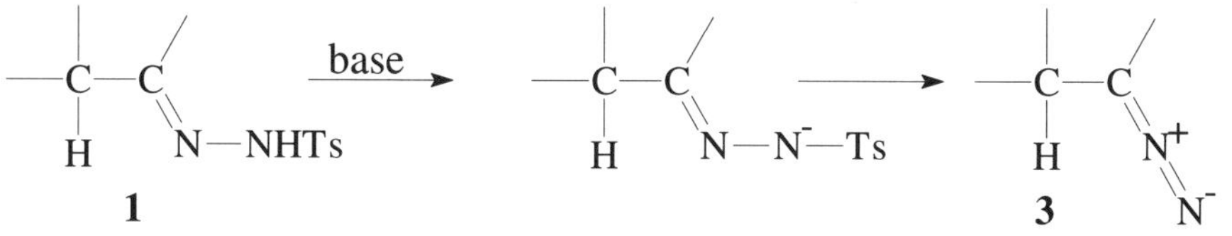

Depending on the reaction conditions, the further reaction can follow either one of two pathways which lead to different products.

In a protic solvent—glycols are often used, with the base being the corresponding sodium glycolate—the reaction proceeds *via* formation of a carbenium ion **5**. The diazo compound **3** can be converted into the diazonium ion **4** through transfer of a proton from the solvent (S—H). Subsequent loss of nitrogen then leads to the carbenium ion **5**:

From **5** the formation of alkene **2** is possible through loss of a proton. However, carbenium ions can easily undergo a *Wagner–Meerwein rearrangement*, and the corresponding rearrangement products may be thus obtained. In case of the Bamford–Stevens reaction under protic conditions, the yield of non-rearranged olefins may be low, which is why this reaction is applied only if other methods (e.g. dehydration of alcohols under acidic conditions) are not practicable.

When an aprotic solvent is used, the reaction proceeds *via* an intermediate carbene **6**. In the absence of a proton donor, a diazonium ion cannot be formed and the diazo compound **3** loses nitrogen to give the carbene **6**:

High boiling ethers such as ethylene glycol dimethyl ether or higher homologs are often used as solvents, and a sodium alkoxide is often used as base. The olefin **2** can be formed by migration of hydrogen. Products from insertion reactions typical for carbenes may be obtained. The 1,2-hydrogen shift generally is the faster process, which is why the aprotic Bamford–Stevens reaction often gives high yields of the desired alkene. Consequently numerous examples have been reported.

A special case is the reaction of the tosylhydrazone **7** of cyclopropane carbaldehyde. It conveniently gives access to bicyclobutane[4] **8**:

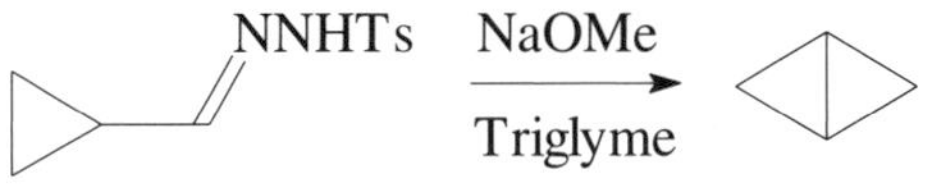

Tosylhydrazones **9** derived from α,β-unsaturated ketones can react *via* vinylcarbenes **10** to yield cyclopropenes[5] **11**:

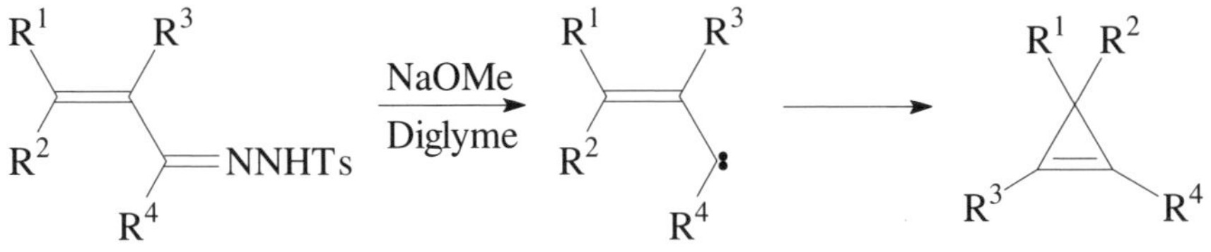

A more promising procedure for the formation of alkenes from tosylhydrazones is represented by the *Shapiro reaction*.[3,6] It differs from the Bamford–Stevens reaction by the use of an organolithium compound (e.g. methyl lithium) as a strongly basic reagent:

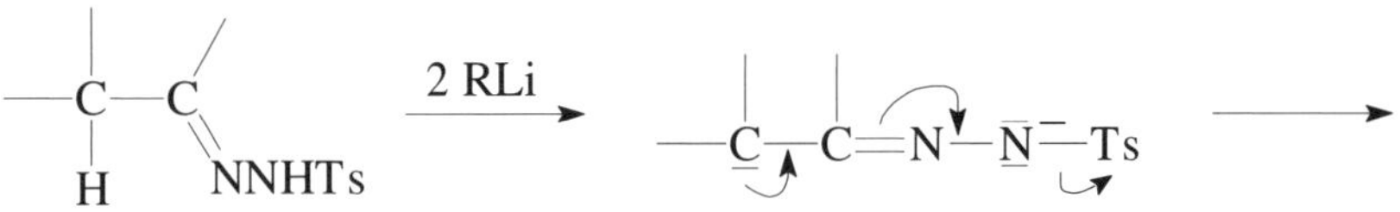

The reaction mechanism has been confirmed by trapping of intermediates **13**, **14** and **15**. Because of the fact that neither a carbene nor a carbenium ion species is involved, generally good yields of non-rearranged alkenes **2** are obtained. Together with the easy preparation and use of tosylhydrazones, this explains well the importance of the Shapiro reaction as a synthetic method.

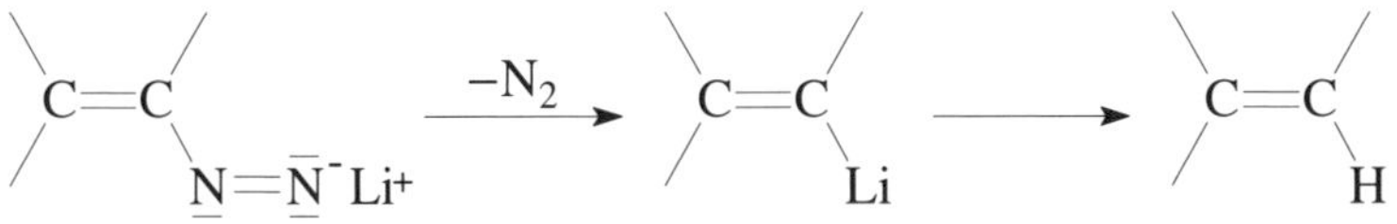

1.   W. R. Bamford, T. S. Stevens, *J. Chem. Soc.* **1952**, 4735–4740.
2.   W. Kirmse, *Carbene Chemistry*, Academic Press, New York, 2nd ed., **1971**, p. 29–34.
3.   R. H. Shapiro, *Org. React.* **1976**, *23*, 405–507;
      A. R. Chamberlin, S. H. Bloom, *Org. React.* **1990**, *39*, 1–83.
4.   H. M. Frey, I. D. R. Stevens, *Proc. Chem. Soc.* **1964**, 144.

5. U. Misslitz, A. de Meijere, *Methoden Org. Chem. (Houben-Weyl)*, **1990**, *Vol. E19b*, p. 675–680.
6. R. M. Adlington, A. G. M. Barrett, *Acc. Chem. Res.* **1983**, *16*, 55–59.

Beckmann Rearrangement

Rearrangement of oximes to give *N*-substituted carboxylic amides

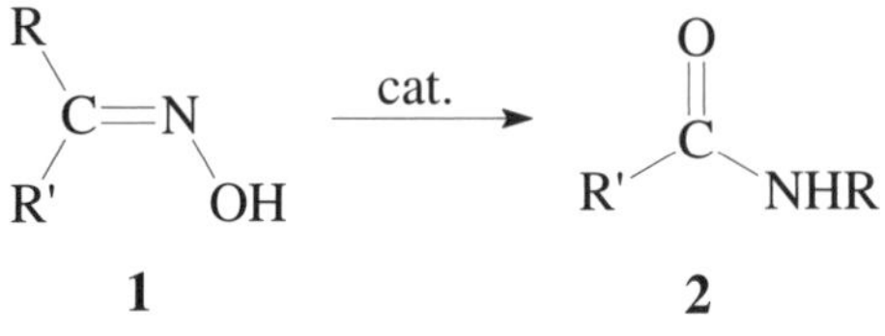

The rearrangement of oximes **1** under the influence of acidic reagents to yield *N*-substituted carboxylic amides **2**, is called the *Beckmann rearrangement*.[1,2] The reaction is usually applied to ketoximes; aldoximes often are less reactive.

Upon treatment with a protic acid, the hydroxy group of the oxime **1** initially is protonated to give an oxonium derivative **3** which can easily lose a water molecule. The migration of the substituent R (together with the bonding electrons) and loss of water proceed simultaneously:[3]

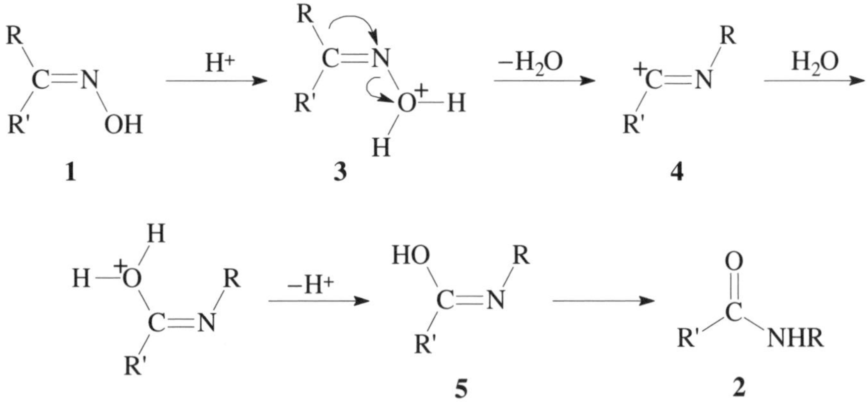

The cationic species **4** thus formed reacts with water to give the iminol **5**, which tautomerizes to a more stable amide tautomer, the *N*-substituted carboxylic amide **2**. Those steps correspond to the formation of amides by the *Schmidt reaction*. A side reaction can give rise to the formation of nitriles.

As reagents concentrated sulfuric acid, hydrochloric acid, liquid sulfur dioxide, thionyl chloride, phosphorus pentachloride and even silica gel[4] can be used. Reagents like phosphorus pentachloride (as well as thionyl chloride and others) first convert the hydroxy group of the oxime **1** into a good leaving group:

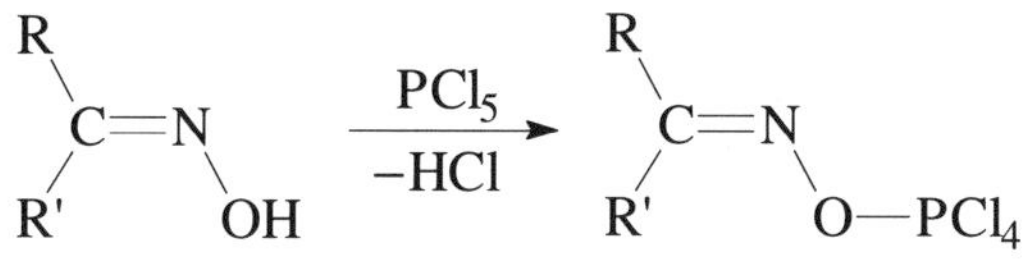

R—C(—R')=N—OH $\xrightarrow[-HCl]{PCl_5}$ R—C(—R')=N—O—PCl_4

1

The stereochemical course of the Beckmann rearrangement often allows for the prediction of the reaction product to be obtained; in general the substituent R *anti* to either the hydroxy or the leaving group will migrate:

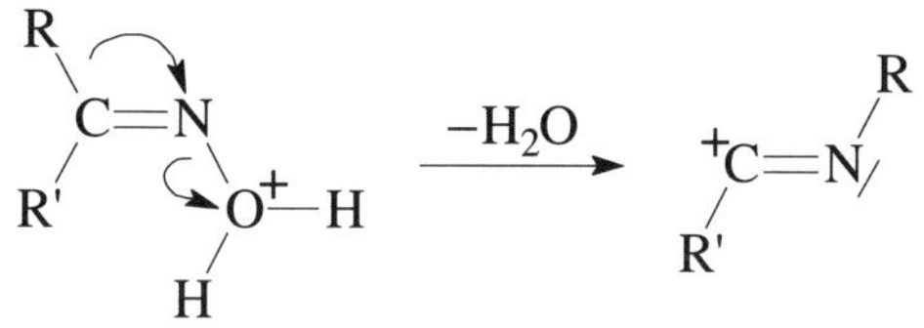

R—C(—R')=N—O⁺(H)—H $\xrightarrow{-H_2O}$ ⁺C(—R')=N—R

In some cases a mixture of the two possible amides may be obtained. This has been rationalized to be a result of partial isomerization of the oxime under the reaction conditions, prior to rearrangement.

With aldoximes (R = H) a migration of hydrogen is seldom found. The Beckmann rearrangement therefore does not give access to *N*-unsubstituted amides.

The reaction with oximes of cyclic ketones leads to formation of lactams (e.g. **6 → 7**) by ring enlargement:

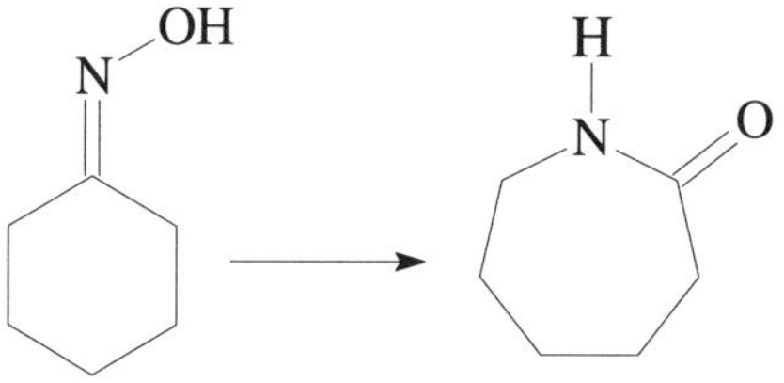

6   **7**

This particular reaction is performed on an industrial scale; ε-caprolactam **7** is used as monomer for polymerization to a polyamide for the production of synthetic fibers.

Substituents R, R' at the starting oxime **1** can be H, alkyl, or aryl.[2,3] The reaction conditions for the Beckmann rearrangement often are quite drastic (e.g. concentrated sulfuric acid at 120 °C), which generally limits the scope to less sensitive substrates. The required oxime can be easily prepared from the respective aldehyde or ketone and hydroxylamine.

1. E. Beckmann, *Ber. Dtsch. Chem. Ges.* **1886**, *19*, 988–993.
2. L. G. Donaruma, W. Z. Heldt, *Org. React.* **1960**, *11*, 1–156;

R. E. Gawley, *Org. React.* **1988**, *35*, 1–420.
D. Schinzer, Y. Bo, *Angew. Chem.* **1991**, *103*, 727; *Angew. Chem. Int. Ed. Engl.* **1991**, *30*, 687; D. Schinzer, E. Langkopf, *Synlett* **1994**, 375.
3. M. I. Vinnik, N. G. Zarakhani, *Russ. Chem. Rev.* **1967**, *36*, 51–64.
4. A. Costa, R. Mestres, J. M. Riego, *Synth. Commun.* **1982**, *12*, 1003–1006.

Benzidine Rearrangement

Rearrangement of hydrazobenzene to yield benzidine

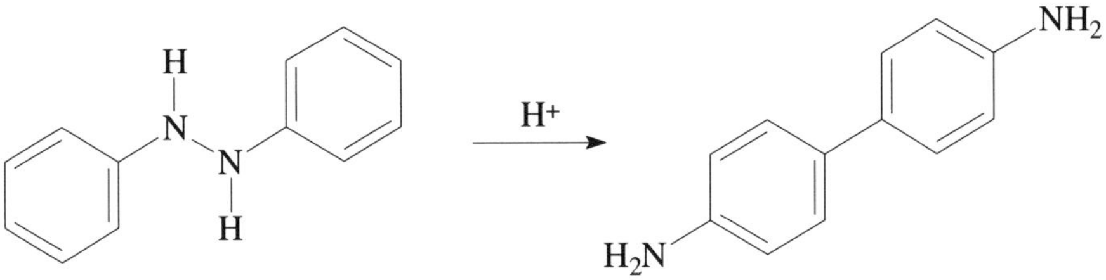

Hydrazobenzene **1** (1,2-diphenyl hydrazine) is converted to benzidine **2** (4,4′-diaminobiphenyl) under acidic conditions.[1,2] This unusual reaction is called the *benzidine rearrangement*,[3,4] and can be observed with substituted diphenyl hydrazines as well.

In accord with the experimental findings a mechanism *via* a [5,5]-sigmatropic rearrangement has been formulated.[5,6] In a first step the hydrazobenzene is protonated to the dicationic species **3**, in which the phenyl groups can arrange in such a way to allow for rearrangement:

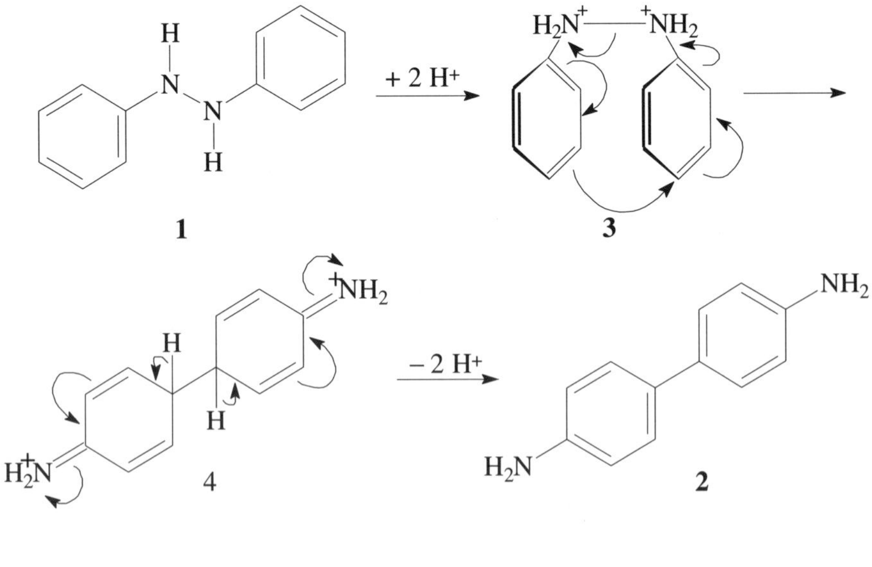

The reaction can be first or second order with respect to the H^+ concentration. In weakly acidic solution it is first order in $[H^+]$, but second order in strongly acidic solution. This indicates that the monoprotonated as well as the diprotonated hydrazobenzene can undergo rearrangement.

The rearranged dicationic species **4**, which has been shown to be an intermediate,[7] leads to the stable benzidine **2** upon deprotonation. It has been demonstrated by crossover experiments that the rearrangement does not proceed *via* a dissociation/recombination process. From the reaction of hydrazobenzene the benzidine is obtained as the major product (up to 70% yield), together with products from side reactions—2,4'-diaminobiphenyl **5** (up to 30% yield) and small amounts of 2,2'-diaminobiphenyl **6** as well as *o*- and *p*-semidine **7** and **8**:

The rearrangement takes place in the presence of strong mineral acids (e.g. hydrochloric or sulfuric acid) in aqueous solution or water–alcohol mixtures at room temperature; in some cases slight warming may be necessary.[3]

The benzidine rearrangement is of interest for mechanistic considerations. The preparative applicability may be limited because of the many side products, together with low yields. Furthermore benzidine is a carcinogenic compound.[8]

1. N. Zinin, *J. Prakt. Chem.* **1845**, *36*, 93–107.
2. P. Jacobsen, *Justus Liebigs Ann. Chem.* **1922**, *428*, 76–121.
3. F. Möller, *Methoden Org. Chem. (Houben-Weyl)* **1957**, *Vol. 11/1*, p. 839–848.
4. R. A. Cox, E. Buncel in *The Chemistry of the Hydrazo, Azo, and Azoxy Groups* (Ed.: S. Patai), Wiley, New York, **1975**, *Vol. 2*, p. 775–807.
5. H. J. Shine, H. Zmuda, K, H, Kwart, A. G. Horgan, C. Collins, B. E. Maxwell, *J. Am. Chem. Soc.* **1981**, *103*, 955–956.
6. H. J. Shine, H. Zmuda, K, H, Kwart, A. G. Horgan, M. Brechbiel, *J. Am. Chem. Soc.* **1982**, *104*, 2501–2509.

7. G. A. Olah, K. Dunne, D. P. Kelly, Y. K. Mo, *J. Am. Chem. Soc.* **1972**, *94*, 7438–7447.
8. Deutsche Forschungsgemeinschaft, *Maximale Arbeitsplatzkonzentration und biologische Arbeitsstofftoleranzwerte*, VCH, Weinheim, **1981**, p. 21.

Benzilic Acid Rearrangement

Rearrangement of 1,2-diketones to give α-hydroxy carboxylic acids

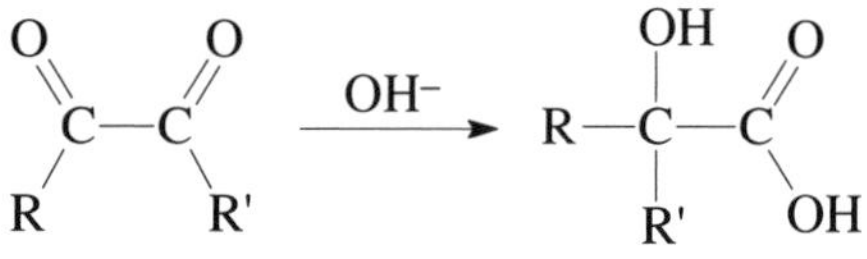

1,2-Diketones **1** can be converted into the salt of an α-hydroxy carboxylic acid upon treatment with alkali hydroxide;[1-3] after acidic workup the free α-hydroxy carboxylic acid **2** is obtained. A well-known example is the rearrangement of benzil (R, R' = phenyl) into benzilic acid (2-hydroxy-2,2-diphenyl acetic acid). The substituents should not bear hydrogens α to the carbonyl group, in order to avoid competitive reactions, e.g. the *aldol reaction*.

The reaction is induced by nucleophilic addition of the hydroxide anion to one of the two carbonyl groups. Then the respective substituent R migrates with the bonding electrons to the adjacent carbon atom (a 1,2-shift). Electron excess at that center is avoided by release of a pair of π-electrons from the carbonyl group to the oxygen:

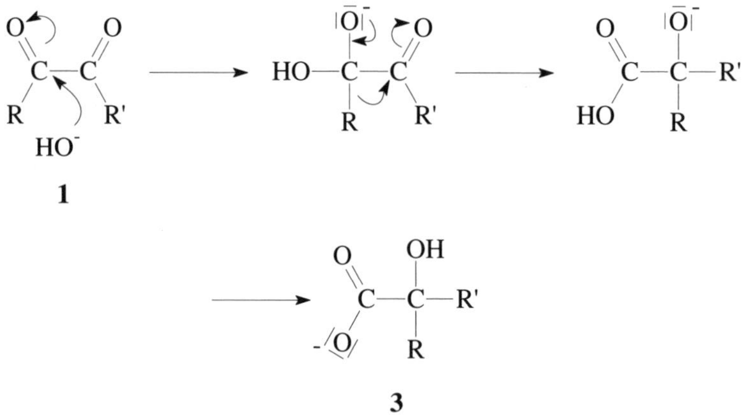

Finally a proton transfer leads to formation of carboxylate anion **3**. Of particular interest is the benzilic acid rearrangement of cyclic diketones such as **4**, since it

leads to a ring contraction:[4]

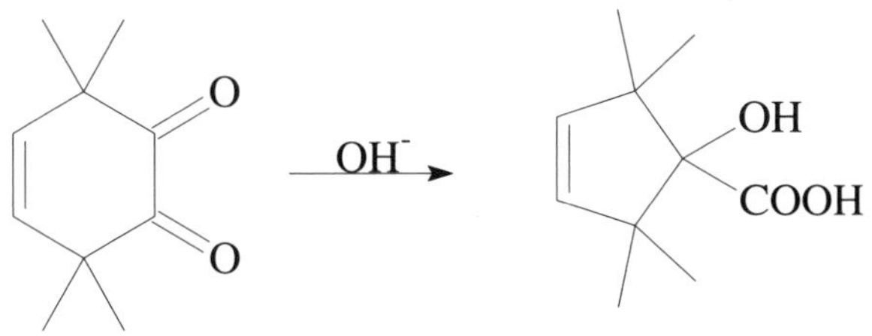

4

A variant is represented by the *benzilic ester rearrangement*,[2,3] where an alkoxide is used as nucleophile. The alkoxide should not be sensitive towards oxidation. The reaction product is the corresponding benzilic acid ester **5**:

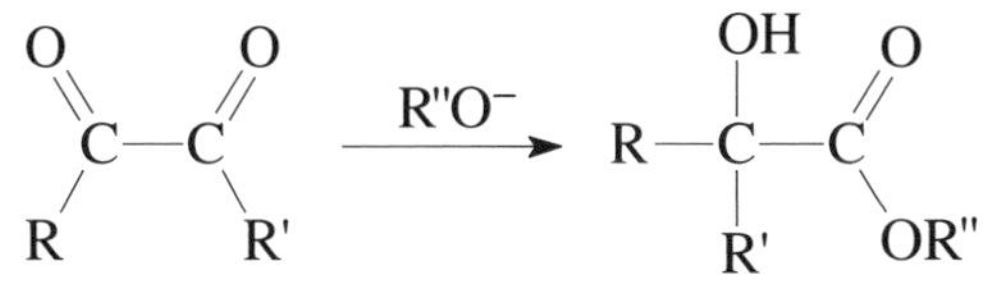

1   **5**

Substrates can be 1,2-diketones with aryl groups as well as some aliphatic substituents, cyclic and heterocyclic diketones. However the benzilic acid rearrangement is of limited preparative importance.

1. N. Zinin, *Justus Liebigs Ann. Chem.* **1839**, *31*, 329–332.
2. S. Selman, J. F. Eastham, *Q. Rev. Chem. Soc.* **1960**, *14*, 221–235.
3. C. J. Collins, J. F. Eastham in *The Chemistry of the Carbonyl Group* (Ed.: S. Patai), Wiley, New York, **1966**, p. 783–787.
4. A. Schaltegger, P. Bigler, *Helv. Chim. Acta* **1986**, *69*, 1666–1670.

Benzoin Condensation

Benzoins from aromatic aldehydes

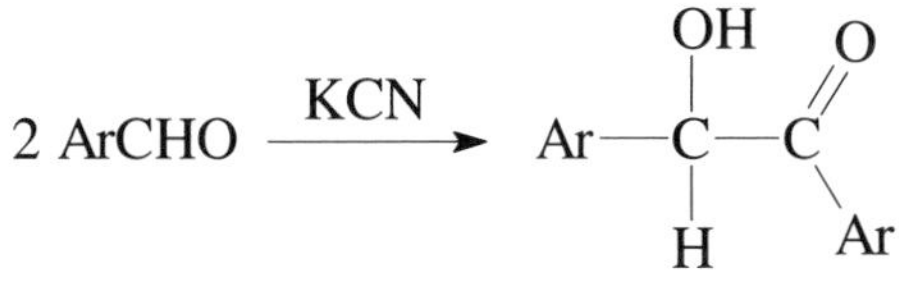

1   **2**

Aromatic aldehydes **1** can undergo a condensation reaction to form α-hydroxy ketones **2** (also called *benzoins*) upon treatment with cyanide anions.[1,2] This reaction, which is called *benzoin condensation*, works by that particular procedure with certain aromatic aldehydes and with glyoxals (RCOCHO).

A cyanide anion as a nucleophile adds to an aldehyde molecule **1**, leading to the anionic species **3**. The acidity of the aldehydic proton is increased by the adjacent cyano group; therefore the tautomeric carbanion species **4** can be formed and then add to another aldehyde molecule. In subsequent steps the product molecule becomes stabilized through loss of the cyanide ion, thus yielding the benzoin **2**:

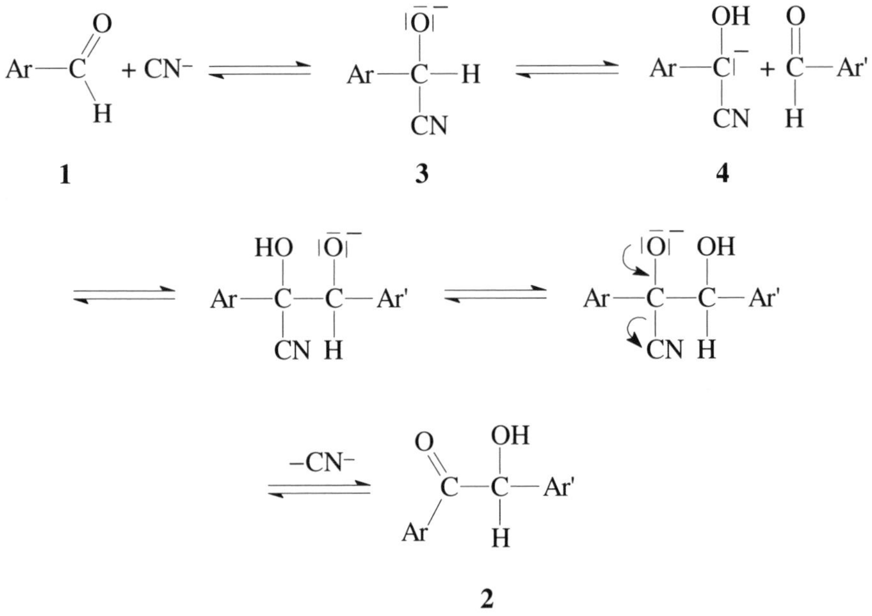

One aldehyde molecule has transferred its aldehyde hydrogen during course of the reaction onto another aldehyde molecule, which is why the reactants are called donor and acceptor (see below).

The cyanide ion plays an important role in this reaction, for it has three functions: in addition to being a good nucleophile, its electron-withdrawing effect allows for the formation of the carbanion species by proton transfer, and it is a good leaving group. These features make the cyanide ion a specific catalyst for the benzoin condensation.

The reaction can also be catalyzed by certain thiazolium salts[3] **5**, in which case it also works with aliphatic substrates. For this modified procedure the following mechanism has been formulated:

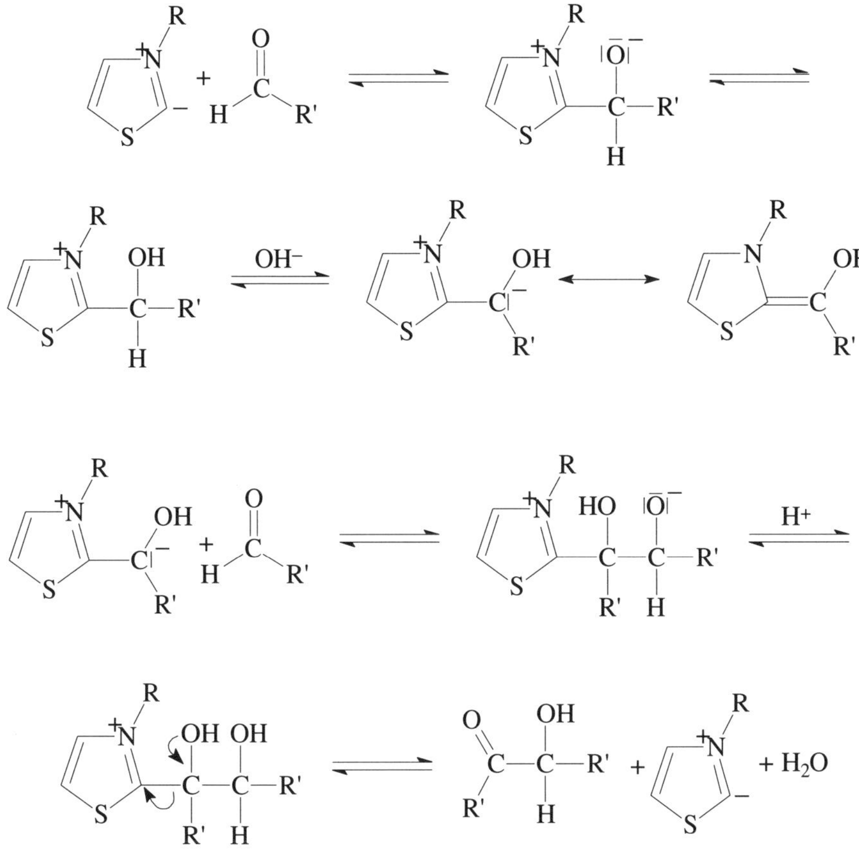

Cross coupling benzoin condensations can be carried out if one aldehyde does not condense with itself, because it either functions only as a donor or acceptor component. *p*-dimethylaminobenzaldehyde acts only as donor and can be condensed, e.g. with benzaldehyde which then acts as acceptor.

1. H. Staudinger, *Ber. Dtsch. Chem. Ges.* **1913**, *46*, 3535–3538.
2. W. S. Ide, J. S. Buck, *Org. React.* **1948**, *4*, 269–304;
   H. Stetter, H. Kuhlmann, *Org. React.* **1991**, *40*, 407–496.
3. H. Stetter, R. Y. Rämsch, H. Kuhlmann, *Synthesis* **1976**, 733–735.

Bergman Cyclization

Cyclization of enediynes

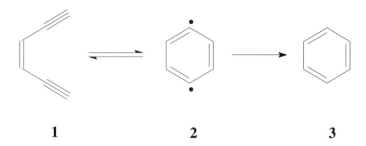

1 **2** **3**

The cycloaromatization of enediynes, having a structure like **1**, proceeds *via* formation of a benzenoid 1,4-diradical **2**, and is commonly called the *Bergman cyclization*.[1,2] It is a relatively recent reaction that has gained importance especially during the last decade. The unusual structural element of enediynes as **1** has been found in natural products (such as calicheamicine and esperamicine)[3] which show a remarkable biological activity.[4,5]

Upon heating the enediyne **1a** rearranges reversibly to the 1,4-benzenediyl diradical **2a**, which in its turn can rearrange to the enediyne **1b** or—in the presence of a hydrogen donor (e.g. cyclohexa-1,4-diene)—react to the aromatic compound[2,8,9] **3a**.

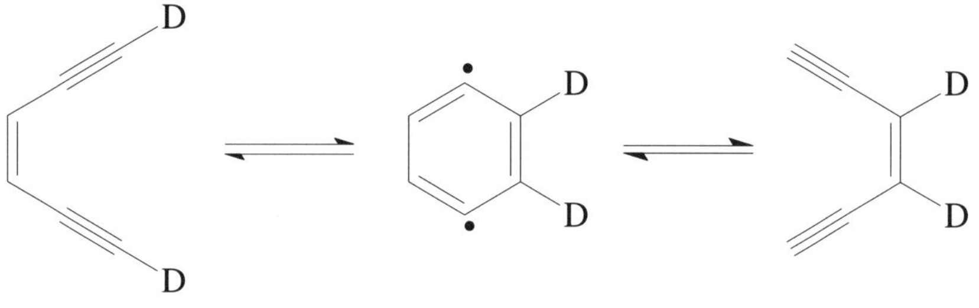

1a **2a** **1b**

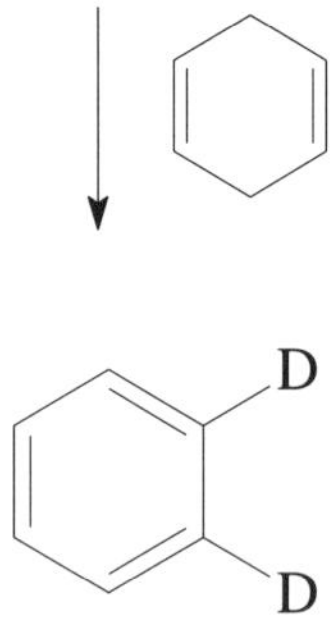

3a

Of great importance for the Bergman cyclization is the distance between the triple bonds. The reaction cannot occur at moderate temperatures if the distance is too large. Optimal reactivity at physiological temperatures is obtained by fitting the enediyne element into a ten-membered ring.[4]

The biological activity of *calicheamicin* **4** (simplified structure) is based on the ability to damage DNA. At the reaction site, initially the distance between the triple bonds is diminished by an addition reaction of a sulfur nucleophile to the enone carbon–carbon double bond, whereupon the Bergman cyclization takes place leading to the benzenoid diradical **5**, which is capable of cleaving double-stranded DNA.[4,5]

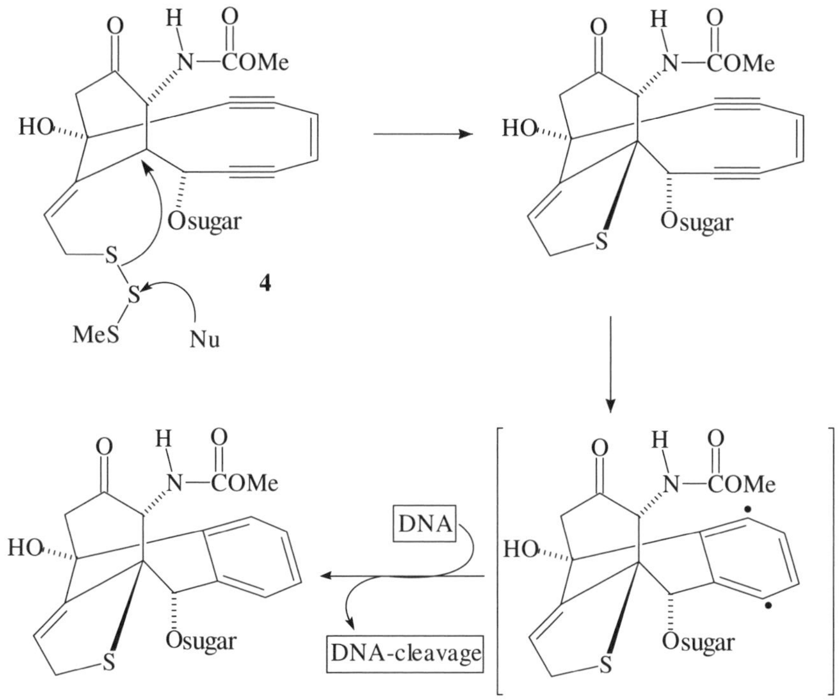

Myers has discovered a related reaction of the natural product *neocarzinostatine*[6,7] **8** (simplified structure). As in the case of the Bergman cyclization a diradical intermediate is generated by a chemical activation step taking place at the reaction site, where it then can cleave DNA. Because of this feature, together with its discriminating affinity towards different DNA strands, neocarzinostatine is regarded as a potential antitumor agent.

The reactive structural element for the *Myers cyclization* is an enyne allene, the heptatrienyne **6**, which reacts to form a diradical species **7**:

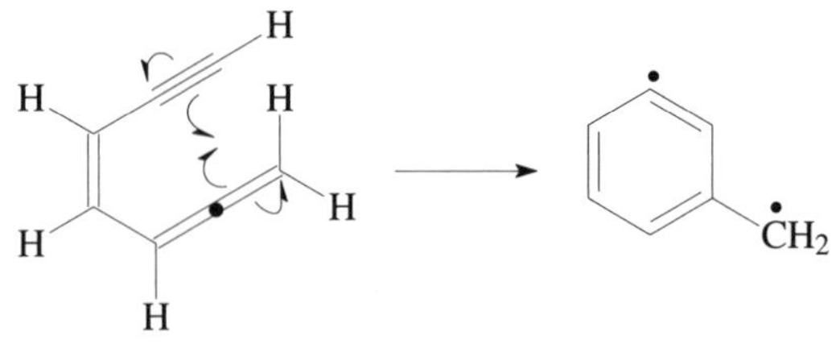

6   **7**

In the initial step, neocarzinostatine **8** (simplified structure) is converted to the cyclization precursor **9**, which contains a cumulated triene unit.[9] The reaction then proceeds *via* the cyclized diradical species **10**, which abstracts hydrogen from a suitable donor to give **11**.

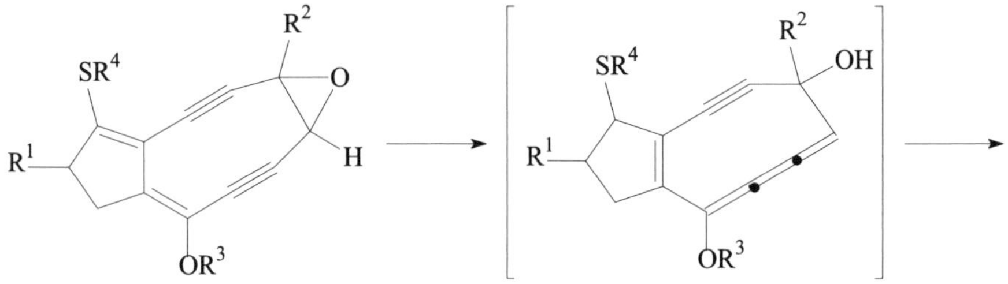

8   **9**

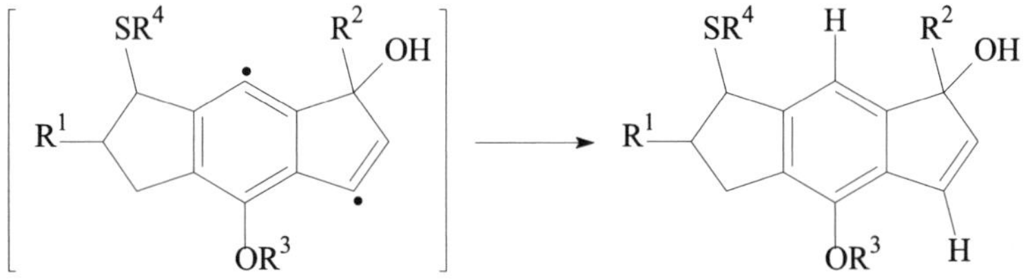

10   **11**

At present the synthetic importance of both the Bergman cyclization and the Myers reaction remains rather small. However, because of the considerable biological activity[4] of the natural products mentioned above, there is great mechanistic interest in these reactions[8,9] in connection with the mode of action of DNA cleavage.

1. R. G. Bergman, R. R. Jones, *J. Am. Chem. Soc.* **1972**, *94*, 660–661.
2. R. G. Bergman, *Acc. Chem. Res.* **1973**, *6*, 25–31.
3. M. D. Lee, T. S. Dunne, C. C. Chang, G. A. Ellestad, M. M. Siegel, G. O. Morton, W. J. McGahren, D. B. Borders, *J. Am. Chem. Soc.* **1987**, *109*, 3466–3468.
4. K. C. Nicolaou, W.-M. Dai, *Angew. Chem.* **1991**, *103*, 1453–1481; *Angew. Chem. Int. Ed. Engl.* **1991**, *30*, 1387.
5. K. C. Nicolaou, G. Zuccarello, C. Riemer, V. A. Estevez, W.-M. Dai, *J. Am. Chem. Soc.* **1992**, *114*, 7360–7371;
   K. C. Nicolaou, *Angew. Chem.* **1993**, *105*, 1462–1471; *ibid. Int. Ed. Engl.* **1993**, *32*, 1377;
   S. A. Hitchcock, S. H. Boyer, M. Y. Chu-Moyer, S. H. Olson, S. J. Danishefsky, *Angew. Chem.* **1994**, *106*, 928–931; *Angew. Chem. Int. Ed. Engl.* **1994**, *33*, 858.
6. A. G. Myers, P. J. Proteau, T. M. Handel, *J. Am. Chem. Soc.* **1988**, *110*, 7212–7214.
7. A. G. Myers, P. S. Dragovich, E. Y. Kuo, *J. Am. Chem. Soc.* **1992**, *114*, 9369–9386.
8. M. E. Maier, *Synlett* **1995**, 13–26;
   H. Lhermitte, D. S. Grierson, *Contemp. Org. Synth.* **1996**, *3*, 41–64.
9. The chemistry of enediynes, enyne allenes and related compounds: J. W. Grissom, G. U. Gunawardena, D. Klingberg, D. Huang, *Tetrahedron* **1996**, *52*, 6453–6518.

Birch Reduction

Partial reduction of aromatic compounds

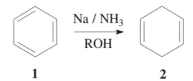

1   **2**

The reduction of aromatic compounds **1** by alkali metals in liquid ammonia in the presence of an alcohol is called the *Birch reduction*, and yields selectively the 1,4-hydrogenated product[1–3] **2**.

Alkali metals in liquid ammonia can transfer an electron to the solvent, leading to so-called solvated electrons. These can add to the aromatic substrate **1** to give a reduced species, the radical anion **3**:

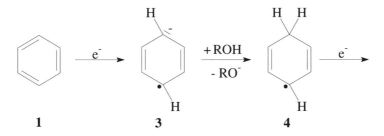

1   **3**   **4**

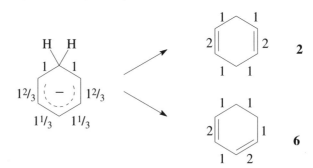

Evidence for the radical anion **3** came from esr spectroscopic experiments, thus supporting this mechanism. The radical anion is protonated by the alcohol to give the radical species **4**, which is further reduced by a solvated electron to give the carbanion **5**. This anion is protonated by the alcohol leading to the 1,4-dihydro product **2**. Thus the alkali metal serves as a source of electrons, while the alcohol serves as a source of protons.

The negative charge of the cyclohexadienyl anion **5** is delocalized over several carbon centers, as is illustrated by the following resonance structures:

At first glance it may be surprising that the 1,4-diene is formed instead of the thermodynamically more stable, conjugated 1,3-diene derivative. An explanation is offered by the *principle of least motion*,[4] which favors those reaction pathways that involve the least change in atomic position and electronic configuration. A description of the bond orders for the carbon–carbon bonds of the carbanionic species **5** and the possible products **2** and **6** by a simplified valence-bond method (1 for a single bond, 2 for a double bond), shows the smaller change when going from **5** to the 1,4-diene **2** ($\Delta = 4/3$) compared to the greater change when going from **5** to the 1,3-diene **2** ($\Delta = 2$).

For the Birch reduction of mono-substituted aromatic substrates the substituents generally influence the course of the reduction process.[5] Electron-donating substituents (e.g. alkyl or alkoxyl groups) lead to products with the substituent located at a double bond carbon center. The reduction of methoxybenzene (anisole) **7** yields 1-methoxycyclohexa-1,4-diene **8**:

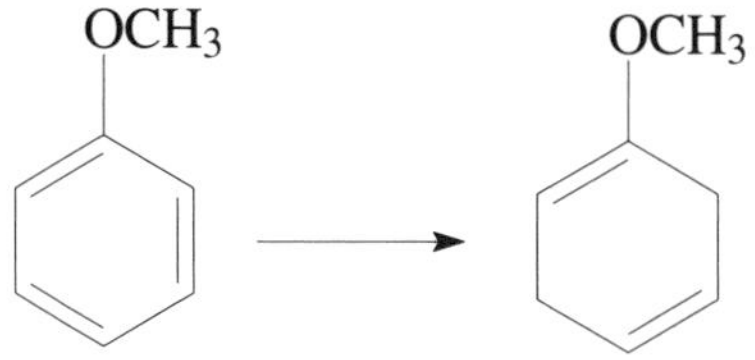

An electron-withdrawing substituent leads to a product where it is bound to a saturated carbon center. Benzoic acid **9** is reduced to the cyclohexa-2,5-diene carboxylic acid **10**:

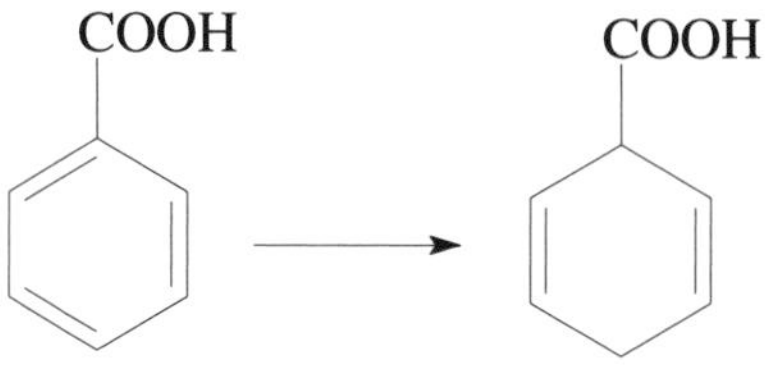

The Birch reduction is the method of choice for the partial reduction of aromatic compounds. The catalytic hydrogenation would lead to fully hydrogenated products. Ordinary olefins are not reduced under Birch conditions, while conjugated olefins will react. Halogen substituents, nitro, aldehyde and keto groups can suffer reduction. In some cases the insufficient solubility of the aromatic substrate in liquid ammonia may cause problems, which can be avoided by use of a co-solvent. The yields are generally good or even high. With polycyclic benzenoid substrates mixtures of isomers may be obtained.

1. A. J. Birch, *J. Chem. Soc.* **1944**, 430–436.
2. P. W. Rabideau, Z. Marcinow, *Org. React.* **1992**, *42*, 1–334.
3. P. W. Rabideau, *Tetrahedron* **1989**, *45*, 1579–1603.
4. J. Hine, *J. Org. Chem.* **1966**, *31*, 1236–1244.
5. H. E. Zimmerman, P. A. Wang, *J. Am. Chem. Soc.* **1990**, *112*, 1280–1281.

Blanc Reaction

Chloromethylation of aromatic compounds

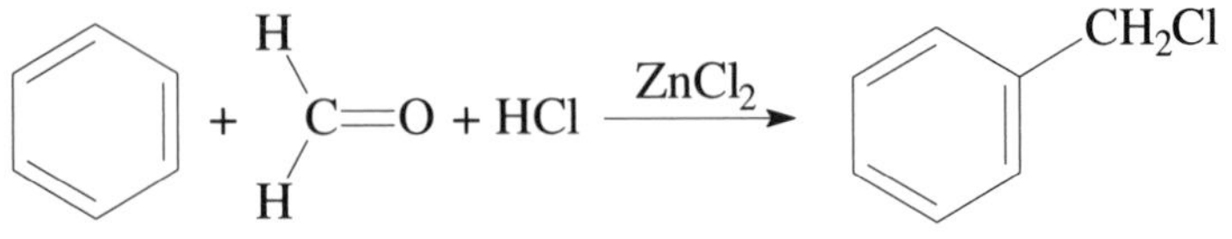

$$\begin{array}{ccc} \mathbf{1} & \mathbf{2} & \mathbf{3} \end{array}$$

The introduction of a chloromethyl group on aromatic compounds (e.g. benzene **1**) by reaction with formaldehyde **2** and gaseous hydrogen chloride in the presence of a catalyst is called the *Blanc reaction*.[1,2]

In a first reaction step the formaldehyde **2** is protonated, which increases its reactivity for the subsequent electrophilic aromatic substitution at the benzene ring. The cationic species **4** thus formed loses a proton to give the aromatic hydroxymethyl derivative **5**, which further reacts with hydrogen chloride to yield the chloromethylated product[3] **3**:

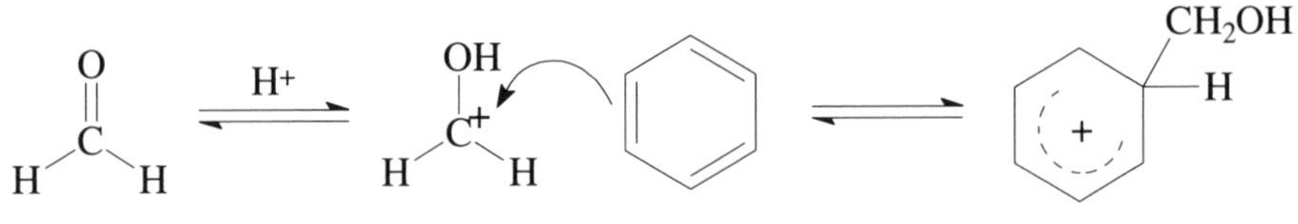

$$\begin{array}{ccc} \mathbf{2} & \mathbf{1} & \mathbf{4} \end{array}$$

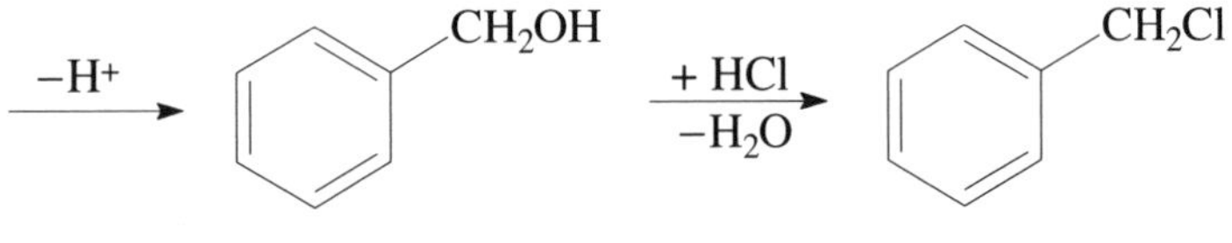

$$\begin{array}{cc} \mathbf{5} & \mathbf{3} \end{array}$$

The rate-determining step is the electrophilic aromatic substitution as in the closely related *Friedel–Crafts reaction*. Both reactions have in common that a Lewis acid catalyst is used. For the Blanc reaction zinc chloride is generally employed,[2] and the formation of the electrophilic species can be formulated as follows:[3]

$$CH_2O + HCl + ZnCl_2 \longrightarrow CH_2OH^+ZnCl_3{}^-$$

Electron-rich aromatic substrates can react without a catalyst present. Modern variants of the Blanc reaction use chloromethyl ether[4] (e.g. $(ClCH_2)_2O$, $ClCH_2OMe$) or methoxyacetyl chloride,[5] since those reagents are more reactive and give higher yields.

The chloromethylation can be generally employed in aromatic chemistry; benzene, naphthaline, anthracene, phenanthrene, biphenyls and many derivatives thereof are appropriate substrates. The benzylic chlorides thus obtained can be further transformed, for example to aromatic aldehydes. Ketones like benzophenone are not reactive enough. In contrast phenols are so reactive that polymeric products are obtained.[2]

An important side reaction is the formation of diaryl methane derivatives $ArCH_2Ar$. Moreover polysubstituted products may be obtained as minor products. Aromatic compounds have been treated with formaldehyde and hydrogen bromide or hydrogen iodide instead of hydrogen chloride. The formaldehyde may be replaced by another aldehyde; the term 'Blanc reaction' however stands for the chloromethylation only.

1. M. G. Blanc, *Bull. Soc. Chim. Fr.* **1923**, *33*, 313–319.
2. R. C. Fuson, C. H. McKeever, *Org. React.* **1942**, *1*, 63–90.
3. L. I. Belenkii, Yu. B. Volkenshtein, I. B. Karmanova, *Russ. Chem. Rev.* 1977, *46*, 891–903.
4. G. A. Olah, D. A. Beal, J. A. Olah, *J. Org. Chem.* **1976**, *41*, 1627–1631.
5. A. McKilloq, F. A. Madjdabadi, D. A. Long, *Tetrahedron Lett.* **1983**, *24*, 1933–1936.

Bucherer Reaction

Interconversion of naphtholes and naphthylamines

An important reaction in the chemistry of naphthalenes is the *Bucherer reaction*,[1-3] i.e. the conversion of naphthols **1** to naphthylamines **2** as well as the reverse reaction. The reaction is carried out in aqueous medium in the presence of catalytic amounts of a sulfite or bisulfite. Apart from very few exceptions it does not apply to benzene derivatives, which limits the scope of that reaction.

Naphthol **1** is initially protonated at a carbon center of high electron density (C-2 or C-4). The cationic species **3** thus formed is stabilized by resonance; it can add a bisulfite anion at C-3. The addition product can tautomerize to give the more stable tetralone sulfonate **4**; the tetralone carbonyl group is then attacked by a nucleophilic amine (e.g. ammonia). Subsequent dehydration leads to the cation **5** which again is stabilized by resonance. Loss of a proton leads to the enamine **6**, which upon loss of the bisulfite leads to the aromatic naphthylamine[4] **2**:

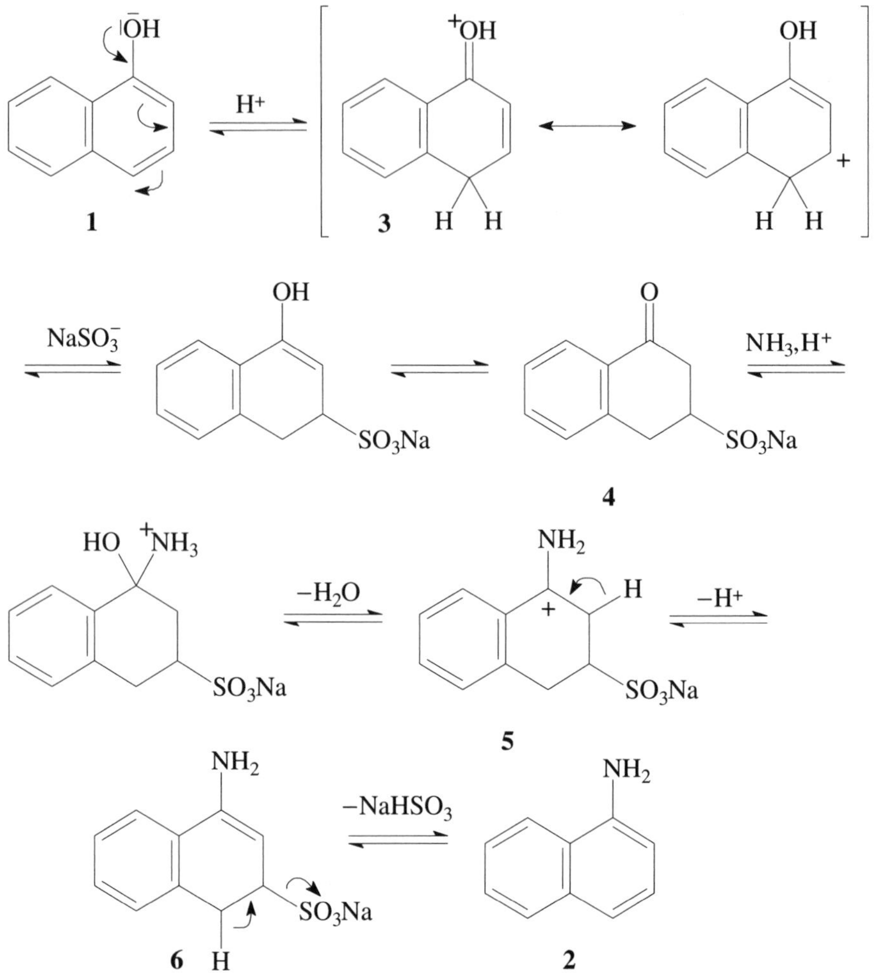

Every step of the Bucherer reaction is reversible, and the reverse sequence is also of synthetic value. The equilibrium can be shifted by varying the concentration of free ammonia.[3]

As mentioned above, the scope of the Bucherer reaction is limited. It works with anthracenes and phenanthrenes, but only very few examples with substituted benzenes are known. Naphthylamines can be converted into the corresponding naphthols, and these can then be further converted into primary, secondary or tertiary naphthylamines (*transamination*). Naphthylamines are of importance for the synthesis of dyes, and derivatives of tetralone sulfonic acid, which are accessible through the Bucherer reaction, are of synthetic value as well.[4]

1. H. T. Bucherer, *J. Prakt. Chem.* **1904**, *69*, 49–91.
2. N. L. Drake, *Org. React.* **1942**, *1*, 105–128.
3. R. Schröter, *Methoden Org. Chem. (Houben-Weyl)* **1957**, *Vol. 11/1*, p. 143–159.
4. H. Seeboth, *Angew. Chem.* **1967**, *79*, 329–340; *Angew. Chem. Int. Ed. Engl.* **1967**, *6*, 307.

C

Cannizzaro Reaction

Disproportionation of aldehydes

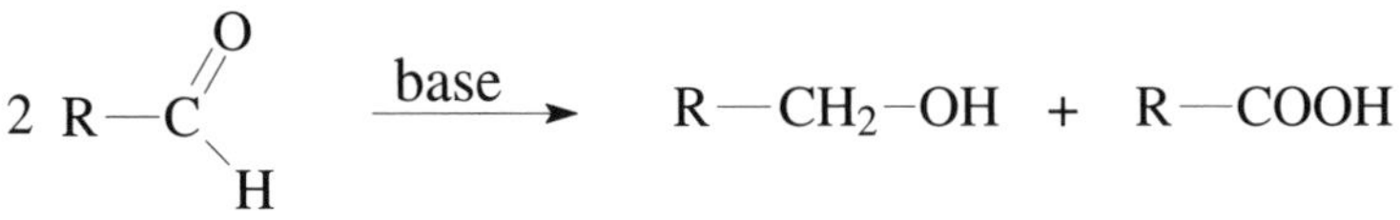

$$2 \ R-\overset{\displaystyle O}{\underset{\displaystyle H}{C}} \quad \xrightarrow{\text{base}} \quad R-CH_2-OH \ + \ R-COOH$$

1   **2**   **3**

Aldehydes **1** that have no α-hydrogen give the *Cannizzaro reaction* upon treatment with a strong base, e.g. an alkali hydroxide.[1,2] In this disproportionation reaction one molecule is reduced to the corresponding alcohol **2**, while a second one is oxidized to the carboxylic acid **3**. With aldehydes that do have α-hydrogens, the *aldol reaction* takes place preferentially.

The key step of the Cannizzaro reaction is a hydride transfer. The reaction is initiated by the nucleophilic addition of a hydroxide anion to the carbonyl group of an aldehyde molecule **1** to give the anion **4**. In a strongly basic medium, the anion **4** can be deprotonated to give the dianionic species **5**:

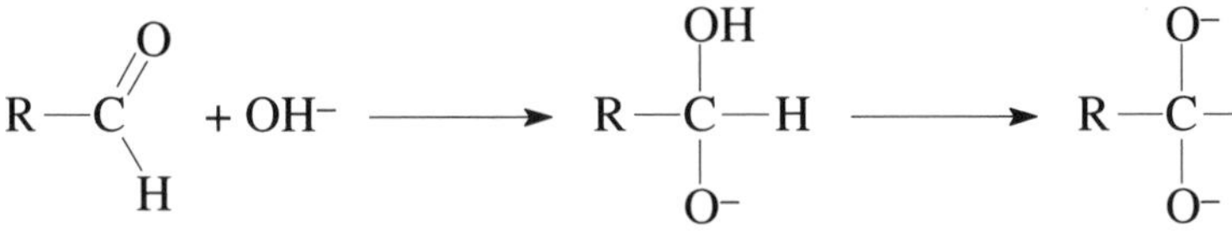

$$R-\overset{\displaystyle O}{\underset{\displaystyle H}{C}} + OH^- \longrightarrow R-\overset{\displaystyle OH}{\underset{\displaystyle O^-}{\overset{\displaystyle |}{\underset{\displaystyle |}{C}}}}-H \longrightarrow R-\overset{\displaystyle O^-}{\underset{\displaystyle O^-}{\overset{\displaystyle |}{\underset{\displaystyle |}{C}}}}-H$$

1   **4**   **5**

The reaction can proceed from both species **4** or **5** respectively. The strong electron-donating effect of one or even two O^--substituents allows for the transfer of a hydride ion H^- onto another aldehyde molecule:

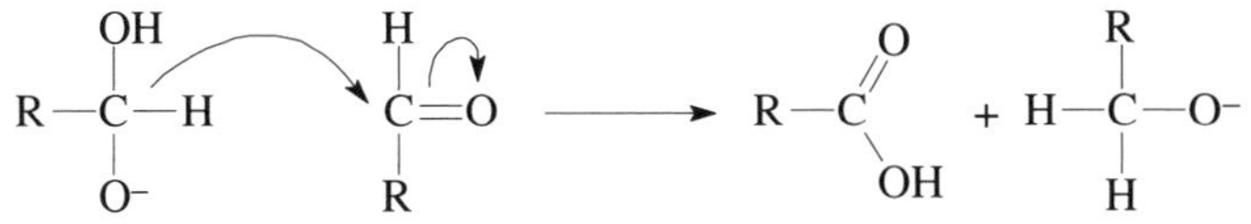

4　　**1**

5　　**1**

This mechanism is supported by the outcome of experiments with D_2O as solvent. The resulting alcohol **2** does not contain carbon-bonded deuterium, indicating that the transferred hydrogen comes from a second substrate molecule, and not from the solvent.

The synthetic importance of the reaction is limited, because as a consequence of the disproportionation, the yield of the alcohol as well as the carboxylic acid is restricted to 50%. However good yields of alcohols can often be obtained when the reaction is carried out in the presence of equimolar amounts of formaldehyde. The formaldehyde is oxidized to formic acid and concomitantly reduces the other aldehyde to the desired alcohol. This variant is called the *crossed Cannizzaro reaction*.

α-Keto aldehydes **6** can be converted to α-hydroxy carboxylic acids **7** by an intramolecular Cannizzaro reaction:

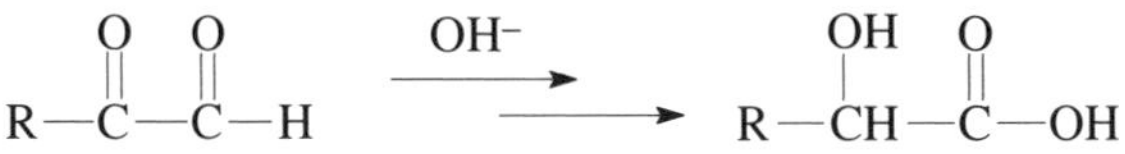

6　　　　　　　**7**

1,4-Dialdehydes **8** have been converted to γ-lactones **9** in the presence of a rhodium phosphine complex as catalyst.[3] The example shown below demonstrates that this reaction works also with aldehydes that contain α-hydrogen atoms.

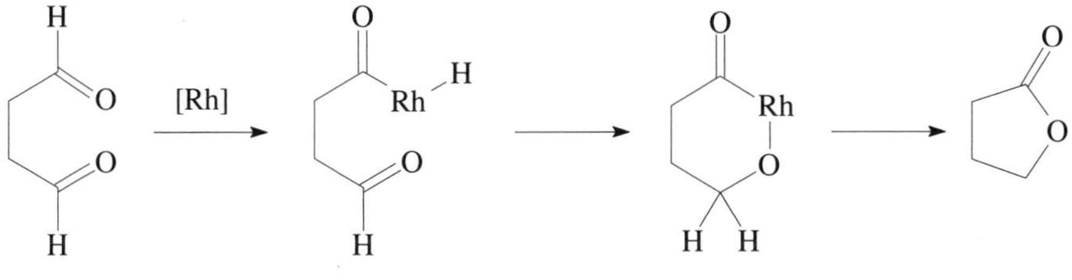

8　　　　　　　　　　　　　　　　　　　　　　　**9**

The applicability of the Cannizzaro reaction may be limited, if the substrate aldehyde can undergo other reactions in the strongly basic medium. For instance an α,α,α-trihalo acetaldehyde reacts according to the *haloform reaction*.

Mechanistically closely related is the *benzilic acid rearrangement*, where an alkyl or aryl group migrates instead of the hydrogen.

1. S. Cannizzaro, *Justus Liebigs Ann. Chem.* **1853**, *88*, 129–130.
2. T. A. Geissman, *Org. React.* **1944**, *2*, 94–113.
3. S. H. Bergens, D. P. Fairlie, B. Bosnich, *Organometallics* **1990**, *9*, 566–571.

Chugaev Reaction

Formation of olefins from xanthates

1   **2**   **3**   **4**

Upon thermolysis of xanthates (xanthogenates) **1** olefins **2** can be obtained, together with gaseous carbon oxysulfide COS **3** and a thiol RSH **4**. This decomposition process is called the *Chugaev reaction*;[1–3] another common transcription for the name of its discoverer is *Tschugaeff*.

The required xanthates **1** can be prepared from alcohols **5** by reaction with carbon disulfide in the presence of sodium hydroxide and subsequent alkylation of the intermediate sodium xanthate **6**. Often methyl iodide is used as the alkylating agent:

$$\text{ROH} + \text{CS}_2 + \text{NaOH} \longrightarrow \text{ROCS}^-\text{Na}^+ \xrightarrow{\text{CH}_3\text{I}} \text{ROCSCH}_3 + \text{NaI}$$

5   **6**   **1**

The thermolysis (pyrolysis) is generally carried out at temperatures ranging from 100–250 °C. Similar to the closely related *ester pyrolysis* the reaction mechanism is of the E_i-type, which involves a six-membered cyclic transition state **7**:

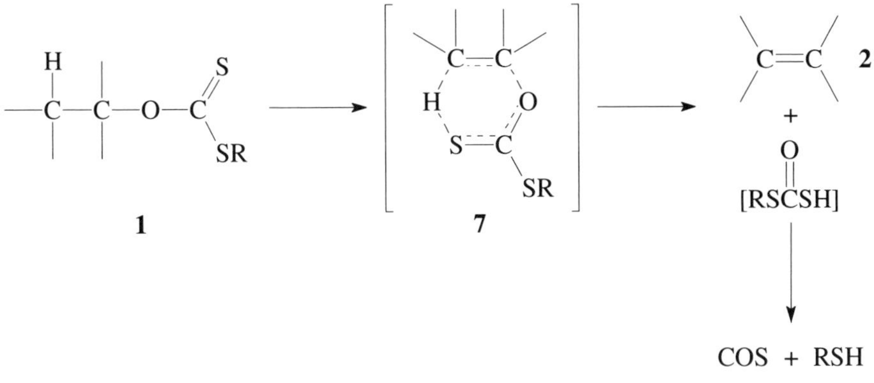

This mechanism of a *syn*-elimination reaction is supported by experimental find-ings with ^{34}S- and ^{13}C-labeled starting materials.[4] The Chugaev reaction is analogous to the ester pyrolysis, but allows for milder reaction conditions—i.e. it occurs at lower temperatures. It is less prone to side reactions, e.g. the formation of rearranged products, and is therefore the preferred method.

The thermolysis of xanthates derived from primary alcohols yields one olefin only. With xanthates from secondary alcohols (acyclic or alicyclic) regioisomeric products as well as E/Z-isomers may be obtained; see below. While acyclic substrates may give rise to a mixture of olefins, the formation of products from alicyclic substrates often is determined by the stereochemical requirements; the β-hydrogen and the xanthate moiety must be *syn* to each other in order to eliminate *via* a cyclic transition state.

An instructive example on how stereochemical features influence the stereo-chemical outcome of the elimination is the pyrolysis of xanthates from *erythro*- and *threo*-1,2-diphenyl-1-propanol. The *erythro*-alcohol **8** is converted into E-methylstilbene **9** only, and the *threo*-alcohol **10** is converted into the corre-sponding Z-isomer **11** only. These results support the assumption of a *syn*-elimination process through a cyclic transition state:[5]

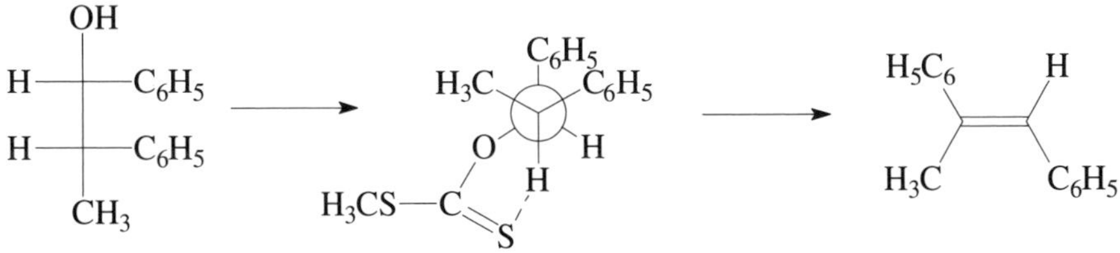

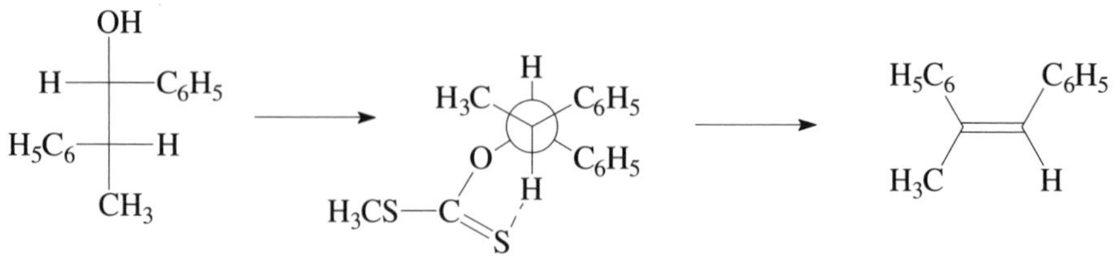

The Chugaev elimination is of synthetic value, because it proceeds without rearrangement of the carbon skeleton.[2] Other non-thermolytic elimination procedures often lead to rearranged products, when applied to the same substrates. However applicability of the Chugaev reaction is limited if the elimination is possible in more than one direction, and if a β-carbon has more than one hydrogen. Complex mixtures of isomeric olefins may then be obtained. For example the thermolysis of xanthate **12**, derived from 3-hexanol yields 28% *E*-hex-3-ene **13**, 13% *Z*-hex-3-ene **14**, 29% *E*-hex-2-ene **15** and 13% *Z*-hex-2-ene[6] **16**:

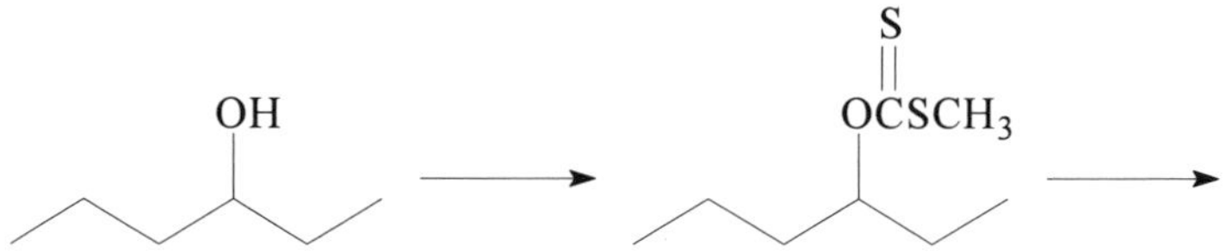

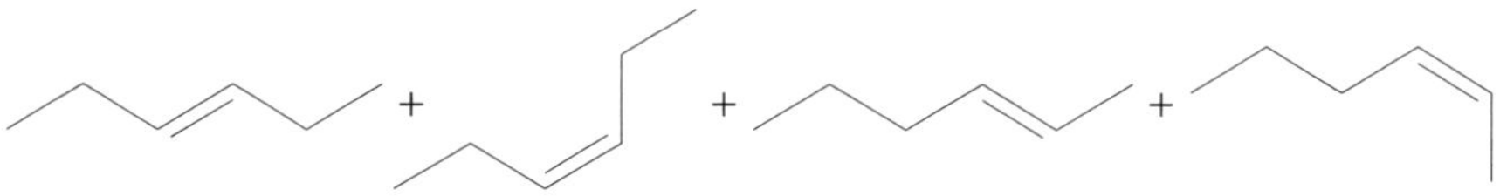

The applicability to alicyclic alcohols may be limited, since the mechanistic course *via* a cyclic transition state requires a suitably positioned hydrogen in order to afford the desired product.

1. L. Tschugaeff, *Ber. Dtsch. Chem. Ges.* **1899**, *32*, 3332–3335.
2. H. R. Nace, *Org. React.* **1962**, *12*, 57–100.
3. C. H. DePuy, R. W. King, *Chem. Rev.* **1960**, *60*, 431–457.
4. R. F. W. Bader, A. N. Bourns, *Can. J. Chem.* **1961**, *39*, 346–358.
5. D. J. Cram, F. A. A. Elhafez, *J. Am. Chem. Soc.* **1952**, *74*, 5828–5835.
6. R. A. Benkeser, J. J. Hazdra, M. L. Burrous, *J. Am. Chem. Soc.* **1959**, *81*, 5374–5379.

Claisen Ester Condensation

Formation of β-keto esters from carboxylic esters

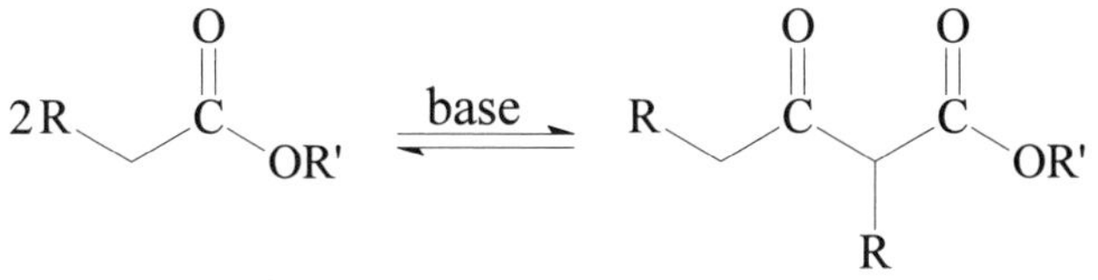

1   **2**

Carboxylic esters **1** that have an α-hydrogen can undergo a condensation reaction upon treatment with a strong base to yield a β-keto ester **2**. This reaction is called the *Claisen ester condensation*[1,2] or *acetoacetic ester condensation*; the corresponding intramolecular reaction is called the *Dieckmann condensation*:[3,4]

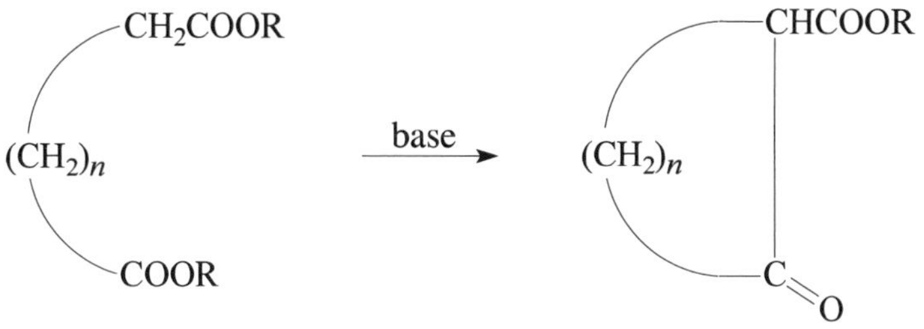

The mechanism involves the formation of anion **3** from ester **1** by reaction with base:

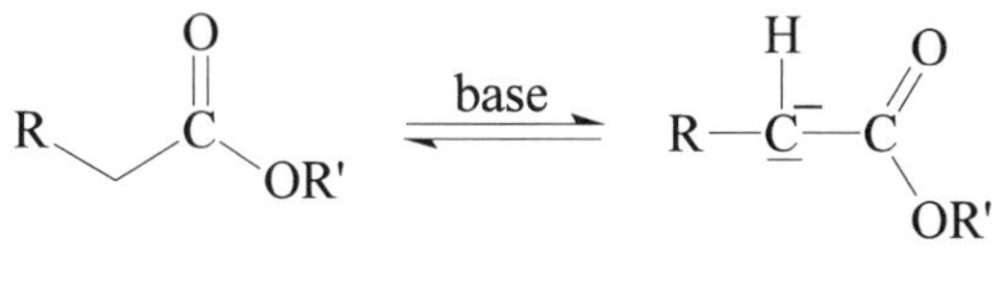

1   **3**

Anion **3** can add to another ester **1**. The resulting anionic species **4** reacts to the stable β-keto ester by loss of an alkoxide anion $R'O^-$ **5**:

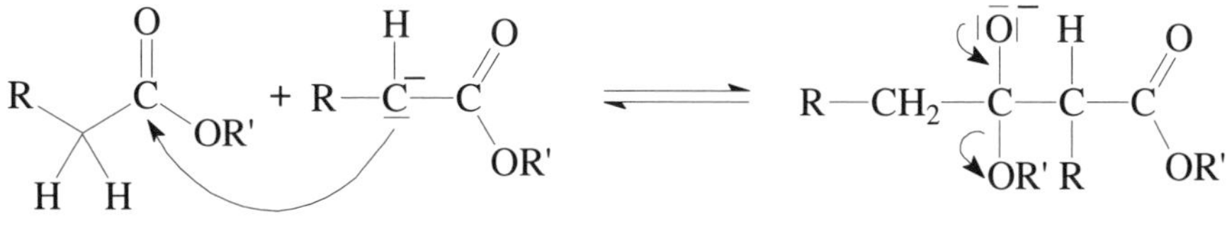

1   **3**   **4**

$$\text{----} \quad \text{R—CH}_2\text{—C—C—C} \quad + \text{RO}^-$$

(structure **2** and **5** with O, H, O; R; OR')

2   **5**

Those steps are reversible reactions, with the equilibrium shifted to the left. However the overall reaction can be carried out in good yield since the β-ketoester **2** is converted into the conjugate base **6** by the lost alkoxide **5**; the ester is more acidic than the alcohol R'OH **7**:

$$\text{R—CH}_2\text{—C—C—C} \quad + \text{RO}^- \quad \rightleftharpoons \quad \text{R—CH}_2\text{—C—C—C} \quad + \text{R'OH}$$

2   **5**   **6**   **7**

This mechanism is supported by the finding, that esters containing just one α-hydrogen do not undergo that reaction, unless much stronger bases are used, since the condensation product **9** cannot be stabilized under the usual reaction conditions and the equilibrium lies to the left:

$$(CH_3)_2CH\text{-C—OC}_2H_5 \quad \overset{\text{NaOEt}}{\underset{\text{EtOH}}{\rightleftharpoons}} \quad (CH_3)_2CH\text{-C—C—C—OC}_2H_5 + EtOH$$

8   **9**

α,α-disubstituted β-ketoesters like **9**, when treated with an alkoxide, can be cleaved into ordinary esters by the reverse of the condensation reaction, the *retro-Claisen reaction*. However the condensation of esters with only one α-hydrogen is possible in moderate yields by using a strong base, e.g lithium diisopropyl amide (LDA).[5]

The most common base used for the Claisen condensation is sodium ethoxide, although for some substrates stronger bases such as sodium amide or sodium hydride may be necessary.

The application to a mixture of two different esters, each with α-hydrogens, is seldom of preparative value, since a mixture of the four possible condensation products will be obtained. If however only one of the starting esters contains α-hydrogens, the *crossed Claisen condensation* often proceeds in moderate to good yields.

The intramolecular condensation reaction of diesters, the *Dieckmann condensation*,[3,4] works best for the formation of 5- to 7-membered rings; larger rings are formed with low yields, and the *acyloin condensation* may then be a faster competitive reaction. With non-symmetric diesters two different products can be formed. The desired product may be obtained regioselectively by a modified procedure using a solid support—e.g with a polystyrene[6] **10**:

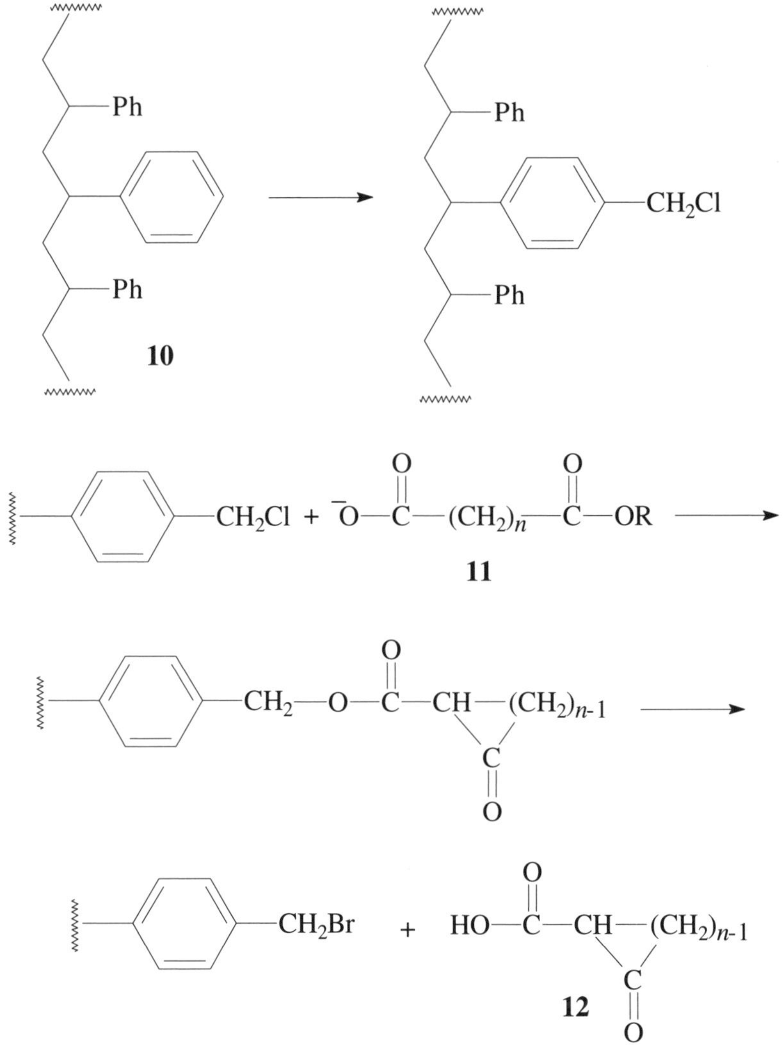

A functional group is introduced to the polystyrene **10** by chloromethylation (*Blanc reaction*) in order to allow for reaction with the substrate **11**. The polymer-bound diester is then treated with base to initiate the Dieckmann condensation.

Finally treatment with HBr leads to cleavage of product **12** from the polymer.

1. L. Claisen, O. Lowman, *Ber. Dtsch. Chem. Ges.* **1887**, *20*, 651–657.
2. C. R. Hauser, B. E. Hudson, *Org. React.* **1942**, *1*, 266–302.
3. W. Dieckmann, *Ber. Dtsch. Chem. Ges.* **1900**, *33*, 2670–2684.
4. J. P. Schaefer, J. J. Bloomfield, *Org. React.* **1967**, *15*, 1–203.
5. M. Hamell, R. Levine, *J. Org. Chem.* **1950**, *15*, 162–168.
6. J. I. Crowley, H. Rapoport, *J. Org. Chem.* **1980**, *45*, 3215–3227.

Claisen Rearrangement

Rearrangement of allyl vinyl ethers or allyl aryl ethers

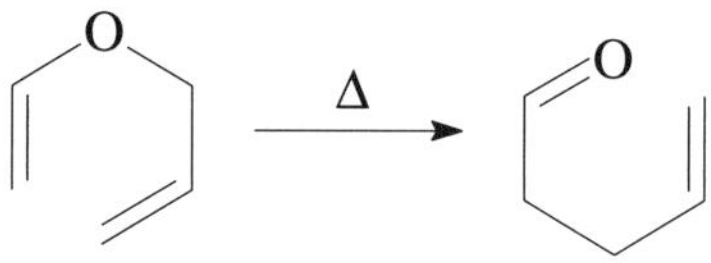

1

The *Claisen rearrangement*[1-3] is a thermal rearrangement of allyl aryl ethers and allyl vinyl ethers respectively. It may be regarded as the *oxa*-version of the closely related *Cope rearrangement*. Claisen has discovered this reaction first on allyl vinyl ethers **1**, and then extended to the rearrangement of allyl aryl ethers[1] **2** to yield *o*-allylphenols **3**:

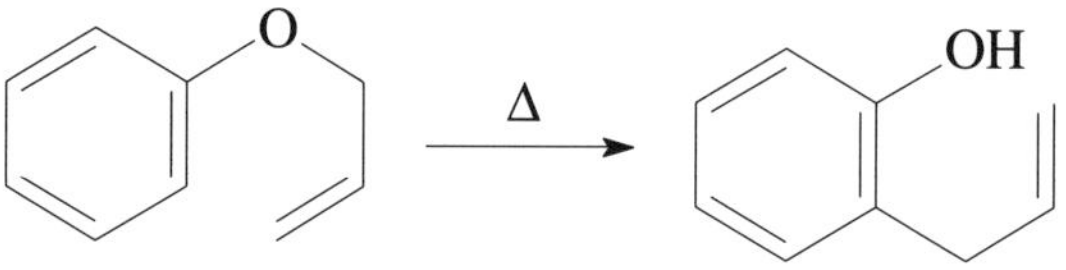

2   **3**

Mechanistically this reaction is described as a concerted pericyclic [3,3] sigmatropic rearrangement. A carbon–oxygen bond is cleaved and a carbon–carbon bond is formed. In a subsequent step the initial product **4** tautomerizes to the stable aromatic allylphenol **3**:

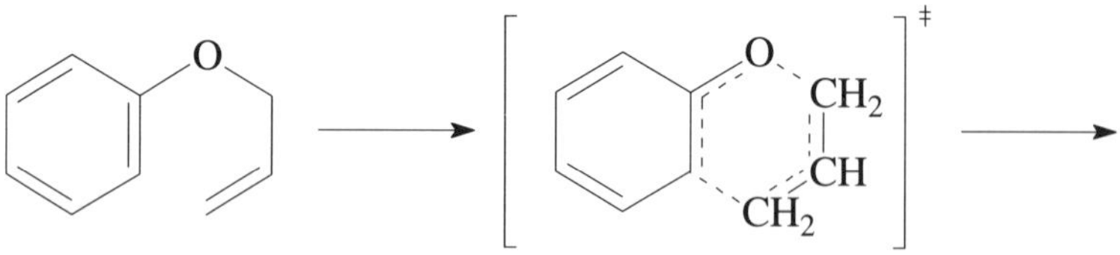

2

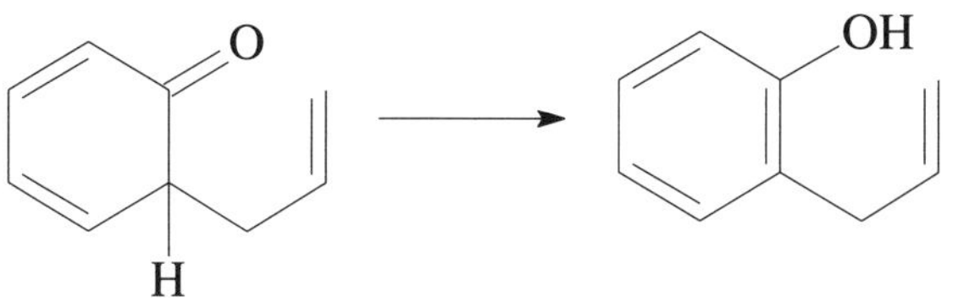

4   3

If both *ortho* positions bear substituents other than hydrogen, the allyl group will further migrate to the *para* position. This reaction is called the *para-Claisen rearrangement*. The formation of the *para*-substituted phenol can be explained by an initial Claisen rearrangement to an *ortho*-allyl intermediate which cannot tautomerize to an aromatic *o*-allylphenol, followed by a *Cope rearrangement* to the *p*-allyl intermediate which can tautomerize to the *p*-allylphenol; e.g. 6:

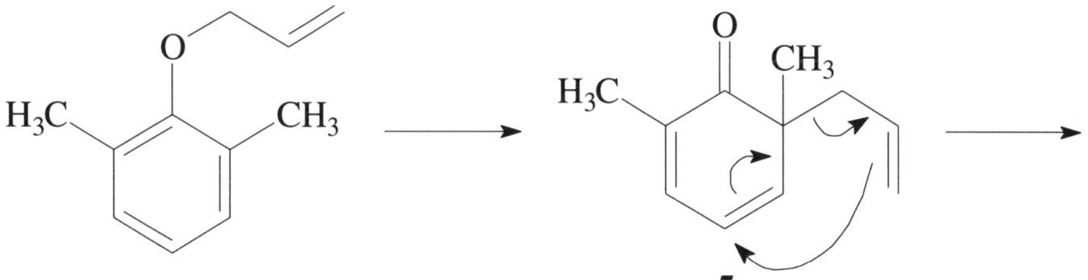

5

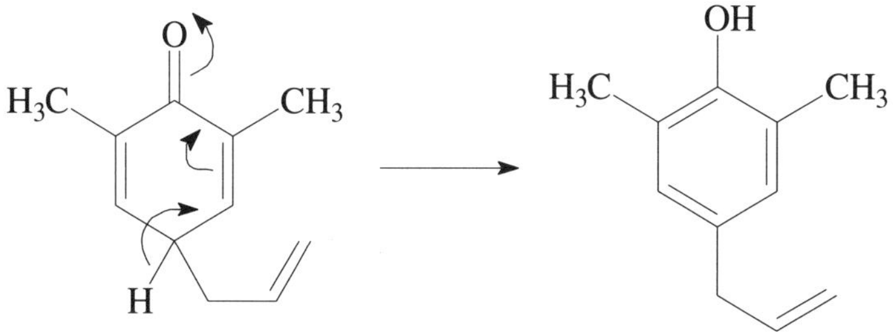

6

Compound 5 can be trapped through a *Diels–Alder reaction* with maleinic anhydride and thus be shown to be an intermediate. Further evidence for a mechanism involving two subsequent allyl conversions has been provided by experiments with ^{14}C-labeled substrates.

If both *ortho*-positions as well as the *para*-position bear substituents other than hydrogen, no Claisen rearrangement product is obtained.

The stereochemical outcome of the reaction is determined by the geometry of the transition state; for the Claisen rearrangement a chairlike conformation is preferred,[4,5] and it proceeds strictly by an intramolecular pathway. It is therefore possible to predict the stereochemical course of the reaction, and thus the configuration of the stereogenic centers to be generated. This potential can be used for the planning of stereoselective syntheses; e.g the synthesis of natural products.[4,6,7]

In order to broaden the scope of the Claisen rearrangement, various analogous reactions have been developed. For instance the *Amino–Claisen rearrangement*,[2] where the oxygen is replaced by a NR group, and the rearrangement of propargyl ethers **7** gives access to penta-3,4-dienals[8] **8**:

7   **8**

An important variant is the rearrangement of silylketene acetals like **10** and **11** which are easily accessible from allyl esters **9**. This so-called *Ireland–Claisen rearrangement*[9–11] is a valuable carbon–carbon bond forming reaction that takes advantage of the fact that the reactants are first connected to each other by an ester linkage as in allyl esters **9**, that are easy prepare.

By proper choice of the reaction conditions, the configuration of the enolate and the silylketene acetal derived thereof can be controlled to a great extent. The stereochemical course of the rearrangement then leads to the conversion of two stereogenic sp[2] centers of specific geometry into stereogenic sp[3] centers with the desired relative configuration predominating. Upon deprotonation of the allyl ester **9** by lithium diisopropylamide (LDA) in tetrahydrofuran solution the *E*-enolate is formed predominantly, and can be silylated (e.g. with *tert*-butyldimethylsilyl chloride TBDMSCl) to give the *E*-configured silylketene acetal **10**. On the other hand in tetrahydrofuran solution containing 23% hexamethylphosphoric triamide (HMPA) the *Z*-configured derivative **11** is formed predominantly. The *E*-derivative **10** upon rearrangement and hydrolytic cleavage of the silyl ester yields the *erythro*-γ,δ-unsaturated carboxylic acid **12**, while the *Z*-derivative **11** yields the corrresponding *threo* product **13**:

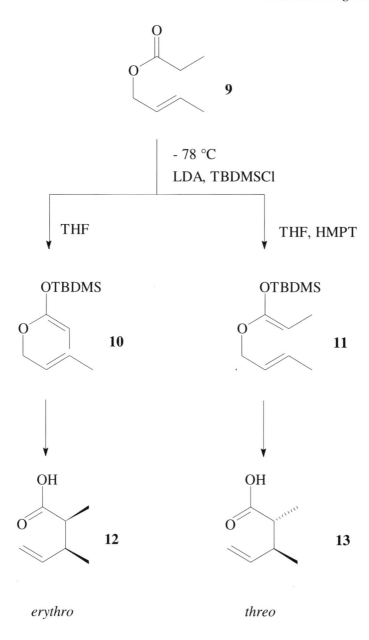

erythro   threo

By employing optically active enol borinates instead of silylketene acetals, the Ireland–Claisen rearrangement has been further developed to an enantioselective reaction.[12]

1. L. Claisen, *Ber. Dtsch. Chem. Ges.* **1912**, *45*, 3157–3166.
2. D. S. Tarbell, *Org. React.* **1944**, *2*, 1–48;
    S. J. Rhoads, N. R. Raulins, *Org. React.* **1975**, *22*, 1–252.
3. B. Ganem, *Angew. Chem.* **1996**, *108*, 1014–1023; *Angew. Chem. Int. Ed. Engl.* **1996**, *35*, 936–945:
    J. J. Gajewski, *Acc. Chem. Res.* **1997**, *30*, 219–225.
4. F. E. Ziegler, *Chem. Rev.* **1988**, *88*, 1423–1452.
5. A. Wunderli, T. Winkler, H.-J. Hansen, *Helv. Chim. Acta* **1977**, *60*, 2436–2459.
6. Y. Hirano, C. Djerassi, *J. Org. Chem.* **1982**, *47*, 2420–2426.
7. S. D. Burke, G. J. Pacofsky, *Tetrahedron Lett.* **1986**, *27*, 445–448.
8. A. Viola, J. J. Collins, N. Filipp, *Tetrahedron* **1981**, *37*, 3785–3791.
9. R. E. Ireland, R. H. Mueller, *J. Am. Chem. Soc.* **1972**, *94*, 5897–5898.
10. R. E. Ireland, R. H. Mueller, A. K. Willard, *J. Am. Chem. Soc.* **1976**, *98*, 2868–2877.
11. A. G. Cameron, D. W. Knight, *J. Chem. Soc., Perkin Trans. 1*, **1986**, 161–167.
12. E. J. Corey, D.-H. Lee, *J. Am. Chem. Soc.* **1991**, *113*, 4026–4028.

Clemmensen Reduction

Reduction of aldehydes and ketones to methylene compounds

By application of the *Clemmensen reduction*,[1,2] aldehydes and ketones **1** can be converted into the corresponding hydrocarbons **2**. As the reducing agent zinc amalgam, together with concentrated hydrochloric acid or gaseous hydrogen chloride, is used.

The reactions of various substrates under various reaction conditions cannot be rationalized by one general mechanism. Alcohols are not considered to be intermediates, since these generally are not reduced under Clemmensen conditions, as has been demonstrated with independently prepared alcohols corresponding to the carbonyl substrates.[3]

The Clemmensen reduction can be formulated to proceed by a sequence of one-electron and proton transfer reactions. It is a heterogenous reaction, taking place at the zinc surface. Initially an electron is transferred from zinc to the carbonyl group of ketone **1**, leading to a radical species **3**, which is presumed to react further to a zinc-carbenoid species[3] **4**:

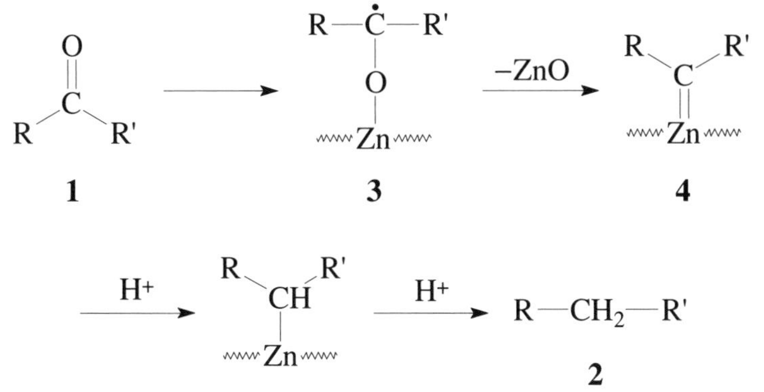

Upon subsequent addition of protons the methylene product **2** is formed.

Alternatively a mechanism involving a α-hydroxyalkylzinc chloride **5** has been formulated:[2]

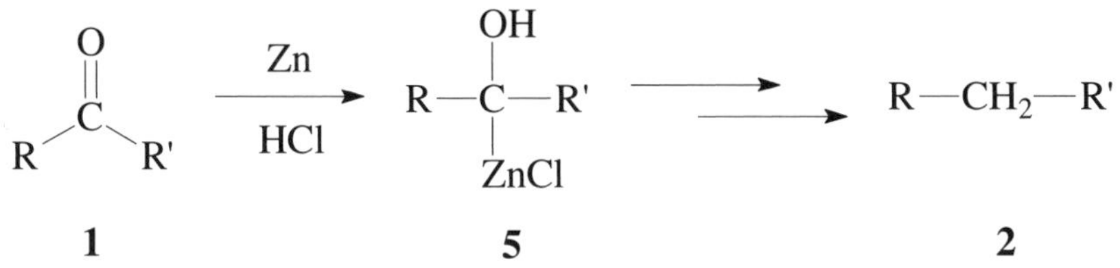

A modified procedure[4] uses activated zinc together with dry gaseous hydrogen chloride in an organic solvent, e.g. acetic acid, as reducing agent. Under those conditions the reaction occurs at lower temperatures as with the original procedure.

Another important synthetic method for the reduction of ketones and aldehydes to the corresponding methylene compounds is the *Wolff–Kishner reduction*. This reaction is carried out under basic conditions, and therefore can be applied for the reduction of acid-sensitive substrates; it can thus be regarded as a complementary method. The experimental procedure for the Clemmensen reduction is simpler: however for starting materials of high molecular weight the Wolff–Kishner reduction is more successful.

1. E. Clemmensen, *Ber. Dtsch. Chem. Ges.* **1913**, *46*, 1837–1843.
2. E. L. Martin, *Org. React.* **1942**, *1*, 155–209;
    E. Vedejs, *Org. React.* **1975**, *22*, 401–422.
3. J. Burdon, R. C. Price, *J. Chem. Soc., Chem. Commun.* **1986**, 893–894.
4. M. L. DiVona, V. Rosnati, *J. Org. Chem.* **1991**, *56*, 4269–4273.

Cope Elimination Reaction

Olefins from amine oxides

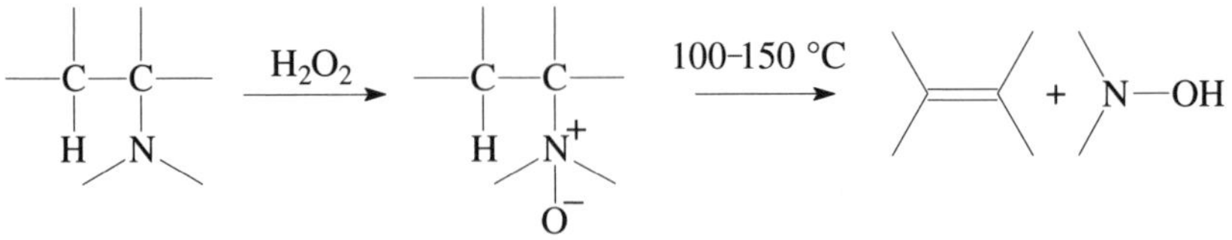

1   **2**   **3**   **4**

Amine oxides **2**, which can be prepared by oxidation of amines **1**, react upon heating to yield an olefin **3** and a hydroxylamine **4**. This reaction is called the *Cope elimination reaction*,[1-3] and as a synthetic method is a valuable alternative to the *Hofmann degradation reaction* of quaternary ammonium salts.

Similar to the *Ester pyrolysis* the mechanism is formulated to proceed by a E_i-mechanism; however in this case *via* a five-membered transition state **5**:

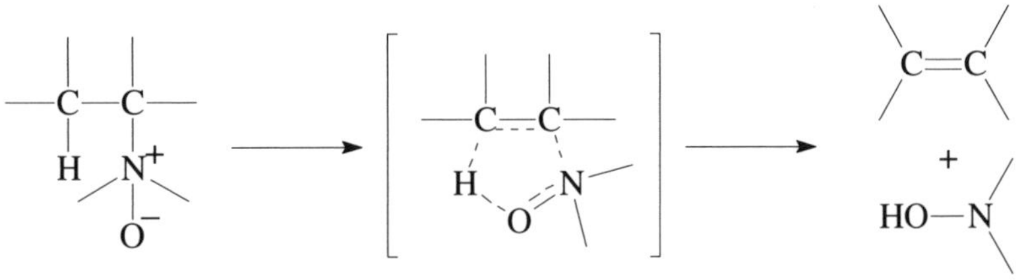

2   **5**

The mechanism is supported by findings from the decomposition reaction of the amine oxides derived from *threo-* and *erythro-2*-amino-3-phenylbutane. The *threo*-amine oxide **6** yields *E*-2-phenylbut-2-ene **7** with a selectivity of 400 : 1, and the *erythro*-derivative **8** yields the *Z*-olefin **9** with a selectivity of 20 : 1:

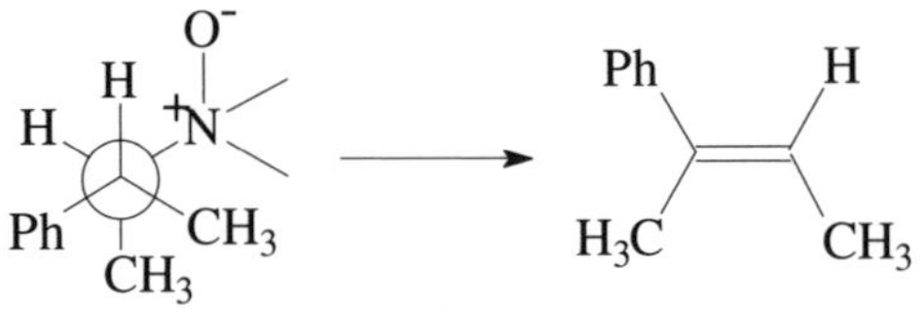

6   **7**

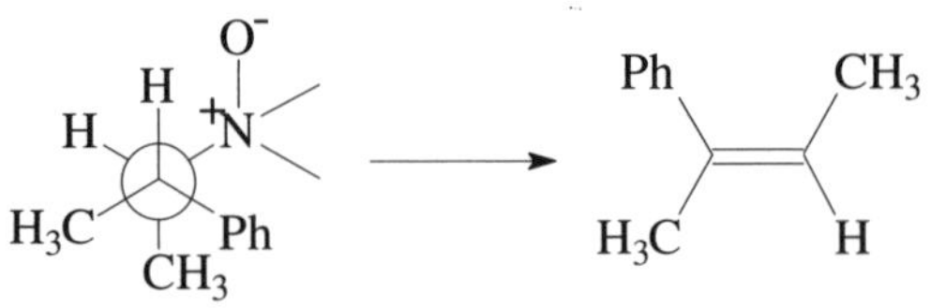

8   **9**

The higher selectivity observed with the *threo*-compound has been rationalized to arise from less steric hindrance in the five-membered transition state.

For the regioselectivity similar rules as for the ester pyrolysis do apply. With simple, alkylsubstituted amine oxides a statistical mixture of regioisomeric olefins is obtained. On the other hand with cycloalkyl amine oxides the regioselectivity is determined by the ability to pass through a planar, five-membered transition state. This has been demonstrated for the elimination reaction of menthyl dimethylamine oxide **10** and neomenthyl dimethylamine oxide **11**:

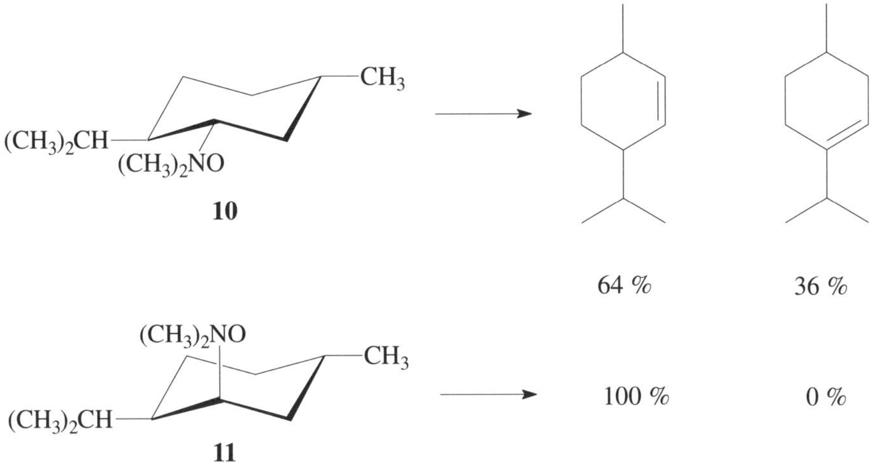

64 %   36 %

100 %   0 %

Certain amine oxides, especially those derived from six-membered heterocyclic amines e.g. N-methylpiperidine oxide, that cannot go through a planar, five-membered transition state, do not undergo the Cope elimination reaction.

In general side reactions are rare. In a few cases an isomerization by shift of the double bond favored by formation of a conjugated system can be observed:

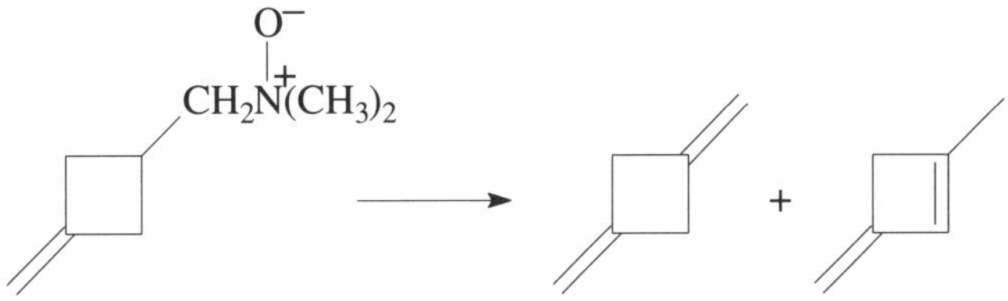

Furthermore the formation of *O*-substituted hydroxylamines **12**, e.g. by migration of an allyl or benzyl substituent, is possible:

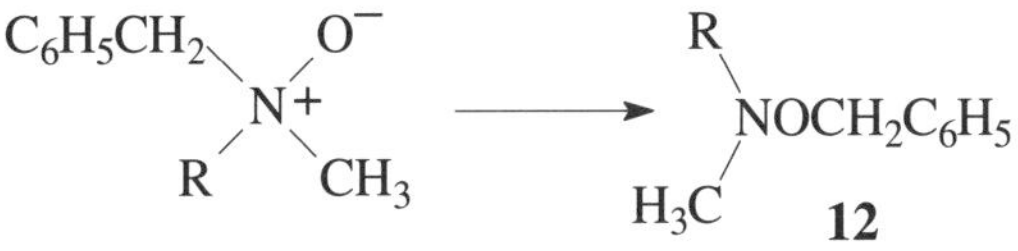

In addition to being a valuable method for the preparation of olefins, the Cope elimination reaction also gives access to *N,N*-disubstituted hydroxylamines.

1. A. C. Cope, T. T. Foster, P. H. Towle, *J. Am. Chem. Soc.* **1949**, *71*, 3929–3935.
2. A. C. Cope, E. R. Trumbull, *Org. React.* **1960**, *11*, 317–493.
3. C. H. DePuy, R. W. King, *Chem. Rev.* **1960**, *60*, 431–457.

Cope Rearrangement

Isomerization of 1,5-dienes

1   **2**

The thermal rearrangement of 1,5-dienes **1** to yield the isomeric 1,5-dienes **2**, is called the *Cope rearrangement*[1,2]—not to be confused with the thermolysis of amine oxides, which is also named after *Arthur C. Cope*.

This reaction proceeds by a concerted, [3,3] sigmatropic rearrangement (cf. the *Claisen rearrangement*) where one carbon–carbon single bond breaks, while the new one is formed. It is a reversible reaction; the thermodynamically more stable isomer is formed preferentially:

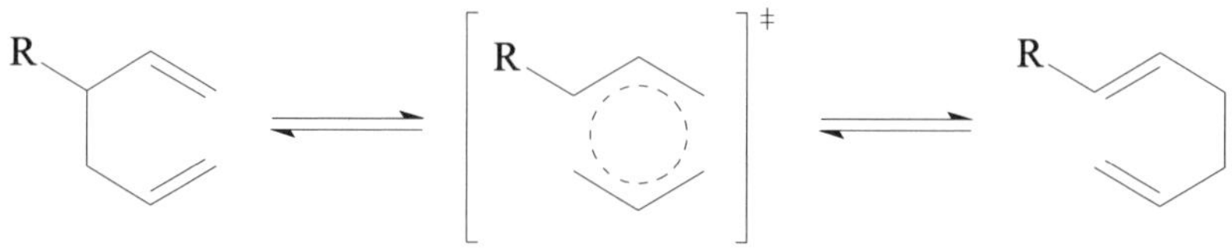

1   **3**   **2**

The diene passes through a six-membered cyclic transition state **3**; preferentially of chair-like conformational geometry:[3]

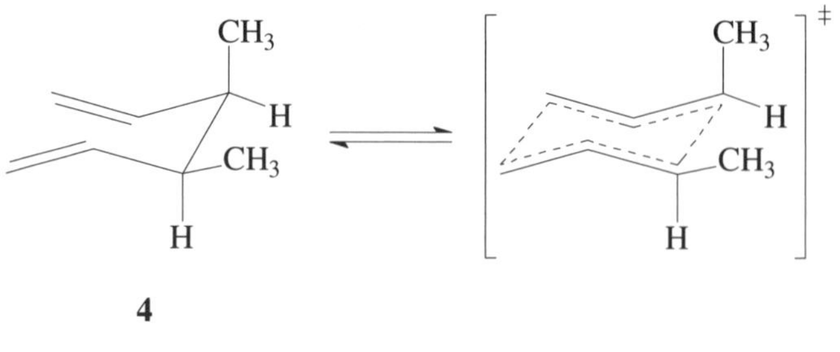

4

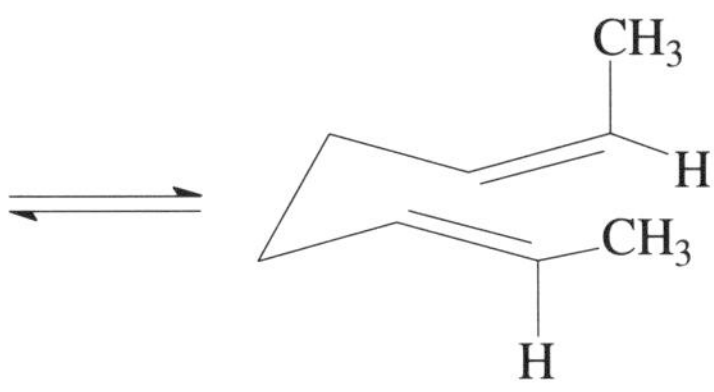

5

Evidence for that stereochemical course comes from the rearrangement of *meso*-3,4-dimethylhexa-1,5-diene **4**, which yields the *E,Z*-configured diene **5** almost quantitatively. With a transition state of boatlike geometry, a *Z,Z*- or *E,E*-configured product would be formed.[4]

With certain donor substituents at C-3 the experimental findings may be rationalized rather by a diradical mechanism,[5] where formation of the new carbon–carbon single bond leads to a diradical species **6**, which further reacts by bond cleavage to give the diene **2**:

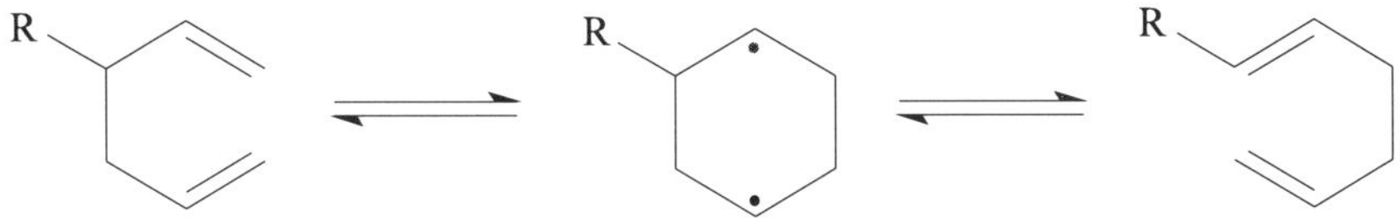

| **1** | **6** | **2** |

The Cope rearrangement of hexa-1,5-diene does not allow for differentiation of starting material and product; this is called a degenerate Cope rearrangement. Another example is the *automerization* of bicyclo[5,1,0]octa-2,5-diene **7**:

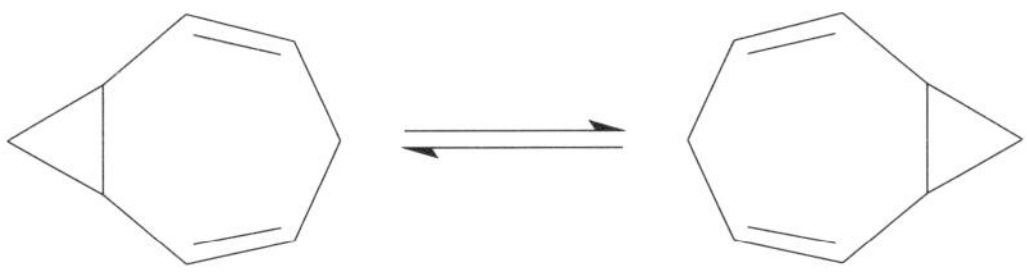

7   **7**

The required temperatures for the Cope rearrangement are generally lower, if the starting material contains a substituent at C-3 or C-4 which can form a conjugated system with one of the new double bonds.

Starting from a 3-hydroxy-1,5-diene **8**, the rearrangement is not reversible, because the Cope product **9** tautomerizes to an aldehyde or ketone **10**, and is thereby removed from equilibrium:

      8   **9**   **10**

This variant is called the *oxy-Cope rearrangement*[6]. The rate of the Cope rearrangement is strongly accelerated with an oxy-anion substituent at C-3;[6] i.e. by use of the corresponding alkoxide instead of alcohol **8**. This *anionic oxy-Cope rearrangement* is especially fast with the potassium alkoxide derivative in the presence of the ionophor 18-crown-6. Appropriate starting materials containing nitrogen or sulfur can also undergo Cope rearrangements;[7] a clear definition as a Cope or Claisen rearrangement then may sometimes be difficult.

In some cases the rearrangement can be catalyzed by transition metal compounds,[7] and thus caused to take place at room temperature. The ordinary, uncatalyzed rearrangement requires temperatures in the range of 150–250 °C.

        11   **12**

The Cope rearrangement is of great importance as a synthetic method;[6] e.g. for the construction of seven- and eight-membered carbocycles from 1,2-divinylcyclopropanes and 1,2-divinylcyclobutanes respectively (e.g. **11** → **12**), and has found wide application in the synthesis of natural products. The second step of the *para-Claisen rearrangement* is also a Cope rearrangement reaction.

1.   A. C. Cope, E. M. Hardy, *J. Am. Chem. Soc.* **1940**, *62*, 441–444.
2.   S. J. Rhoads, N. R. Raulins, *Org. React.* **1975**, *22*, 1–252.
3.   R. Hoffmann, R. B. Woodward, *J. Am. Chem. Soc.* **1965**, *87*, 4389–4390.

4. W. v. E. Doering, W. R. Roth, *Tetrahedron* **1962**, *18*, 67–74.
5. M. Dollinger, W. Henning, W. Kirmse, *Chem. Ber.* **1982**, *115*, 2309–2325.
6. D. A. Evans, A. M. Golob, *J. Am. Chem. Soc.* **1975**, *97*, 4765–4766;
   L. A. Paquette, *Angew. Chem.* **1990**, *102*, 642–660; *Angew. Chem. Int. Ed. Engl.* **1990**, *29*, 609.
7. R. P. Lutz, *Chem. Rev.* **1984**, *84*, 205–247.

Corey–Winter Fragmentation

Olefins from vicinal diols

By application of the *Corey–Winter reaction*,[1,2] vicinal diols **1** can be converted into olefins **3**. The key step is the cleavage of cyclic thionocarbonates **2** (1,3-dioxolanyl-2-thiones) upon treatment with trivalent phosphorus compounds. The required cyclic thionocarbonate **2** can be prepared from a 1,2-diol **1** and thiophosgene **4** in the presence of 4-dimethylaminopyridine (DMAP):

In addition there are certain other methods for the preparation such compounds.[2]
Upon heating of the thionocarbonate **2** with a trivalent phosphorus compound e.g. trimethyl phosphite, a *syn*-elimination reaction takes place to yield the olefin **3**. A nucleophilic addition of the phosphorus to sulfur leads to the zwitterionic species **6**, which is likely to react to the phosphorus ylide **7** *via* cyclization and subsequent desulfurization. An alternative pathway for the formation of **7** *via* a 2-carbena-1,3-dioxolane **8** has been formulated. From the ylide **7** the olefin **3** is formed stereospecifically by a concerted 1,3-dipolar cycloreversion (see *1,3-dipolar cycloaddition*), together with the unstable phosphorus compound **9**, which decomposes into carbon dioxide and R_3P. The latter is finally obtained as R_3PS:

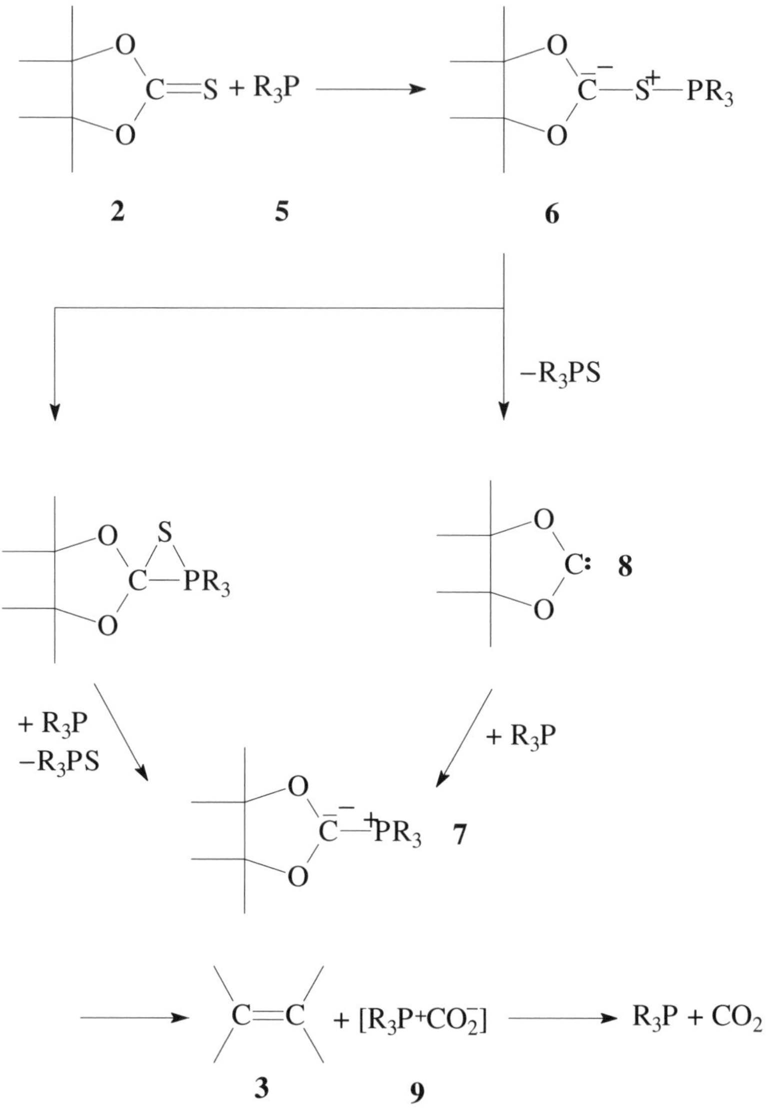

The Corey–Winter reaction provides a useful method for the preparation of olefins that are not accessible by other routes. For instance it may be used for the synthesis of sterically crowded targets, since the initial attack of phosphorus at the sulfur takes place quite distantly from sterically demanding groups that might be present in the substrate molecule. Moreover the required vicinal diols are easily accessible, e.g. by the carbon–carbon bond forming *acyloin ester condensation* followed by a reductive step. By such a route the twistene **10** has been synthesized:[3]

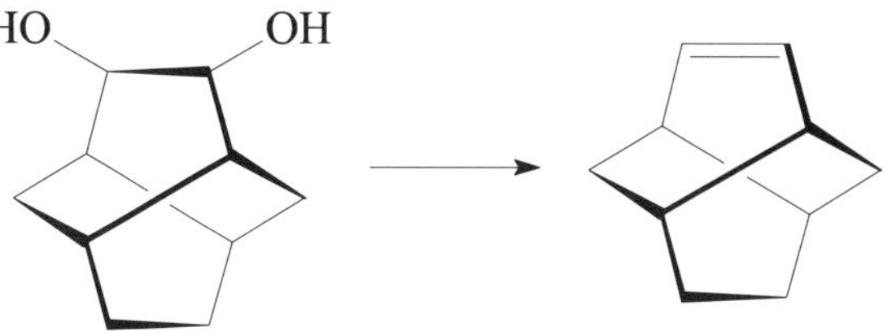

10

Furthermore highly strained compounds such as bicyclo[3.2.1]oct-1-ene **11**, containing a double bond to a bridgehead carbon atom, have been prepared; however this strained olefin could be identified only as its Diels–Alder product from subsequent reaction with an added diene.[4]

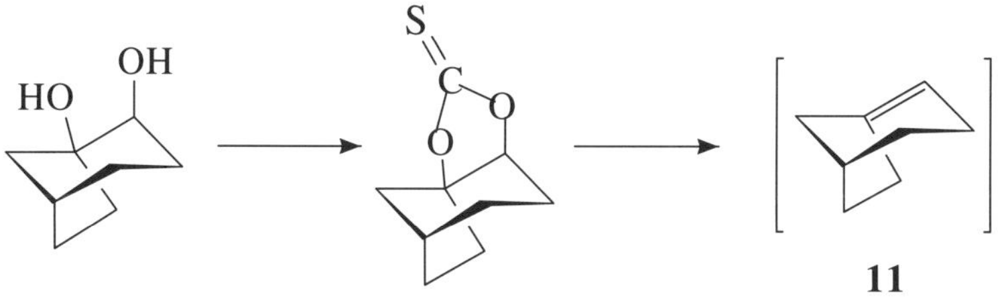

11

According to *Bredt's rule* such olefins of small ring size are unstable; ordinary elimination reactions usually yield an isomeric olefin where a bridgehead carbon does not participate in the double bond.

1. E. J. Corey, R. A. E. Winter, *J. Am. Chem. Soc.* **1963**, *85*, 2677–2678.
2. E. Block, *Org. React.* **1984**, *30*, 457–566.
3. M. Tichy, J. Sicher, *Tetrahedron Lett.* **1969**, 4609–4613.
4. J. A. Chong, J. R. Wiseman, *J. Am. Chem. Soc.* **1972**, *94*, 8627–8629.

Curtius Reaction

Isocyanates from acyl azides

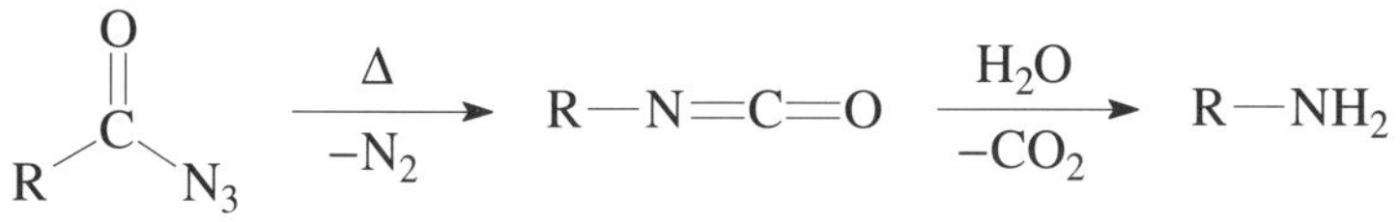

The thermal decomposition of an acyl azide **1** to yield an isocyanate **2** by loss of N_2, is called the *Curtius reaction*[1,2] or *Curtius rearrangement*. It is closely related to the *Lossen reaction* as well as the *Hofmann rearrangement*, and like these allows for the preparation of amines **3** *via* intermediate formation of an isocyanate. The Curtius reaction can thus be applied to convert carboxylic acids into primary amines.

The required acyl azide **1** can be prepared from the corresponding acyl chloride **4** and azide ion (e.g. with sodium azide) or alternatively from an acylhydrazine **5** by treatment with nitrous acid:

4   **1**

5   **1**

Loss of N_2 and migration of the group R is likely to be a concerted process,[3,4] since evidence for a free acyl nitrene RCON in the thermal reaction has not been found:[4]

1   **2**

The Curtius rearrangement can be catalyzed by Lewis acids or protic acids, but good yields are often obtained also without a catalyst. From reaction in an inert solvent (e.g. benzene, chloroform) in the absence of water, the isocyanate can be isolated, while in aqueous solution the amine is formed. Highly reactive acyl azides may suffer loss of nitrogen and rearrange already during preparation in aqueous solution. The isocyanate then cannot be isolated because it immediately reacts with water to yield the corresponding amine.

An isocyanate **2** formed by a Curtius rearrangement can undergo various subsequent reactions, depending on the reaction conditions. In aqueous solution the isocyanate reacts with water to give a carbaminic acid **6**, which immediately decarboxylates to yield an amine **3**. When alcohol is used as solvent, the isocyanate reacts to a carbamate **7**:

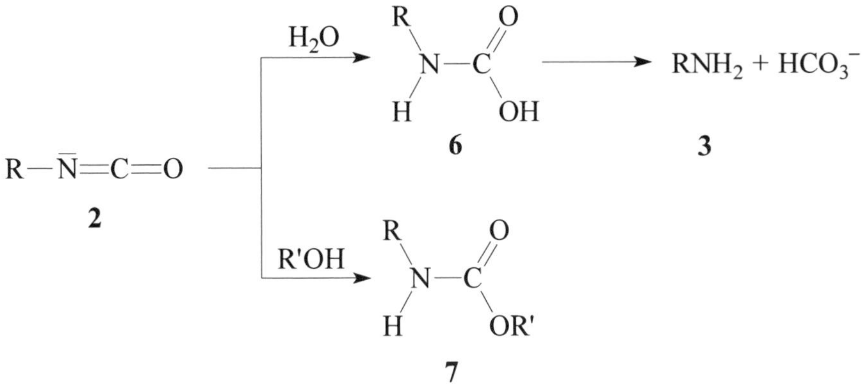

Acyl azides can undergo photolytic cleavage and rearrangement upon irradiation at room temperature or below. In that case acyl nitrenes **8** have been identified by trapping reactions and might be reactive intermediates in the *photo Curtius rearrangement*. However there is also evidence that the formation of isocyanates upon irradiation proceeds by a concerted reaction as in the case of the thermal procedure, and that the acyl nitrenes are formed by an alternative and competing pathway:[3,4]

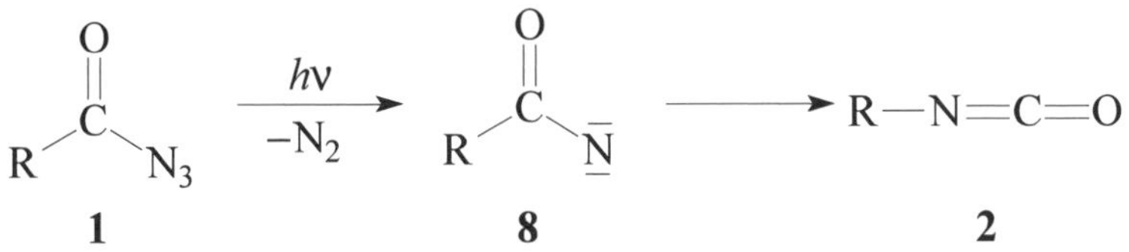

The Curtius rearrangement is a useful method for the preparation of isocyanates as well as of products derived thereof.[5] The substituent R can be alkyl, cycloalkyl, aryl, a heterocyclic or unsaturated group; most functional groups do not interfere.

1. T. Curtius, *Ber. Dtsch. Chem. Ges.* **1890**, *23*, 3023–3041.
2. P. A. S. Smith, *Org. React.* **1946**, *3*, 337–449.
3. A. Rauk, P. Alewood, *Can. J. Chem.* **1977**, *55*, 1498–1510.
4. W. Lwowski, *Angew. Chem.* **1967**, *79*, 922–931; *Angew. Chem. Int. Ed. Engl.* **1967**, *6*, 897;
    W. Lwowski in: *Azides and Nitrenes* (Ed.: E. F. V. Scriven), Academic Press, Orlando, **1984**, p. 205–246.
5. N. A. LeBel, R. M. Cherluck, E. A. Curtis, *Synthesis* **1973**, 678–679.

1,3-Dipolar Cycloaddition

Five-membered heterocycles through a cycloaddition reaction

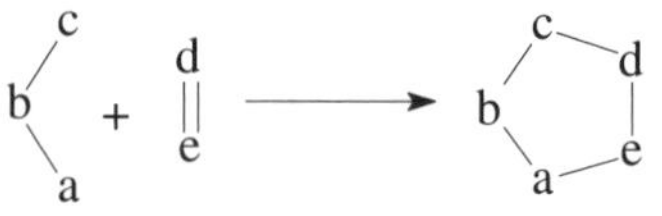

1   **2**   **3**

Huisgen[1,2] has reported in 1963 about a systematic treatment of the *1,3-dipolar cycloaddition reaction*[3–5] as a general principle for the construction of five-membered heterocycles. This reaction is the addition of a 1,3-dipolar species **1** to a multiple bond, e. g. a double bond **2**; the resulting product is a heterocyclic compound **3**. The 1,3-dipolar species can consist of carbon, nitrogen and oxygen atoms (seldom sulfur) in various combinations, and has four non-dienic π-electrons. The 1,3-dipolar cycloaddition is thus a $4\pi + 2\pi$ cycloaddition reaction, as is the *Diels–Alder reaction.*

Mechanistically the 1,3-dipolar cycloaddition reaction very likely is a concerted one-step process *via* a cyclic transition state. The transition state is less symmetric and more polar as for a Diels–Alder reaction; however the symmetry of the frontier orbitals is similar. In order to describe the bonding of the 1,3-dipolar compound, e.g. diazomethane **4**, several Lewis structures can be drawn that are resonance structures:

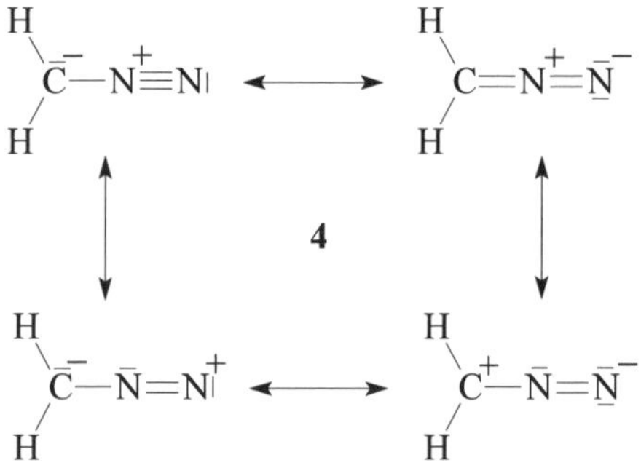

4

The cycloaddition reaction of diazomethane **4** and an olefin, e.g. methyl acrylate **5**, leads to a dihydropyrazole derivative **6**:

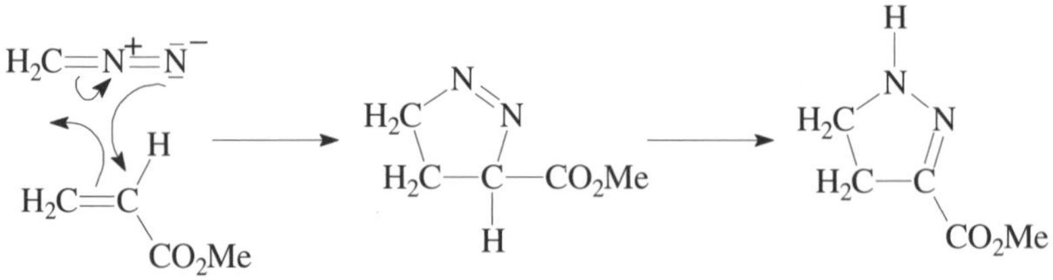

5   **6**

The shifting of electrons as shown in the scheme should be taken as a simplified depiction only. A more thorough understanding follows from consideration of the frontier orbitals and their coefficients;[6] this may then permit a prediction of the regiochemical course of the cycloaddition.

A well-known example for a 1,3-dipolar compound is ozone. The reaction of ozone with an olefin is a 1,3-dipolar cycloaddition (see *ozonolysis*).

Further examples of 1,3-dipolar compounds:[4]

Diazoalkanes	$\diagdown \!\!\! \overset{+}{\underset{\diagup}{C}} \!-\! \overline{N} \!=\! \overline{\underline{N}}^{-}$
Azides	$-\overset{+}{\underline{N}} \!-\! \underline{N} \!=\! \overline{\underline{N}}^{-}$
Nitriloxides	$-\overset{+}{C} \!\equiv\! N \!-\! \overline{\underline{O}}|^{-}$
Azomethinylides	$R_2\overset{-}{\underline{C}} \!-\! \overline{N} \!-\! \overset{+}{C}R''_2$
	$\qquad\quad |\ R'$
Ozone	$|\overline{\underline{O}}^{-} \!-\! \overline{O} \!-\! \overline{O}^{+}$
Nitrones	$|\overline{\underline{O}}^{-} \!-\! \overline{N} \!-\! \overset{+}{C}R'_2$
	$\qquad\quad |\ R$

Dipolar compounds often are highly reactive, and therefore have to be generated *in situ*.

The 2π component **2**, the so-called *dipolarophile* (analogously to the dienophile of the Diels–Alder reaction) can be an alkene or alkyne or a heteroatom derivative thereof. Generally those substrates will be reactive as dipolarophiles, that also are good dienophiles.

An interesting perspective for synthesis is offered by the reaction sequence cycloaddition/cycloreversion.[7,8] It often does not lead to the initial reactants, but to a different pair of dipole and dipolarophile instead:

By analogy to the *alkene metathesis*, this reaction sequence is called *1,3-dipol metathesis*.

Strained bicyclic compounds can be obtained e.g. when cyclopropenes are used as dipolarophiles. Reaction of 3,3-dimethylcyclopropene **7** with diazomethane **4** gives the heterobicyclic cycloaddition product **8** in 85% yield:[9]

7   4   8

The importance of the 1,3-dipolar cycloaddition reaction for the synthesis of five-membered heterocycles arises from the many possible dipole/dipolarophile combinations. Five-membered heterocycles are often found as structural subunits of natural products. Furthermore an intramolecular variant[10] makes possible the formation of more complex structures from relatively simple starting materials. For example the tricyclic compound **10** is formed from **9** by an intramolecular cycloaddition in 80% yield:[11]

9   10

In many cases a change of the solvent has little effect on the 1,3-dipolar cycloaddition; but similar to the Diels–Alder reaction, the intermolecular 1,3-dipolar cycloaddition exhibits a large negative volume of activation, and therefore a rate enhancement is often observed on application of high pressure.[12] Inert solvents such as benzene, toluene or xylene are often used, or even no solvent at all. Depending on the reactivity of the starting materials the further reaction conditions can range from a few minutes at room temperature to several days at 100 °C or higher.

1. R. Huisgen, *Angew. Chem.* **1963**, *75*, 604–637; *Angew. Chem. Int. Ed. Engl.* **1963**, *14*, 565.
2. R. Huisgen, *Angew. Chem.* **1963**, *75*, 742–754; *Angew. Chem. Int. Ed. Engl.* **1963**, *14*, 633.
3. R. Huisgen in *1,3-Dipolar Cycloaddition Chemistry*, (Ed.: A. Padwa), Wiley, New York, **1984**, *Vol. 1*, p. 1–176.

4. W. Carruthers, *Cycloaddition Reactions in Organic Synthesis*, Pergamon Press, Oxford, **1990**, p. 269–331.
5. P. N. Confalone, E. M. Huie, *Org. React.* **1988**, *36*, 1–173.
6. I. Fleming, *Frontier Orbitals and Organic Chemical Reactions*, Wiley, London, **1976**, p. 148–161.
7. B. Stanovnik, *Tetrahedron* **1991**, *47*, 2925–2945.
8. G. Bianchi, C. De Micheli, R. Gandolfi, *Angew. Chem.* **1979**, *91*, 781–798; *Angew. Chem. Int. Ed. Engl.* **1979**, *18*, 673.
9. M. L. Deem, *Synthesis* **1982**, 701–716.
10. A. Padwa, *Angew. Chem.* **1976**, *88*, 131–144; *Angew. Chem. Int. Ed. Engl.* **1976**, *15*, 123.
11. W. Kirmse, H. Dietrich, *Chem. Ber.* **1967**, *100*, 2710–2718.
12. K. Matsumoto, A. Sera, *Synthesis* **1985**, 999–1027.

[2 +2] Cycloaddition

Photochemical dimerization of alkenes

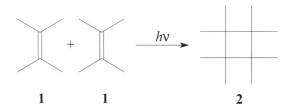

        1   **1**   **2**

According to the *Woodward–Hofmann rules*[1] the thermal $[2\pi + 2\pi]$ cycloaddition reaction of alkenes **1** in a suprafacial manner is symmetry-forbidden, and is observed in special cases only. In contrast the photochemical $[2\pi + 2\pi]$ cycloaddition is symmetry-allowed, and is a useful method for the synthesis of cyclobutane derivatives[2,3] **2**.

For the mechanistic course, two pathways have to be considered: the direct activation of an alkene through absorption of light upon irradiation and the activation mediated through a photosensitizer. For simple alkenes the former process can be difficult to realize experimentally, since these substrates absorb in the far UV (i.e. beyond 200 nm). A photosensitizer (e.g. an aldehyde or ketone) may then be added that absorbs light of higher (longer) wavelength than the alkene. A photosensitizer molecule in an excited state can transfer its excess energy to an alkene molecule which is thus brought to a triplet excited state—a diradical—while the sensitizer drops to its ground state. The excited alkene can react with a second alkene molecule by cycloaddition to yield the dimer.

When buta-1,3-diene **3** is irradiated in the presence of a *photosensitizer*[2] (e.g. benzophenone), the isomeric divinylcyclobutanes **6** and **7** are formed *via* the intermediate diradical species **4** and **5** respectively; in addition the [4 + 2] cycloaddition product 4-vinylcyclohexene (see *Diels–Alder reaction*) is obtained as a side product:

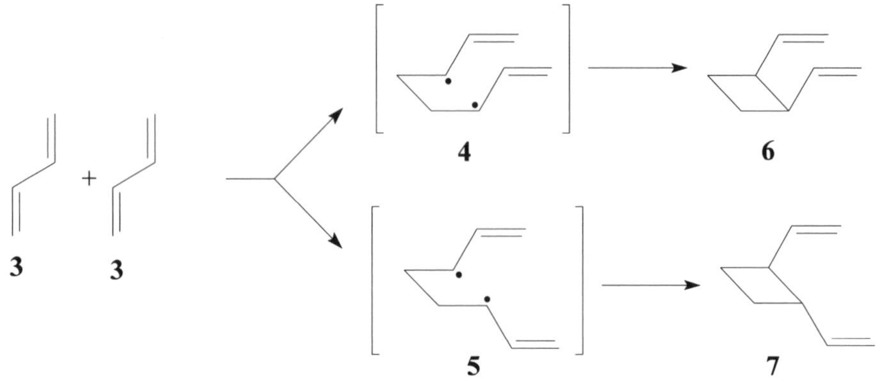

Irradiation of Z-but-2-ene **8** initiates a cyclodimerization reaction, even without a photosensitizer.[4] This cycloaddition proceeds from a singlet state and is likely to be a concerted, one-step reaction. Bond formation occurs suprafacial with respect to both reactants, whereupon only the tetramethylcyclobutanes **9** and **10** can be formed:

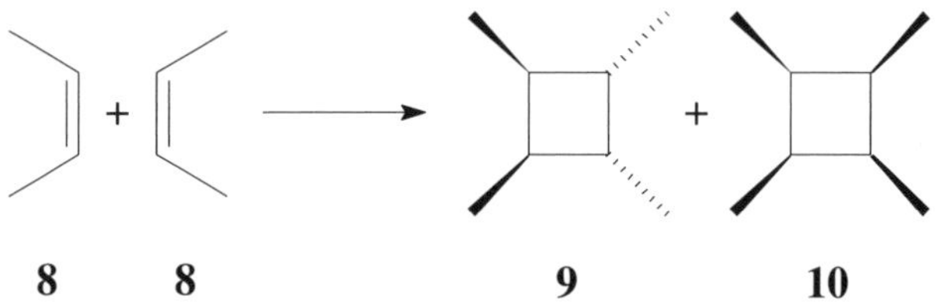

Two different alkenes can be brought to reaction to give a [2 + 2] cycloaddition product. If one of the reactants is an α,β-unsaturated ketone[5] **11**, this will be easier to bring to an excited state than an ordinary alkene or an enol ether e.g. **12**. Consequently the excited carbonyl compound reacts with the ground state enol ether. By a competing reaction pathway, the *Paterno–Büchi reaction* of the α,β-unsaturated ketone may lead to formation of an oxetane,[3] which however shall not be taken into account here:

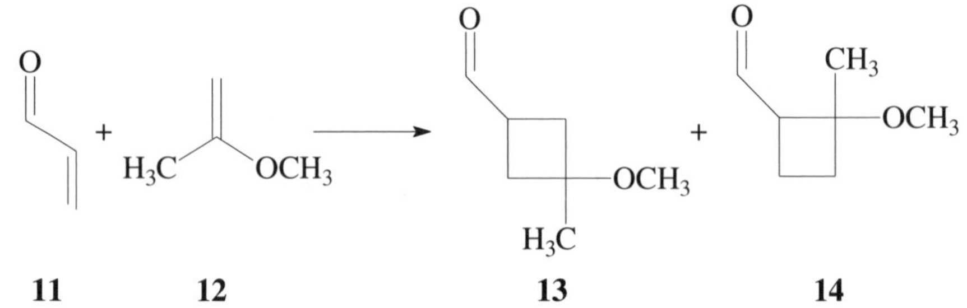

Regio- and stereoisomeric cycloaddition products might be formed in such reactions.[6] In this case the regioisomer **13** is formed as the major product in 98.5% yield.[5]

The intramolecular variant leads to formation of more than one ring; an interesting example is the formation of an intermediate in the synthesis of tetraasterane **16** by *Musso* and coworkers[8] from 3,6-dihydrophthalic anhydride **15** by two subsequent [2 + 2] cycloaddition reactions, an intermolecular step followed by an intramolecular one:

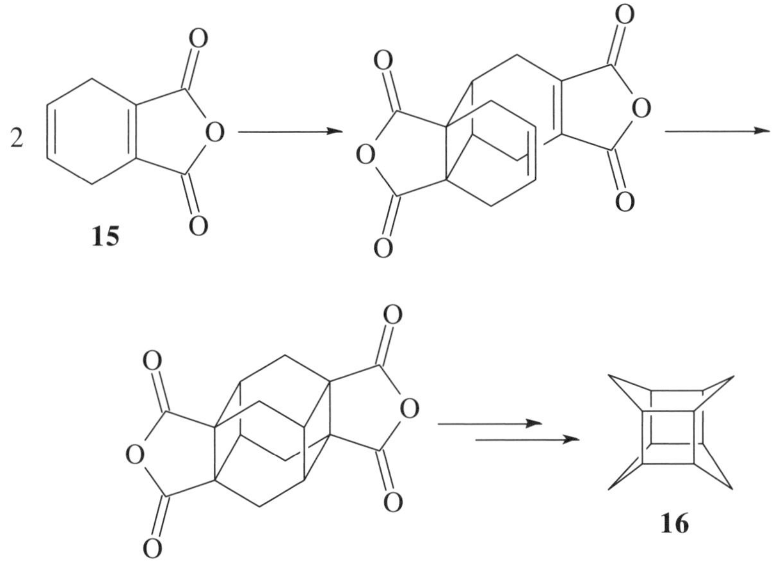

The thermal [2 + 2] cycloaddition[9] is limited to certain activated alkenes. For instance tetrafluoroethylene, tetrachloroethylene, allenes e.g. **17**, ketenes and enamines can form cyclic dimers or react with other alkenes:

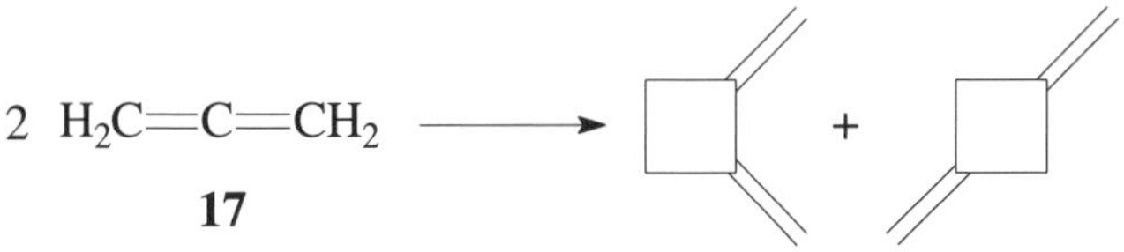

$$2\ H_2C{=}C{=}CH_2 \longrightarrow \qquad + \qquad$$

17

From a preparative point of view, the photochemical [2 + 2] cycloaddition is the most important of the photochemical reactions; especially the cycloaddition involving enones.[5] The [2 + 2] cycloaddition is the method of choice for the construction of cyclobutane derivatives as well as cyclobutane units within larger target molecules.

1. R. B. Woodward, R. Hoffmann, *Angew. Chem.* **1969**, *81*, 797–869; *Angew. Chem. Int. Ed. Engl.* **1969**, *8*, 781–853;
   R. B. Woodward, R. Hoffmann, *The Conservation of Orbital Symmetry*, Academic Press, New York, **1970**.
2. N. J. Turro, *Modern Molecular Photochemistry*, Benjamin/Cunnings Publishing Co., London, **1978**, p. 419–465.
3. J. Ninomiya, T. Naito, *Photochemical Synthesis*, Academic Press, New York, **1989**, p. 59–109.
4. Y. Yamazaki, R. J. Cvetanovic, *J. Am. Chem. Soc.* **1969**, *91*, 520–522.
5. M. Demuth, G. Mikhail, *Synthesis* **1989**, 145–162.
6. K. Y. Burstein, E. P. Serebryakov, *Tetrahedron* **1978**, *34*, 3233–3238.
7. M. T. Crimmins, *Chem. Rev.* **1988**, *88*, 1453–1473.
8. H.-G. Fritz, H.-M. Hutmacher, H. Musso, G. Ahlgren, B. Akermark, R. Karlsson, *Chem. Ber.* **1976**, *109*, 3781–3792.
9. J. D. Roberts, C. M. Shorts, *Org. React.* **1962**, *12*, 1–56.

D

Darzens Glycidic Ester Condensation

α,β-Epoxycarboxylic esters from aldehydes or ketones and α-halo esters

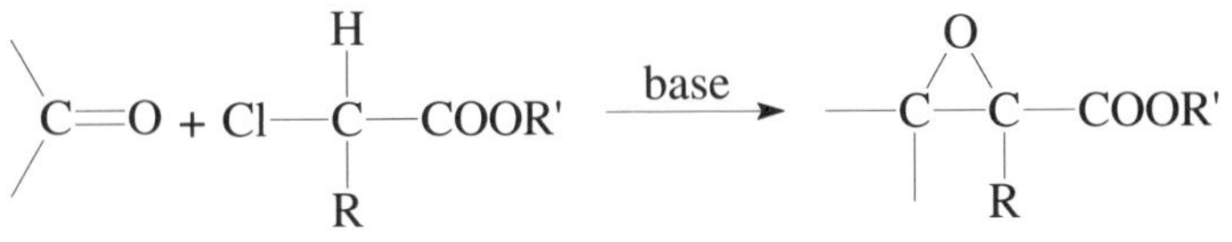

$$
\underset{\textbf{1}}{\diagdown\!\!\underset{\diagup}{C}\!\!=\!\!O} + \underset{\textbf{2}}{Cl\!-\!\overset{\overset{\displaystyle H}{|}}{\underset{\underset{\displaystyle R}{|}}{C}}\!-\!COOR'} \xrightarrow{\text{base}} \underset{\textbf{3}}{-\overset{\overset{\displaystyle O}{\diagup\diagdown}}{C}\!-\!\overset{}{\underset{\underset{\displaystyle R}{|}}{C}}\!-\!COOR'}
$$

An α,β-epoxycarboxylic ester (also called *glycidic ester*) **3** is formed upon reaction of a α-halo ester **2** with an aldehyde or ketone **1** in the presence of a base such as sodium ethoxide or sodium amide.[1-3] Mechanistically it is a Knoevenagel-type reaction of the aldehyde or ketone **1** with the deprotonated α-halo ester to the α-halo alkoxide **4**, followed by an intramolecular nucleophilic substitution reaction to give the epoxide **3**:

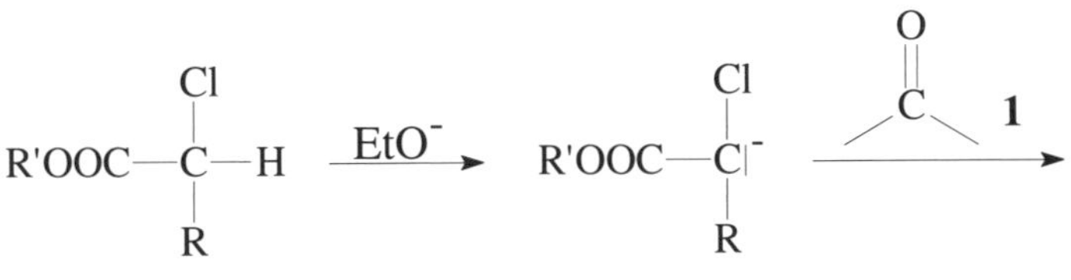

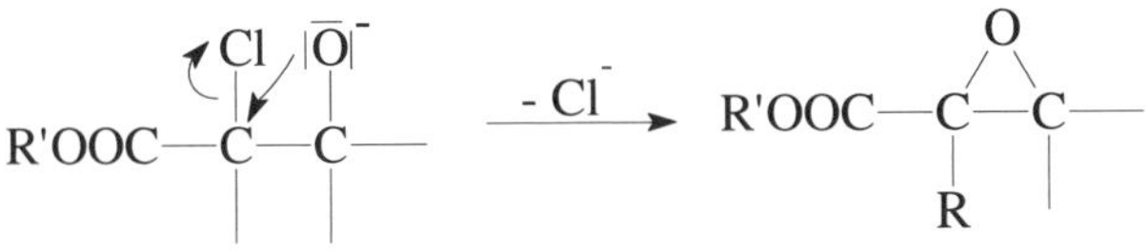

Generally the intermediate **4** is not isolated; however this can be done, thus supporting the formulated mechanism.

Good yields are usually obtained with aromatic aldehydes or ketones. Aliphatic aldehydes are poor substrates for the ordinary procedure, but react much better if the halo ester is first deprotonated with lithium diisopropylamide (LDA) in tetrahydrofuran at $-78\,^{\circ}$C, prior to addition of the aldehyde.

Instead of α-halo esters, related reactants can be used e.g. the α-halo derivatives of ketones, nitriles, sulfones and N,N-disubstituted amides. The Darzens condensation is also of some importance as a synthetic method because a glycidic acid can be converted into the next higher homolog of the original aldehyde, or into a branched aldehyde (e.g. **5**) if the original carbonyl substrate was a ketone:

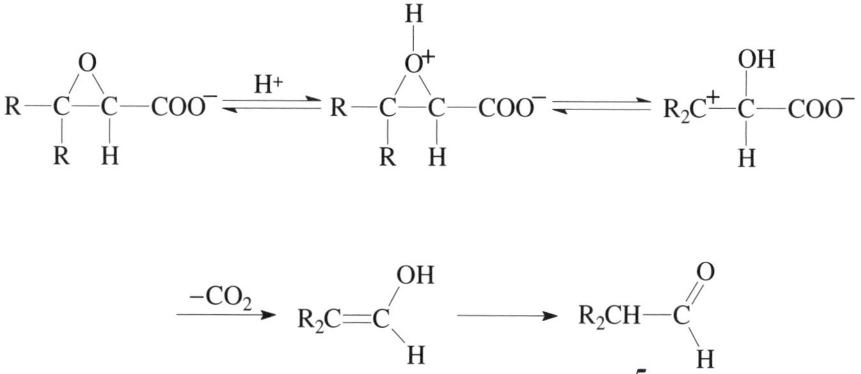

By reaction of an imine **6** with a α-halo ester **2**, an aziridine derivative **7** can be obtained:[4]

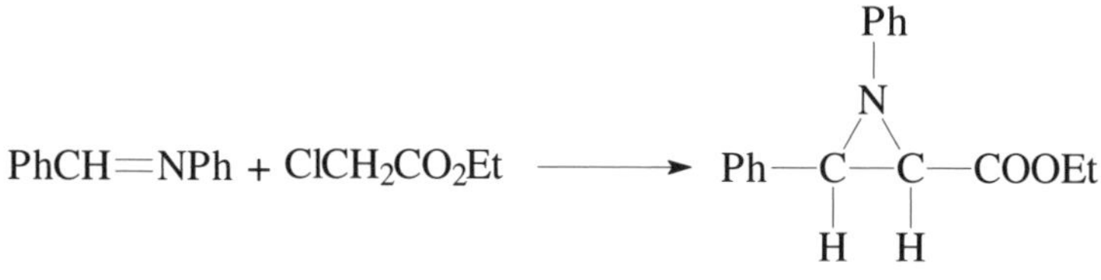

Although this variant often gives yields of less than 50%, it is a general method for the preparation of aziridines, especially of aziridinecarboxylic esters such as **7**.

1. E. Erlenmeyer, *Justus Liebigs Ann. Chem.* **1892**, *271*, 137–163.
2. M. S. Newman, B. J. Magerlein, *Org. React.* **1949**, *5*, 413–440.
3. G. Berti, *Top. Stereochem.* **1973**, *7*, 210–218.
4. J. A. Deyrup, *J. Org. Chem.* **1969**, *34*, 2724–2727.

Delépine Reaction

Primary amines through reaction of alkyl halides with hexamethylenetetramine

$$R\!-\!CH_2\!-\!X + (CH_2)_6N_4 \longrightarrow R\!-\!CH_2\!-\!NH_2$$

1　　　　　**2**　　　　　　　**3**

Reaction of alkyl halides **1** with hexamethylenetetramine **2** (trivial name: *urotropine*) followed by a hydrolysis step, leads to formation of primary amines **3** free of higher substituted amines. This method is called the *Delépine reaction*;[1,2] a comparable method is the *Gabriel synthesis*.

An alkyl halide **1** reacts with hexamethylenetetramine **2** to the quaternary ammonium salt **4**, which crystallizes from the reaction mixture:

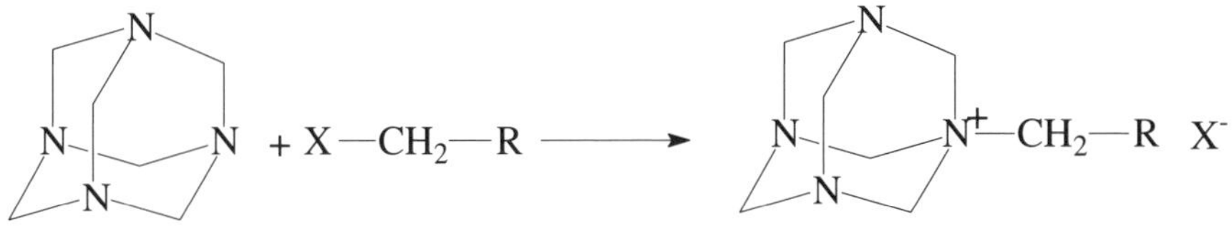

2　　　　　　**1**　　　　　　　**4**

Upon heating with a mixture of concentrated hydrochloric acid and ethanol under reflux, the hexaminium salt **4** is cleaved into the primary amine and formaldehyde. The latter can further react with ethanol under the acidic conditions to give formaldehyde diethylacetal:

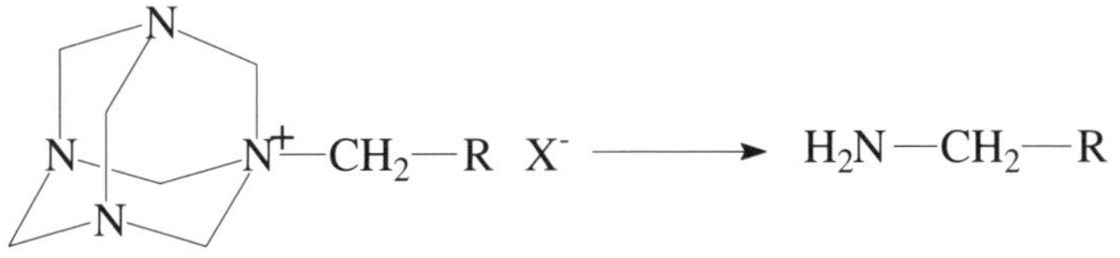

4

The Delépine reaction is a useful synthetic method, since it permits the selective preparation of primary amines from simple starting materials under simple reaction conditions and with short reaction time.

1. M. Delépine, *Bull. Soc. Chim. Fr.* **1895**, *13*, 352–361.
2. N. Blazevic, D. Kolbah, B. Belin, V. Sunjic, F. Kafjez, *Synthesis* **1979**, 161–176.

Diazo Coupling

Coupling reaction of diazonium ions with electron-rich aromatic compounds

$$ArN_2^+ + Ar'H \longrightarrow Ar-N{=}N-Ar'$$

$$\mathbf{1} \qquad \mathbf{2} \qquad\qquad \mathbf{3}$$

Arenediazonium ions **1** can undergo a coupling reaction with electron-rich aromatic compounds **2** like aryl amines and phenols to yield azo compounds[1,2] **3**. The substitution reaction at the aromatic system **2** usually takes place *para* to the activating group; probably for steric reasons. If the para position is already occupied by a substituent, the new substitution takes place *ortho* to the activating group.

Arenediazonium ions are stable in acidic or slightly alkaline solution; in moderate to strong alkaline medium they are converted into diazohydroxides **4**:

$$ArN_2^+ \xrightarrow{\ OH^-\ } Ar-N{=}N-OH$$

$$\mathbf{1} \qquad\qquad \mathbf{4}$$

The optimal pH-value for the coupling reaction depends on the reactant. Phenols are predominantly coupled in slightly alkaline solution, in order to first convert an otherwise unreactive phenol into the reactive phenoxide anion. The reaction mechanism can be formulated as electrophilic aromatic substitution taking place at the electron-rich aromatic substrate, with the arenediazonium ion being the electrophile:

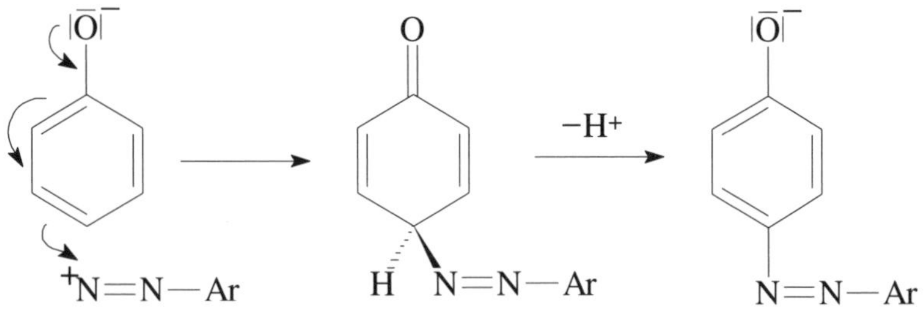

3a

For aryl amines the reaction mixture should be slightly acidic or neutral, in order to have a high concentration of free amine as well as arenediazonium ions. Aryl ammonium species—$ArNH_3^+$—are unreactive. The coupling of the diazonium species with aromatic amines proceeds by an analogous mechanism:

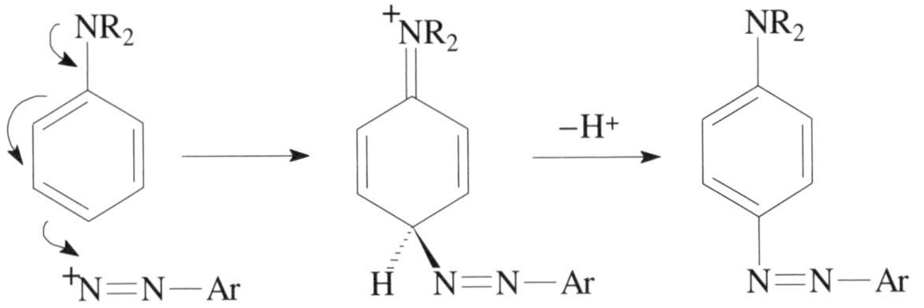

3b

With primary and secondary aryl amines a reaction at the amino nitrogen can occur, leading to formation of an aryl triazene **5**:

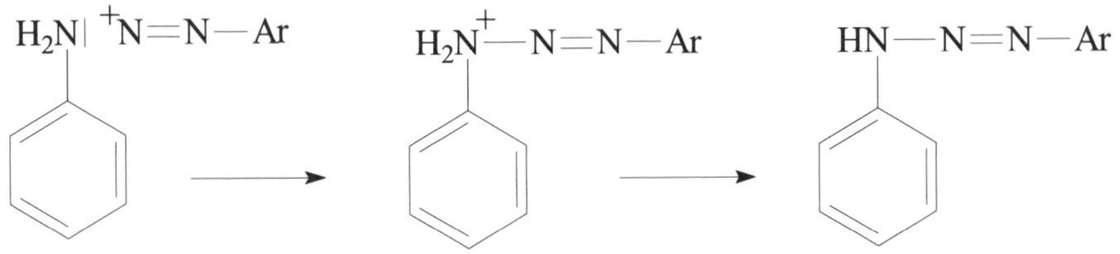

5

The N-azo compound **5** thus obtained can isomerize by an intermolecular process to give the C-azo derivative:[3–5]

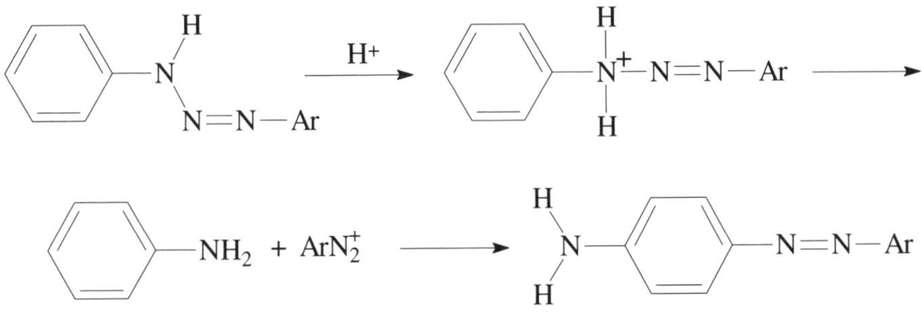

3b

If the para position is not already occupied, this isomerization generally leads to the para isomer. The desired C-azo product can be obtained in one laboratory step.[5]

Arenediazonium ions are relatively weak electrophiles, and therefore react only with electron-rich aromatic substrates like aryl amines and phenols. Aromatic compounds like anisole, mesitylene, acylated anilines or phenolic esters are ordinarily not reactive enough to be suitable substrates; however they may be coupled to activated arenediazonium ions. An electron-withdrawing substituent in the para position of the arenediazonium system increases the electrophilicity of the diazo group by concentrating the positive charge at the terminal nitrogen:

Certain aliphatic diazonium species such as bridgehead diazonium ions and cyclopropanediazonium ions, where the usual loss of N_2 would lead to very unstable carbocations, have been coupled to aromatic substrates.[1]

The opposite case—reaction of an arenediazonium species with an aliphatic substrate[6]—is possible if a sufficiently acidic C−H bond is present; e.g. with β-keto esters and malonic esters. The reaction mechanism is likely to be of the $S_E 1$-type; an electrophilic substitution at aliphatic carbon:

$$Z-CH_2-Z' \xrightarrow{\text{base}} Z-\overset{-}{\underset{H}{C}}-Z' \xrightarrow{+ ArN_2^+}$$

$$Z-\overset{Z'}{\underset{H}{C}}-N=N-Ar \longrightarrow Z-\overset{Z'}{C}=N-NH-Ar$$

(Z, Z′ = COOR, CHO, COR, $CONR_2$, COO−, CN, NO_2, SOR, SO_2R, SO_2OR, SO_2NR_2)

The diazo coupling with C−H acidic aliphatic substrates is a feature of the *Japp–Klingemann reaction*.

Suitably substituted azo compounds constitute an important class of dyes—the azo dyes. Some derivatives such as *p*-dimethylaminoazobenzene-*p*′-sulfonic acid Na-salt (trivial name: *methyl orange*) are used as pH-indicators.

1. I. Szele, H. Zollinger, *Top. Curr. Chem.* **1983**, *112*, 1–66.
2. A. F. Hegarty in *The Chemistry of the Diazonium and Diazo Groups, Vol. 2* (Ed.: S. Patai), Wiley, New York, **1978**, p. 545–551.
3. H. J. Shine, *Aromatic Rearrangements*, American Elsevier, New York, **1967**, p. 212–221.
4. J. R. Penton, H. Zollinger, *Helv. Chim. Acta* **1981**, *64*, 1717–1727 and 1728–1738.
5. R. P. Kelly, J. R. Penton, H. Zollinger, *Helv. Chim. Acta* **1982**, *65*, 122–132.
6. S. M. Parmerter, *Org. React.* **1959**, *10*, 1–142.

Diazotization

Diazonium salts from primary aromatic amines

$$Ar-NH_2 + HONO \longrightarrow ArN^{\pm}\!\!\equiv\!\!N|$$

$$\textbf{1} \qquad\qquad \textbf{2} \qquad\qquad\qquad \textbf{3}$$

The nitrosation of primary aromatic amines **1** with nitrous acid **2** and a subsequent dehydration step lead to the formation of diazonium ions **3**.[1-3] The unstable nitrous acid can for example be prepared by reaction of sodium nitrite with aqueous hydrochloric acid.

The reactive species for the transfer of the nitrosyl cation NO^+ is not the nitrous acid **2** but rather N_2O_3 **4** which is formed in weakly acidic solution. Other possible nitrosating agents are NOCl or $H_2NO_2{}^+$, or even free NO^+ in strong acidic solution. The initially formed N_2O_3 **4** reacts with the free amine **1**:

$$Ar-\overset{\overset{\displaystyle H}{|}}{\underset{\underset{\displaystyle H}{|}}{N}}| + \overset{\displaystyle \overline{N}=O}{\underset{\displaystyle \overset{|}{O}}{}} \longrightarrow Ar-\overset{\overset{\displaystyle H}{|}}{\underset{\underset{\displaystyle H}{|}}{N}}{}^{+}-\overline{N}=O + NO_2^-$$

$$\underset{\displaystyle \overline{N}=O}{}$$

$$\textbf{1} \qquad \textbf{4} \qquad\qquad\qquad \textbf{5}$$

Although the diazotization reaction takes place in acidic solution, it is the free amine that reacts,[1] and not the ammonium salt $ArNH_3{}^+$ X^-. Even in acidic solution there is a small amount of free amine present, since aromatic amines are relatively weak bases.

$$Ar-\overset{\overset{\displaystyle H}{|}}{\underset{\underset{\displaystyle H}{|}}{N}}{}^{+}-\overline{N}=O \xrightarrow{-H^+} Ar-\overset{\overset{\displaystyle }{}}{\underset{\underset{\displaystyle H}{|}}{\overline{N}}}-\overline{N}=O \rightleftharpoons Ar-\overline{N}=\overline{N}-OH$$

$$\textbf{5} \qquad\qquad\qquad \textbf{6} \qquad\qquad\qquad \textbf{7}$$

$$\xrightarrow{H^+} Ar-\overline{N}=\overline{N}-\overset{+}{\underset{\underset{\displaystyle H}{|}}{O}}H \xrightarrow{-H_2O} Ar-N^{\pm}\!\!\equiv\!\!N|$$

$$\textbf{3}$$

The cation **5** loses a proton to give the more stable nitrosoamine **6**, which is in equilibrium with the tautomeric diazohydroxide **7**. Protonation of the hydroxy group in **7** and subsequent loss of H_2O leads to the diazonium ion **3**.

Aliphatic primary amines also undergo the diazotization reaction in weakly acidic solution; however the resulting aliphatic diazonium ions are generally unstable, and easily decompose into nitrogen and highly reactive carbenium ions. The arenediazonium ions are stabilized by resonance with the aromatic ring:

$$\text{(structures of benzenediazonium resonance forms)}$$

However even arenediazonium ions generally are stable only at temperatures below 5 °C; usually they are prepared prior to the desired transformation, and used as reactants without intermediate isolation[5]. They can be stabilized more effectively through complexation by crown ethers.[4,5]

Most functional groups do not interfere with the diazotization reaction. Since aliphatic amines are stronger bases and therefore completely protonated at a pH < 3, it is possible that an aromatic amino group is converted into a diazonium group, while an aliphatic amino group present in the same substrate molecule is protected as ammonium ion and does not react.[6]

Instead of a diazonium salt, a *diazo compound* is obtained from reaction of a primary aliphatic amine **8** that has an electron-withdrawing substituent at the α-carbon (e.g. Z = COOR, CN, CHO, COR) as well as an α-hydrogen:[7]

$$\underset{\textbf{8}}{Z-CH_2-NH_2} + \underset{\textbf{2}}{HONO} \longrightarrow \underset{\textbf{9}}{Z-CH=\overset{+}{N}\!\!=\!\!\overset{-}{N}}$$

Diazonium salts are important intermediates in organic synthesis, e.g. for the *Sandmeyer reaction*. The most important use is the coupling reaction with phenols or aromatic amines to yield azo dyes (see *Diazo coupling*).

1. B. C. Challis, A. R. Butler in *The Chemistry of the Amino Group* (Ed.: S. Patai), Wiley, New York, **1968**, p. 305–320.
2. K. Schank in *The Chemistry of the Diazonium and Diazo Groups* (Ed.: S. Patai), Wiley, New York, **1978**, Vol. 2, p. 645–657.
3. J. H. Ridd, *Q. Rev. Chem. Soc.* **1961**, *15*, 418–441.
4. S. H. Korzeniowski, A. Leopold, J. R. Beadle, M. F. Ahern, W. A. Sheppard, R. K. Khanna, G. W. Gokel, *J. Org. Chem.* **1981**, *46*, 2153–2159.
5. R. A. Bartsch in *The Chemistry of Functional Groups, Supp. C* (Eds.: S. Patai, Z. Rappoport), Wiley, New York, **1983**, Vol. 1, p. 889–915.
6. N. Kornblum, D. C. Iffland, *J. Am. Chem. Soc.*, **1949**, *71*, 2137–2143.
7. M. Regitz in *The Chemistry of the Diazonium and Diazo Groups* (Ed.: S. Patai), Wiley, New York, **1978**, Vol. 2, p. 659–708.

Diels–Alder Reaction

[4 + 2] Cycloaddition of diene and dienophile

$$\textbf{1} \qquad \textbf{2} \qquad\qquad \textbf{3}$$

The *Diels–Alder reaction*,[1–4] is a cycloaddition reaction of a conjugated diene with a double or triple bond (the *dienophile*); it is one of the most important reactions in organic chemistry. For instance an electron-rich diene **1** reacts with an electron-poor dienophile **2** (e.g. an alkene bearing an electron-withdrawing substituent Z) to yield the unsaturated six-membered ring product **3**. An illustrative example is the reaction of butadiene **1** with maleic anhydride **4**:

The Diels–Alder reaction is of wide scope. Not all the atoms involved in ring formation have to be carbon atoms; the *hetero-Diels–Alder reaction* involving one or more heteroatom centers can be used for the synthesis of six-membered heterocycles.[5] The reverse of the Diels–Alder reaction—the *retro-Diels–Alder reaction*[6,7]—also is of interest as a synthetic method. Moreover and most importantly the usefulness of the Diels–Alder reaction is based on its *syn*-stereospecificity, with respect to the dienophile as well as the diene, and its predictable regio- and *endo*-selectivities.[8–10]

For a discussion of the mechanistic course of the reaction, many aspects have to be taken into account.[9,11] The *cisoid* conformation of the diene **1**, which is in equilibrium with the thermodynamically more favored *transoid* conformation, is a prerequisite for the cycloaddition step. Favored by a fixed cisoid geometry are those substrates where the diene is fitted into a ring, e.g. cyclopentadiene **5**. This particular compound is so reactive that it dimerizes easily at room temperature by undergoing a Diels–Alder reaction:

Since the equilibrium lies to the left at higher temperatures, cyclopentadiene can be obtained by thermolytic cleavage of the dimer and distilling the monomer prior to use (*cracking distillation*).

In case of a Z,Z-configurated diene **6**, the transoid conformation is favored, because of unfavorable steric interactions of substituents at C-1 and C-4 in the cisoid conformation:

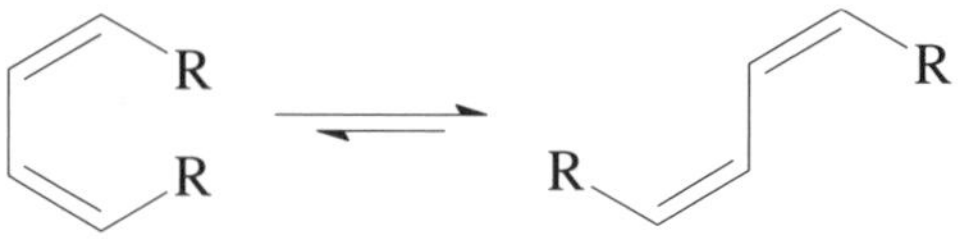

6

Mechanistically the observed stereospecificity can be rationalized by a concerted, pericyclic reaction. In a one-step cycloaddition reaction the dienophile **8** adds 1,4 to the diene **7** *via* a six-membered cyclic, aromatic transition state **9**, where three π-bonds are broken and one π- and two σ-bonds are formed. The arrangement of the substituents relative to each other at the stereogenic centers of the reactants is retained in the product **10**, as a result of the stereospecific *syn*-addition.

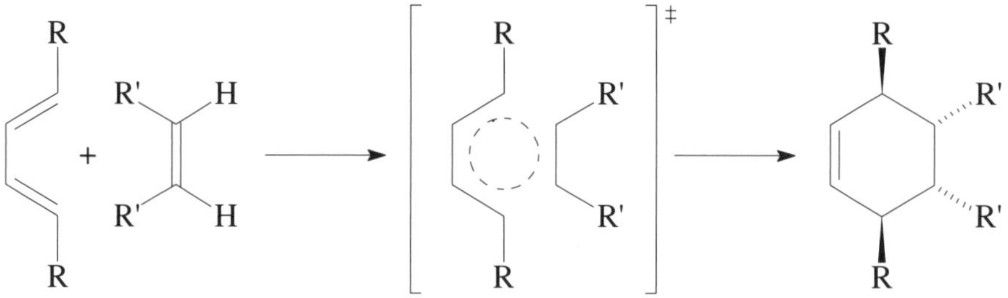

7   **8**   **9**   **10**

An explanation for the finding that concerted [4 + 2] cycloadditions take place thermally, while concerted *[2 + 2] cycloadditions* occur under photochemical conditions, is given through the *principle of conservation of orbital symmetry*[12]. According to the *Woodward–Hofmann rules*[12] derived thereof, a concerted, pericyclic [4 + 2] cycloaddition reaction from the ground state is symmetry-allowed.

From reaction of an unsymmetrically substituted diene **11a, b** and dienophile **12**, different regioisomeric products **13a, b** and **14a, b** can be formed. The so-called '*ortho*' and the '*para*' isomer **13a**, resp. **13b**, is formed preferentially.

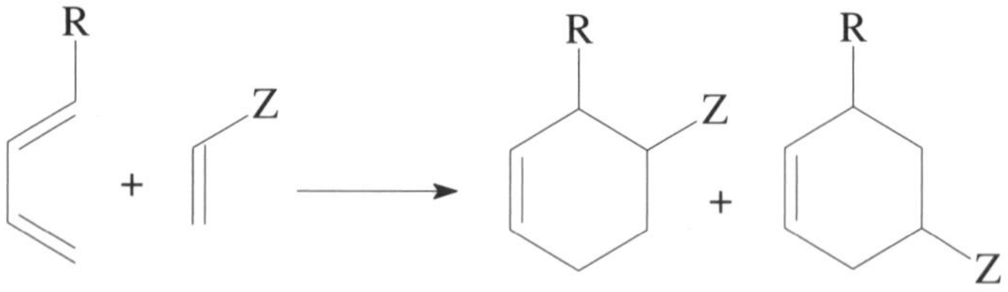

11a   **12**   **13a**   **14a**

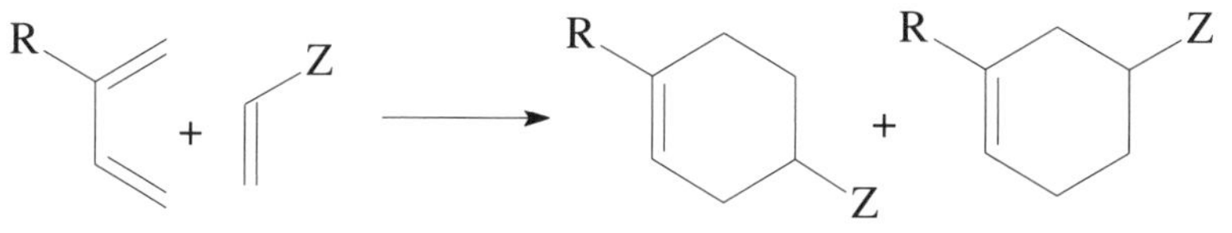

11b   12   **13b**   **14b**

The observed regioselectivity[9] can be explained by taking into account the frontier orbital coefficients of the reactants.[13]

The Diels–Alder reaction of a diene with a substituted olefinic dienophile, e.g. **2**, **4**, **8**, or **12**, can go through two geometrically different transition states. With a diene that bears a substituent as a stereochemical marker (any substituent other than hydrogen; deuterium will suffice[27]) at C-1 (e.g. **11a**) or substituents at C-1 and C-4 (e.g. **5**, **6**, **7**), the two different transition states lead to diastereomeric products, which differ in the relative configuration at the stereogenic centers connected by the newly formed σ-bonds. The respective transition state as well as the resulting product is termed with the prefix *endo* or *exo*. For example, when cyclopentadiene **5** is treated with acrylic acid **15**, the *endo*-product **16** and the *exo*-product **17** can be formed. Formation of the *endo*-product **16** is kinetically favored by secondary orbital interactions (*endo rule* or *Alder rule*).[11,13] Under kinetically controlled conditions it is the major product, and the thermodynamically more stable *exo*-product **17** is formed in minor amounts only.

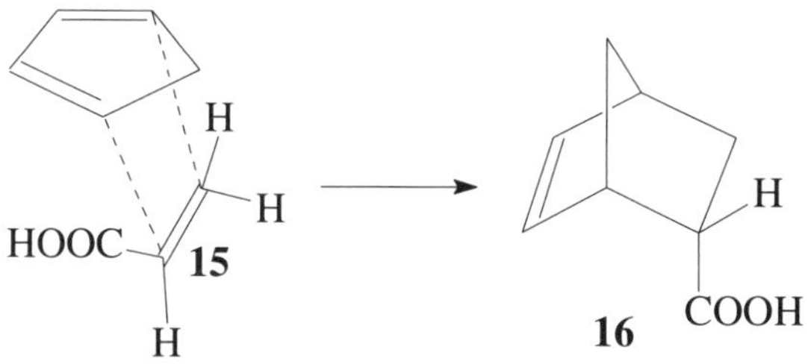

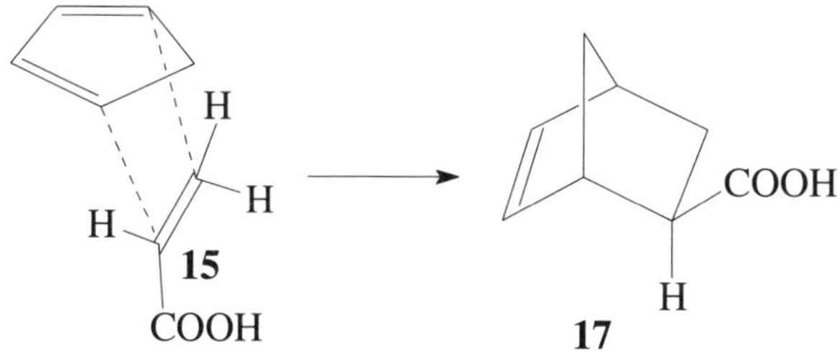

For most Diels–Alder reactions a concerted mechanism as described above, is generally accepted. In some cases, the kinetic data may suggest the intermediacy of a diradical intermediate[9] **18**:

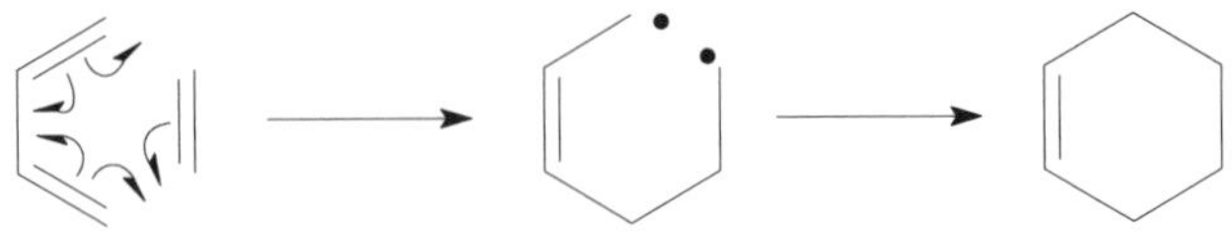

18

Furthermore in certain cases Diels–Alder reactions may proceed by an ionic mechanism.[14]

For the ordinary Diels–Alder reaction the dienophile preferentially is of the electron-poor type; electron-withdrawing substituents have a rate enhancing effect. Ethylene and simple alkenes are less reactive. Substituent Z in **2** can be e.g. CHO, COR, COOH, COOR, CN, Ar, NO$_2$, halogen, C=C. Good dienophiles are for example maleic anhydride, acrolein, acrylonitrile, dehydrobenzene, tetracyanoethylene (TCNE),[15] acetylene dicarboxylic esters. The diene preferentially is of the electron-rich type; thus it should not bear an electron-withdrawing substituent.

Because of their strong aromatic character, benzene and naphthalene are very unreactive as dienes; however anthracene **19** reacts with highly reactive dienophiles, such as dehydrobenzene (benzyne) **20**:

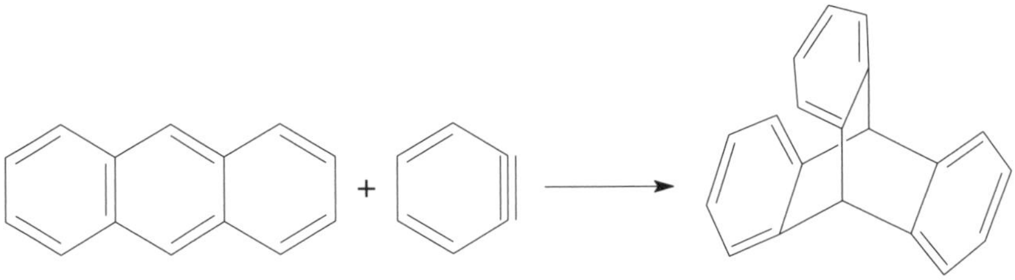

19   **20**

There are Diels–Alder reactions known where the electronic conditions outlined above are just reversed. Such reactions are called Diels–Alder reactions with *inverse electron demand*.[9] For example[16] the electron-poor diene hexachlorocyclopentadiene **21** reacts with the electron-rich styrene **22**:

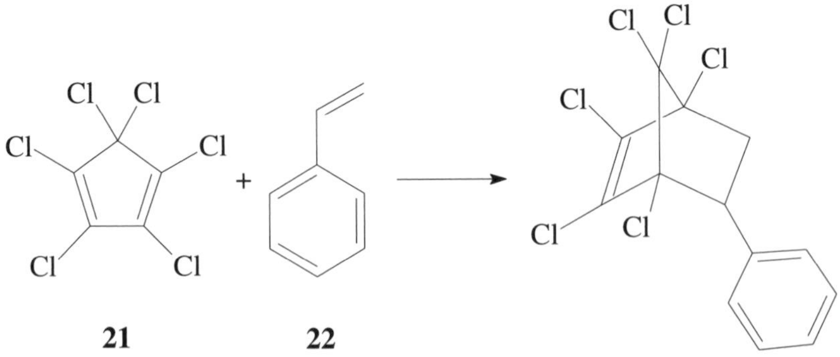

21   **22**

In many cases the solvent has little effect on the Diels–Alder reaction, which is an additional argument for a concerted mechanism. The intermolecular Diels–Alder reaction exhibits a large negative volume of activation, together with a large negative volume of reaction, and therefore the application of high pressure can lead to rate enhancement and improved *endo*-selectivity.[17,18] The strong acceleration of the Diels–Alder reaction and the enhanced *endo*-selectivity observed when using a special solvent system, consisting of a five-molar solution of lithium perchlorate in diethyl ether,[19] is attributed to an internal pressure effect, rather than to an ordinary solvent effect.

The use of catalysts for a Diels–Alder reaction is often not necessary, since in many cases the product is obtained in high yield in a reasonable reaction time. In order to increase the regioselectivity and stereoselectivity (e.g. to obtain a particular *endo*- or *exo*-product), Lewis acids as catalysts (e.g. $TiCl_4$, $AlCl_3$, BF_3-etherate) have been successfully employed.[4] The usefulness of strong Lewis acids as catalysts may however be limited, because they may also catalyze polymerization reactions of the reactants. Chiral Lewis acid catalysts are used for catalytic enantioselective Diels–Alder reactions.[20]

The Diels–Alder reaction with triple bond dienophiles gives access to cyclohexa-1,4-diene derivatives. Further reaction of a reactive intermediate thus produced or a subsequent oxidation step can then lead to a six-membered ring aromatic target molecule.

An example is the synthesis of substituted [2.2]paracyclophanes as reported by *Hopf et al.*[21] When hexa-1,2,4,5-tetraene **23** is treated with dimethyl acetylenedicarboxylate **24** (an electron-poor acetylenic dienophile), the initially formed reactive intermediate **25** dimerizes to yield the [2.2]paracyclophane **26**:

Numerous examples of intramolecular Diels–Alder reactions have been reported;[22,23] especially from application in the synthesis of natural products, where stereoselectivity is of particular importance; e.g. syntheses of steroids.[24]

A *domino reaction*,[26] in this case consisting of an inter- and an intramolecular Diels–Alder reaction, is a key step in the synthesis of the hydrocarbon pagodane **30**, reported by *Prinzbach et al.*[25] When the *bis*-diene **27** is treated with maleic anhydride **4**, an initial intermolecular reaction leads to the intermediate product **28**, which cannot be isolated, but rather reacts intramolecularly to give the pagodane precursor **29**:

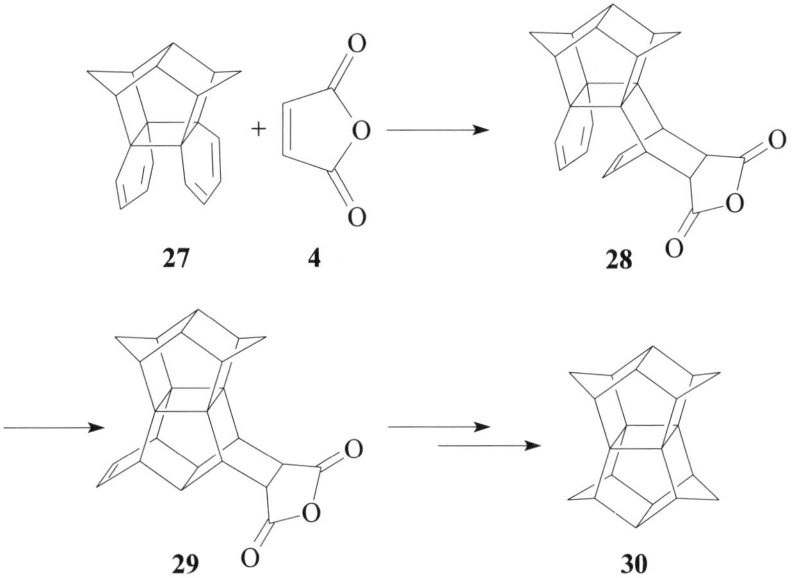

The versatility of the Diels–Alder reaction becomes especially obvious, when considering the hetero-variants.[5] One or more of the carbon centers involved can be replaced by hetero atoms like nitrogen, oxygen and sulfur. An illustrating example is the formation of the bicyclic compound **31**, by an intramolecular *hetero-Diels–Alder reaction*:[5]

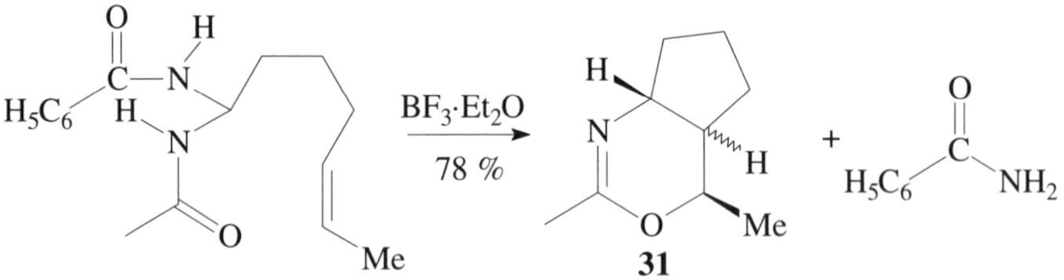

The Diels–Alder reaction is the most important method for the construction of six-membered rings. For example it can be used as a step in a benzo-anellation procedure. The experimental procedure is simple, and yields are generally good; side reactions play only a minor role.

1. O. Diels, K. Alder, *Justus Liebigs Ann. Chem.* **1928**, *460*, 98–122.
2. J. A. Norton, *Chem. Rev.* **1942**, *31*, 319–523.
3. J. G. Martin, R. K. Hill, *Chem. Rev.* **1961**, *61*, 537–562.
4. W. Carruthers, *Cycloaddition Reactions in Organic Synthesis*, Pergamon Press, Oxford, **1990**, p. 1–208.
5. D. L. Boger, S. M. Weinreb, *Hetero-Diels-Alder Methodology in Organic Synthesis*, Academic Press, New York, **1987**; S. M. Weinreb, P. M. Scola, *Chem. Rev.* **1989**, *89*, 1525–1534.
6. A. Ichihara, *Synthesis* **1987**, 207–222.
7. M.-C. Lasne, J.-L. Ripoll, *Synthesis* **1985**, 121–143; M. E. Bunnage, K. C. Nicolaou, *Chem. Eur. J.* **1997**, *3*, 187–192.
8. D. Craig, *Chem. Soc. Rev.* **1987**, *16*, 187–238.
9. J. Sauer, R. Sustmann, *Angew. Chem.* **1980**, *92*, 773–801; *Angew. Chem. Int. Ed. Engl.* **1980**, *19*, 779.
10. W. Oppolzer, *Angew. Chem.* **1984**, *96*, 840–854; *Angew. Chem. Int. Ed. Engl.* **1984**, *23*, 876.
11. J. Sauer, *Angew. Chem.* **1967**, *79*, 76–94; *Angew. Chem. Int. Ed. Engl.* **1967**, *6*, 16.
12. R. B. Woodward, R. Hoffmann, *Angew. Chem.* **1969**, *81*, 797–869; *Angew. Chem. Int. Ed. Engl.* **1969**, *8*, 781–853; R. B. Woodward, R. Hoffmann, *The Conservation of Orbital Symmetry*, Academic Press, New York, **1970**.
13. I. Fleming, *Frontier Orbitals and Organic Chemical Reactions*, Wiley, London, **1976**, p. 110–142 and 161–168.
14. P. G. Gassman, D. B. Gorman, *J. Am. Chem. Soc.* **1990**, *112*, 8624–8626.
15. A. J. Fatiadi, *Synthesis* **1987**, 749–789.
16. J. Sauer, H. Wiest, *Angew. Chem.* **1962**, *74*, 353; *Angew. Chem. Int. Ed. Engl.* **1962**, *1*, 268.
17. K. Matsumoto, A. Sera, *Synthesis* **1985**, 999–1027.
18. L. F. Tietze, T. Hübsch, J. Oelze, C. Ott, W. Tost, G. Wörner, M. Buback, *Chem. Ber.* **1992**, *125*, 2249–2258.
19. P. A. Grieco, *Aldrichimica Acta*, **1991**, *24(3)*, 59–66; H. Waldmann, *Angew. Chem.* **1991**, *103*, 1335–1337; *Angew. Chem. Int. Ed. Engl.* **1991**, *30*, 1306.
20. H. B. Kagan, O. Riant, *Chem. Rev.* **1992**, *92*, 1007–1019; Y. Hayashi, J. J. Rohde, E. J. Corey, *J. Am. Chem. Soc.* **1996**, *118*, 5502–5503.
21. H. Hopf, G. Weber, K. Menke, *Chem. Ber.* **1980**, *113*, 531–541.
22. E. Ciganek, *Org. React.* **1984**, *32*, 1–374.
23. G. Brieger, J. N. Bennett, *Chem. Rev.* **1980**, *80*, 63–97.
24. R. L. Funk, K. P. C. Vollhardt, *Chem. Soc. Rev.* **1980**, *9*, 41–62.
25. W.-D. Fessner, C. Grund, H. Prinzbach, *Tetrahedron Lett.* **1989**, *30*, 3133–3136.
26. L. F. Tietze, U. Beifuss, *Angew. Chem.* **1993**, *105*, 137–170; *Angew. Chem. Int. Ed. Engl.* **1993**, *32*, 131.
27. L. M. Stephenson, D. E. Smith, S. P. Current, *J. Org. Chem.* **1982**, *47*, 4170.

Di-π-Methane Rearrangement

Photochemical rearrangement of 1,4-dienes to vinylcyclopropanes

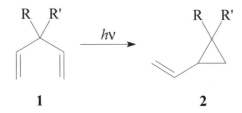

1   **2**

The photochemical isomerization of 1,4-dienes **1**, bearing substituents at C-3, leads to vinyl-cyclopropanes **2**, and is called the *di-π-methane rearrangement*.[1,2,3] This reaction produces possible substrates for the *vinylcyclopropane rearrangement*.

A mechanism has been formulated[2] that would involve formation of diradical species **3** and **4**, which however might not be real intermediates. At least one substituent at C-3 is required in order to stabilize the radical **4**, and thereby facilitate the cleavage of the C-2/C-3 bond:

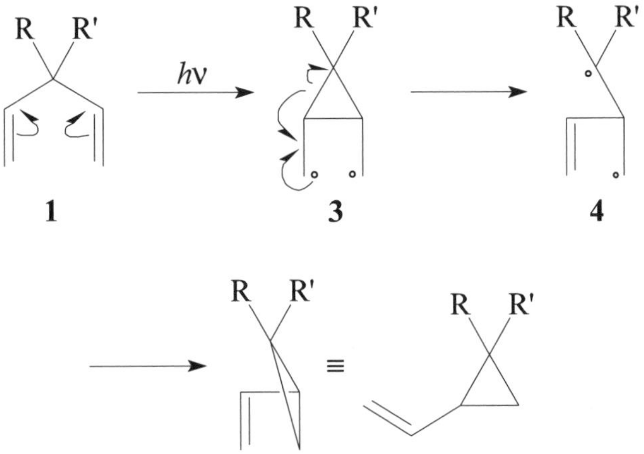

1   **3**   **4**

2

The rearrangement proceeds from the S_1-state of the 1,4-diene **1**. The T_1-state would allow for different reactions like double bond isomerization. Rigid systems like cyclic dienes, where E/Z-isomerization of a double bond is hindered for steric reasons, can react through the T_1-state. When the rearrangement proceeds from the S_1-state, it proves to be stereospecific at C-1 and C-5; no E/Z-isomerization is observed. Z-1,1-Diphenyl-3,3-dimethyl-1,4-hexadiene **5** rearranges to the Z-configured vinylcyclopropane **6**.[4] In this case the reaction also is regiospecific. Only the vinylcyclopropane **6** is formed, but not the alternative product **7**.[4]

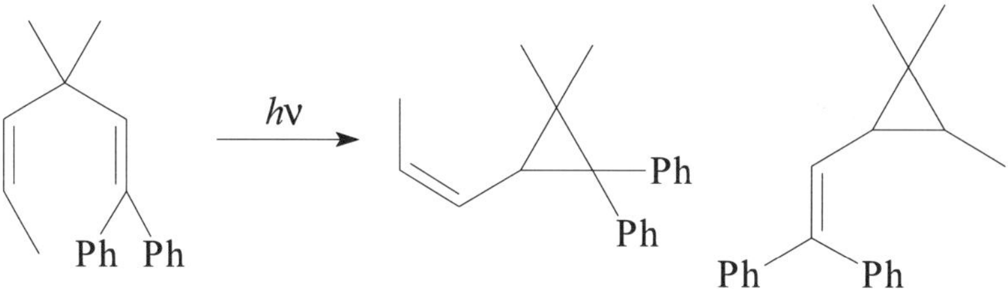

5   **6**   **7**

However, from substrates where the substituents at C-1 and C-5 are not that different in structure, a mixture of regioisomers may be obtained.

The di-π-methane rearrangement is a fairly recent reaction. One of the first examples has been reported in 1966 by Zimmerman and Grunewald[1] with the isomerization of barrelene **8** to semibullvalene **9**. This rearrangement reaction occurs in the presence of acetone as photosensitizer, and proceeds from the T_1-state.[5]

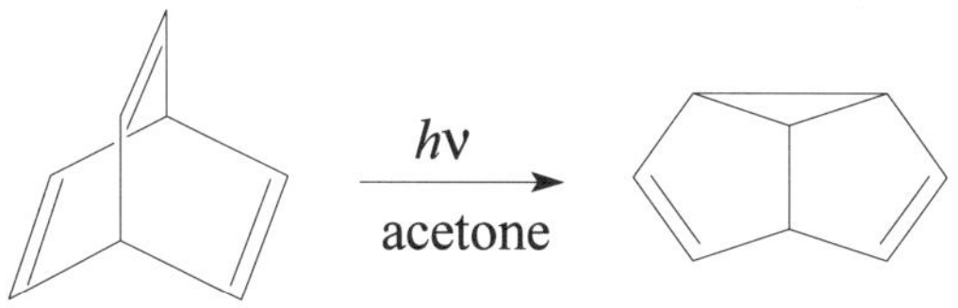

8   **9**

A related reaction is the *oxa-di-π-methane rearrangement*,[2,6] where one of the C=C double bonds is replaced by a C=O double bond. The substrates are thus β,γ-unsaturated ketones. The rearrangement proceeds from the triplet state. This *oxa*-variant gives access to highly strained molecules containing small rings, as has been demonstrated by irradiation of norborn-5-ene-2-one **10**:

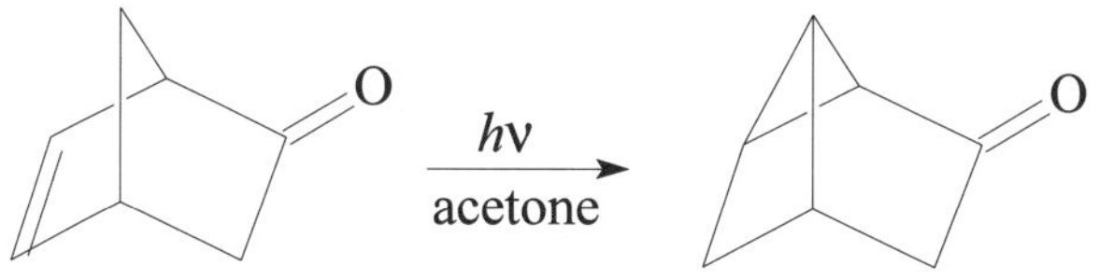

10

Yields of the di-π-methane rearrangement reaction strongly depend on substrate structure, and are ranging from poor to nearly quantitative. Acetone and acetophenone have been used as photosensitizers.[4,5]

1. H. E. Zimmerman, G. L. Grunewald, *J. Am. Chem. Soc.* **1966**, *88*, 183–184.
2. S. S. Hixson, P. S. Mariano, H. E. Zimmerman, *Chem. Rev.* **1973**, *73*, 531–551.
3. D. Döpp, H. E. Zimmerman, *Methoden Org. Chem. (Houben-Weyl)* **1975**, *Vol. 4/5a*, p. 413–432.
   H. E. Zimmerman, D. Armesto, *Chem. Rev*, **1996**, *96*, 3065–3112.
4. L. A. Paquette, E. Bay, A. Yeh Ku, N. G. Rondan, K. N. Houk, *J. Org. Chem.* **1982**, *47*, 422–428.
5. H. E. Zimmerman, *Angew. Chem.* **1969**, *81*, 45–55; *Angew. Chem. Int. Ed. Engl.* **1969**, *8*, 1.
6. M. Demuth, G. Mikhail, *Synthesis* **1989**, 145–162.

Dötz Reaction

Benzo-anellation *via* chromium carbene complexes

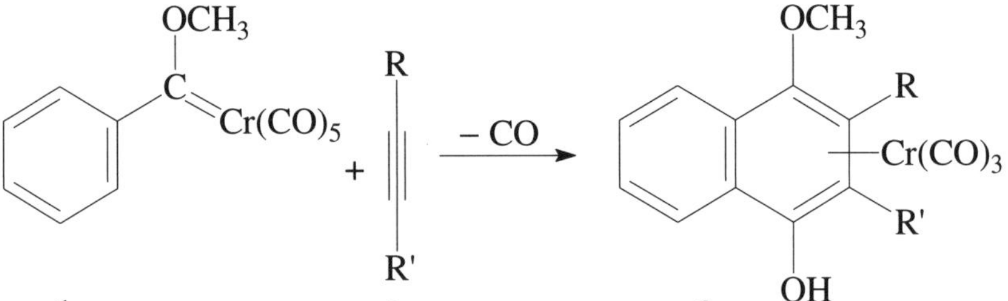

The coupling reaction of an α,β-unsaturated chromium carbene complex, e.g. **1**, and an alkyne **2**, through coordination to the chromium center, is called the *Dötz reaction*.[1–3] The initial product is the chromium tricarbonyl complex of a hydroquinone derivative **3**, which can easily be converted to a free hydroquinone or quinone.

By a photochemically induced elimination of CO, a chromium carbene complex with a free coordination site is generated. That species can coordinate to an alkyne, to give the alkyne–chromium carbonyl complex **4**. The next step is likely to be a cycloaddition reaction leading to a four-membered ring compound **5**. A subsequent electrocyclic ring opening and the insertion of CO leads to the vinylketene complex **6**:[3–5]

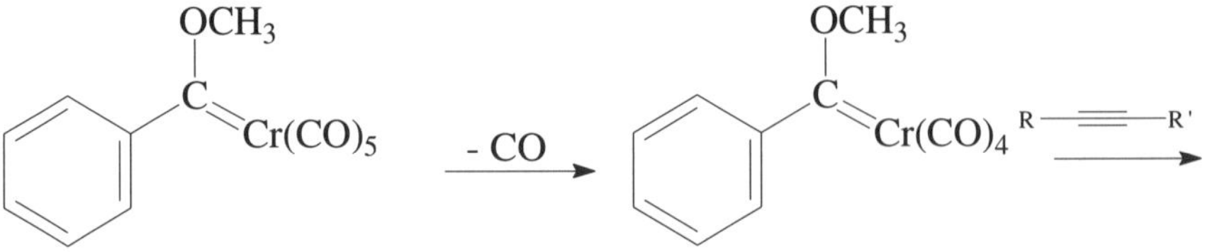

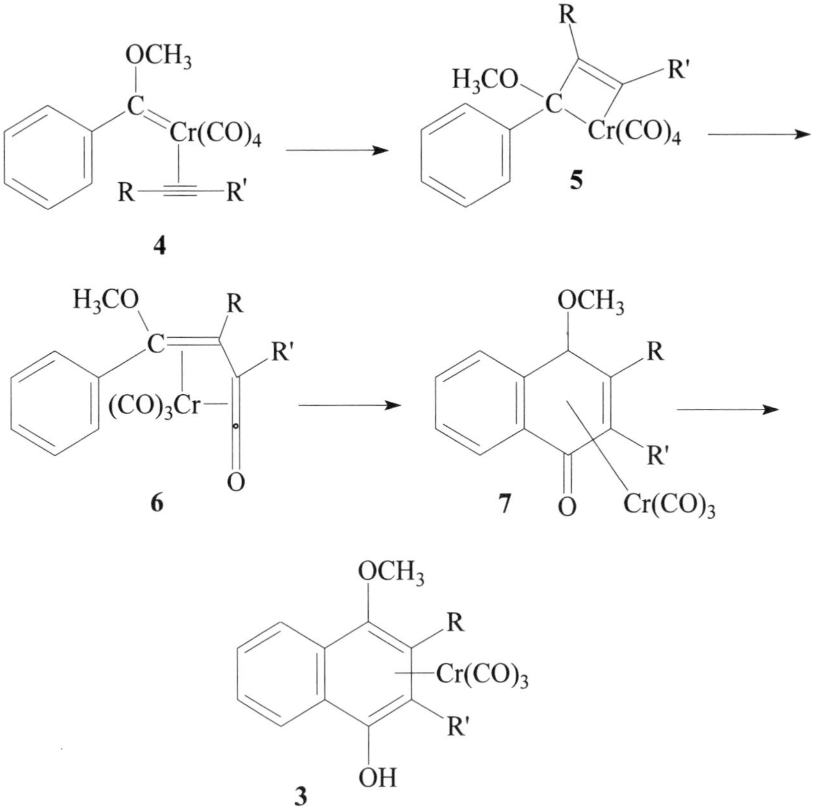

An electrocyclic ring closure then leads to a cyclohexadienone complex **7**, which upon migration of a proton, yields the chromium tricarbonyl-hydroquinone complex **3**.

The regioselectivity of the reaction with unsymmetrical alkynes is poor. Mixtures of isomers are obtained with alkyl substituted acetylenes, if the alkyl groups do not differ much in size. A solution to this problem has been reported by *Semmelhack et al.*[6] The reactants are connected by a −OCH$_2$CH$_2$O-tether, which can later be removed; the coupling step thus becomes intramolecular:

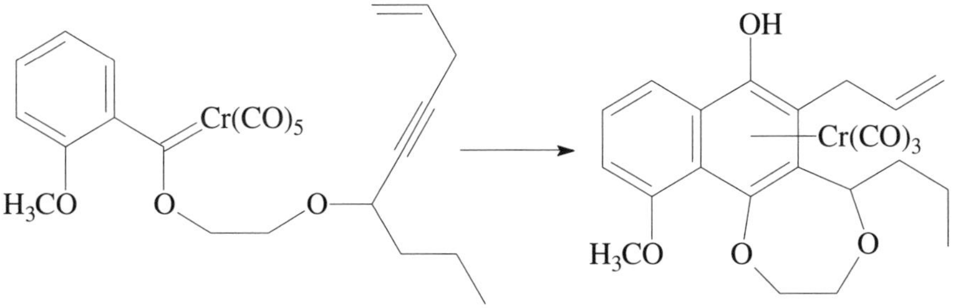

The synthetic value of the Dötz reaction has for example been demonstrated by the synthesis of vitamin $K_{1(20)}$ **10** (simplified structure). This natural product has been prepared synthetically from the chromium carbene complex **8** and the alkyne **9** in two steps; the second step being the oxidative decomplexation to yield the free product **10**:[7]

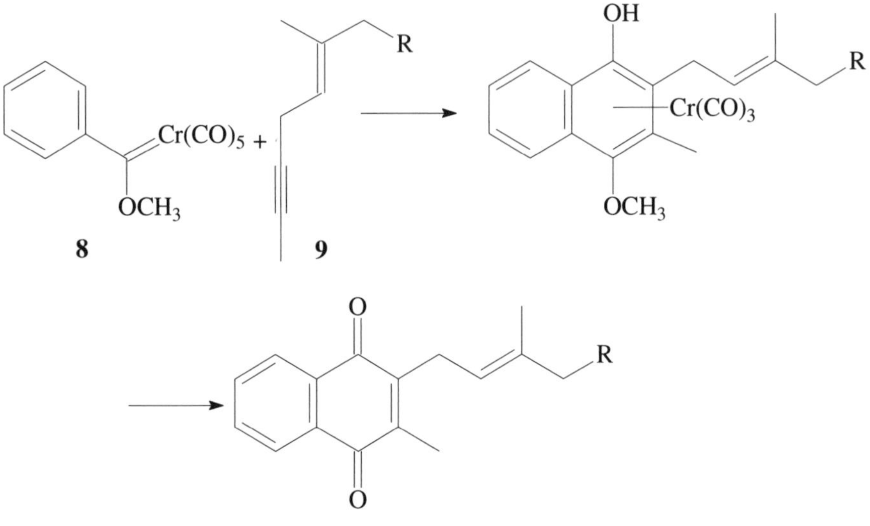

Chromium carbene complexes like **13**, which are called *Fischer carbene complexes*,[3] can conveniently be prepared from chromium hexacarbonyl **11** and an organolithium compound **12**, followed by an O-alkylation step:

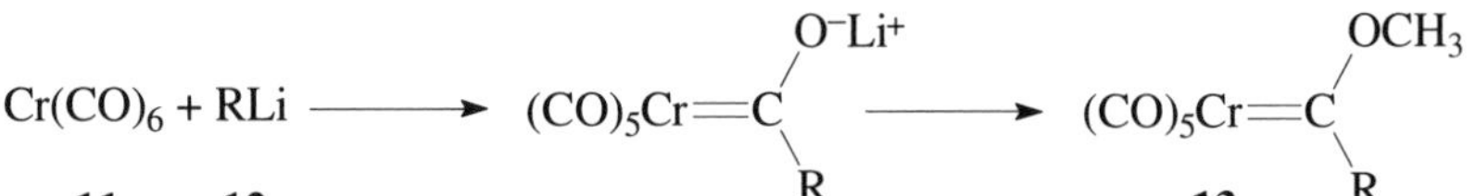

The unsaturated substituent in the carbene complex **1** often is aromatic or heteroaromatic, but can also be olefinic. The reaction conditions of the Dötz procedure are mild; various functional groups are tolerated. Yields are often high. The use of chromium hexacarbonyl is disadvantageous, since this compound is considered to be carcinogenic;[8] however to date it cannot be replaced by a less toxic compound. Of particular interest is the benzo-anellation procedure for the synthesis of anthracyclinones, which are potentially cytostatic agents.[9]

1. K. H. Dötz, *Angew. Chem.* **1975**, *87*, 672–673; *Angew. Chem. Int. Ed. Engl.* **1975**, *14*, 644.
2. N. E. Schore, *Chem. Rev.* **1988**, *88*, 1081–1119.

3. J. Mulzer, H.-U. Reissig, H.-J. Altenbach, M. Braun, K. Krohn, *Organic Synthesis Highlights*, VCH, Weinheim, **1991**, p. 186–191.
4. K. S. Chan, G. A. Peterson, T. A. Brandvold, K. L. Faron, C. A. Challener, C. Hyldahl, W. D. Wulff, *J. Organomet. Chem.* **1987**, *334*, 9–56.
5. J. S. McCallum, F.-A. Kunng, S. R. Gilbertson, W. D. Wulff, *Organometallics* **1988**, 7, 2346–2360.
6. M. F. Semmelhack, J. J. Bozell, L. Keller, T. Sato, E. J. Spiess, W. D. Wulff, A. Zask, *Tetrahedron* **1985**, *41*, 5803–5812.
7. K. H. Dötz, *Angew. Chem.* **1984**, *96*, 573–594; *Angew. Chem. Int. Ed. Engl.* **1984**, *23*, 587.
8. L. Roth, *Krebserzeugende Stoffe*, Wissenschaftl. Verlagsgesellschaft, Stuttgart, **1983**, p. 16.
9. K. H. Dötz, M. Popall, *Chem. Ber.* **1988**, *121*, 665–672.

E

Elbs Reaction

Oxidation of phenols by peroxodisulfate

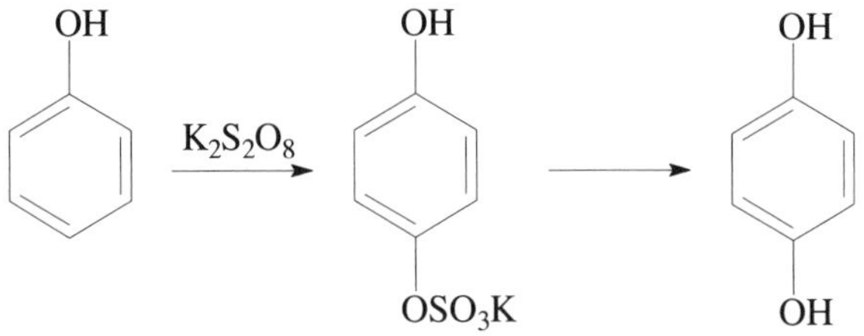

The hydroxylation of a phenol **1** upon treatment with a peroxodisulfate in alkaline solution, to yield a 1,2- or 1,4-dihydroxybenzene **3**, is called the *Elbs reaction*.[1,2]

The phenol is deprotonated by base to give a phenolate anion **4**, that is stabilized by resonance, and which is activated at the *ortho* or the *para* position towards reaction with an electrophilic agent:

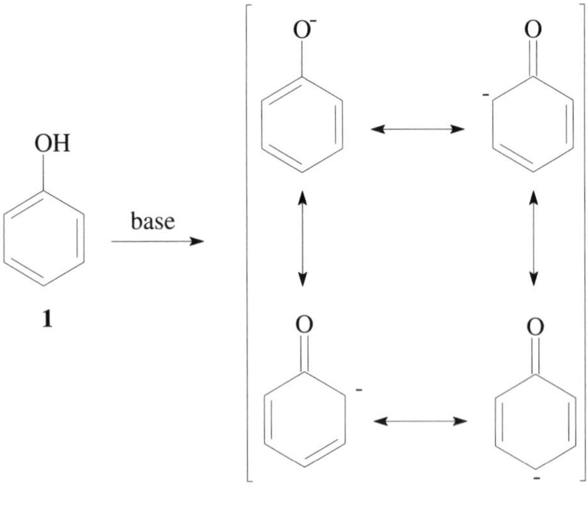

Reaction with the electrophilic peroxodisulfate occurs preferentially at the *para* position, leading to formation of a cyclohexadienone derivative **5**, which loses a proton to give the aromatic compound **6**. Subsequent hydrolysis of the sulfate **6** yields 1,4-dihydroxybenzene **3**:

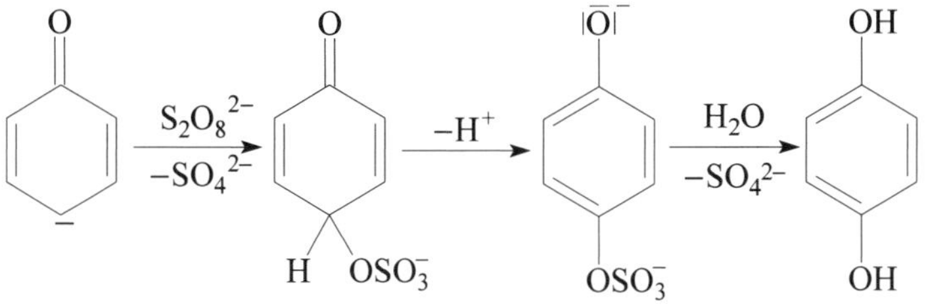

The main product of the Elbs reaction is the 1,4-dihydroxybenzene (hydro-quinone). If the *para* position is already occupied by a substituent, the reaction occurs at an *ortho* position, leading to a catechol derivative; although the yields are not as good as for a hydroquinone. Better yields of catechols 7 can be obtained by a copper-catalyzed oxidation of phenols with molecular oxygen:[3]

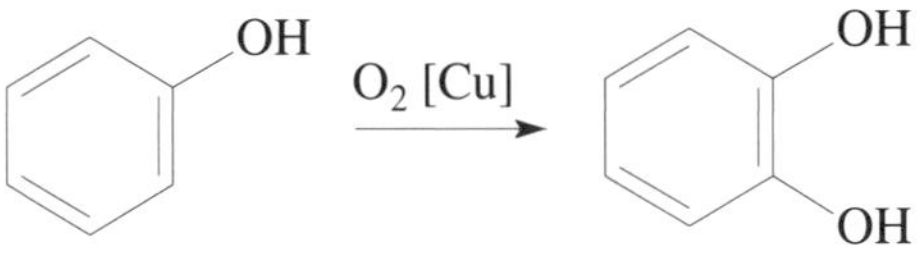

This reaction, that can be viewed as a modified Elbs reaction, often gives good yields, while the ordinary procedure often gives catechols in less then 50% yield.

Nevertheless the Elbs reaction is a valuable method for the preparation of dihy-droxybenzenes. The experimental procedure is simple, and the reaction conditions are mild; a variety of functional groups is tolerated.

1. K. Elbs, *J. Prakt. Chem.* **1893**, 48, 179–185.
2. E. J. Behrman, *Org. React.* **1988**, *35*, 421–511.
3. P. Capdevielle, M. Maumy, *Tetrahedron Lett.* **1982**, *23*, 1573–1576 and 1577–1580.

Ene Reaction

Addition of a double bond to an alkene with allylic hydrogen

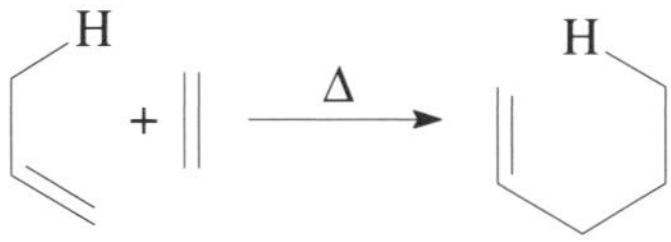

The *ene reaction* as a reaction principle has been first recognized and systematically investigated by *Alder*.[1] It is a thermal addition reaction of a double bond species **2**—the *enophile*—and an alkene **1**—the *ene*—that has at least one allylic hydrogen.[2] The intramolecular variant[3] is of greater synthetic importance than is the intermolecular reaction.

Just like the *Diels Alder reaction* or the 1,5-sigmatropic hydrogen shift, the ene reaction is believed to proceed *via* a six-membered aromatic transition state.

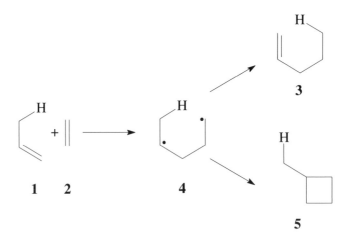

1   2   3

The overall reaction includes allylic transposition of a double bond, migration of the allylic hydrogen and formation of a bond between ene and enophile. Experimental findings suggest a concerted mechanism. Alternatively a diradical species **4** might be formed as intermediate; however such a species should also give rise to formation of a cyclobutane derivative **5** as a side-product. If such a by-product is not observed, one might exclude the diradical pathway:

A primary allylic hydrogen at the ene **1** is especially reactive; a secondary hydrogen migrates less facile, and a tertiary one is even less reactive. The enophile unit should be of an electron-poor nature; it can consist of a carbon–carbon double or triple bond, a carbonyl group or an azo group. Mixtures of regioisomeric products may be obtained with substituted enophiles. The acrylic ester **6** reacts with propene **1** to give the regioisomers **7** (88%) and **8** (12%):

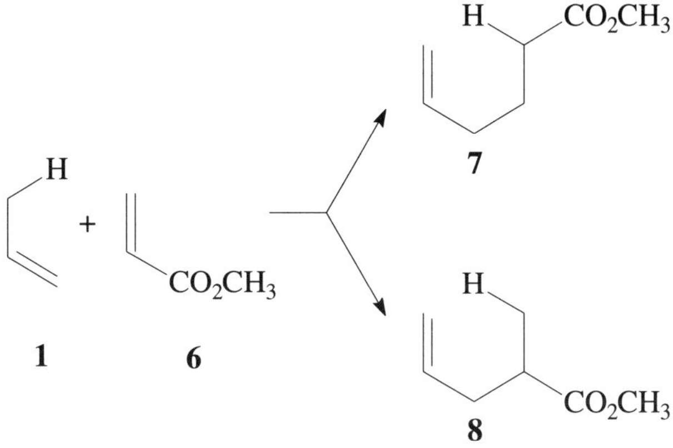

For the intramolecular variant, synthetically valuable applications have been developed during the last decade.[4-6] Three types of intramolecular ene reactions are formulated—depending on the structure of the starting material:[3]

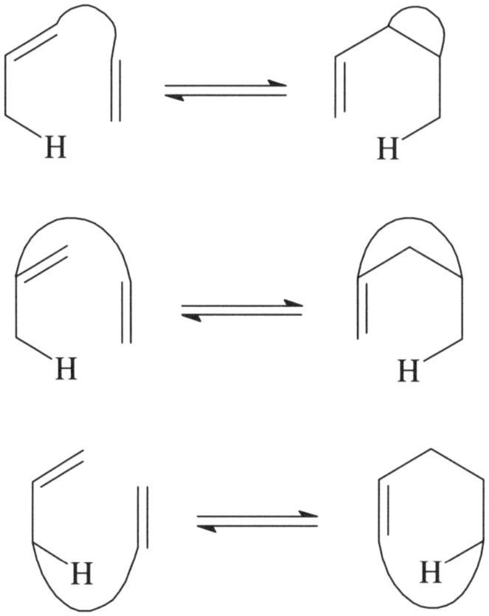

A modern variant is the intramolecular *magnesium-ene reaction*, e.g. the reaction of the alkene-allylic-Grignard compound **9** to give the five-membered ring product **10**. This reaction proceeds regio- and stereoselectively, and is a key step in a synthesis of the sesquiterpenoid 6-protoilludene:[6]

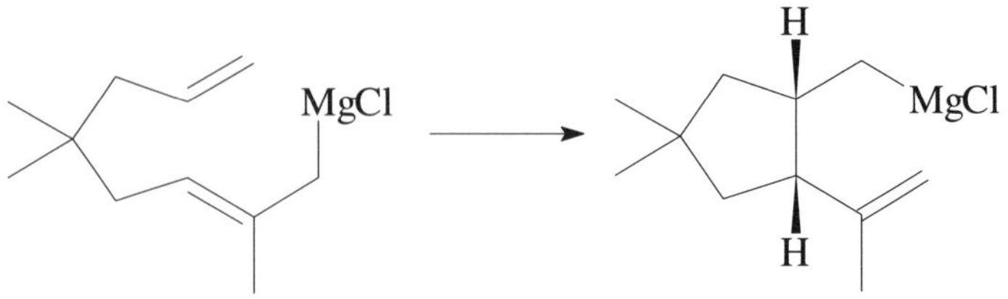

9   **10**

The *retro-ene reaction* also is of synthetic importance. While the application
of high pressure facilitates the ene reaction, the retro-ene reaction is favored
at higher temperatures.[2] Furthermore small-ring strain can shift the equilibrium towards the side of the dienes. The vinylcyclopropane **11** rearranges by
a synchronous process to the open-chain diene **12**. Formally this process is the
reverse of an intramolecular ene reaction:

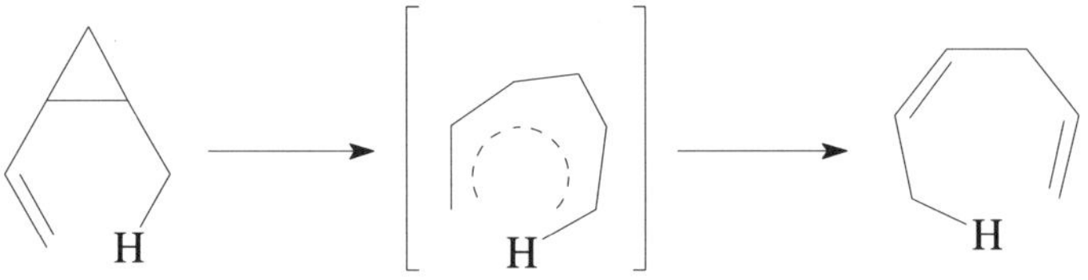

11   **12**

β-Hydroxyalkenes are especially suitable starting materials for the retro-ene reaction; since a stable carbonyl compound is then released as product. The retro-ene
reaction of β-hydroxyalkynes, e.g. **13** → **14**, can be used for the preparation of
allenes:[7,8]

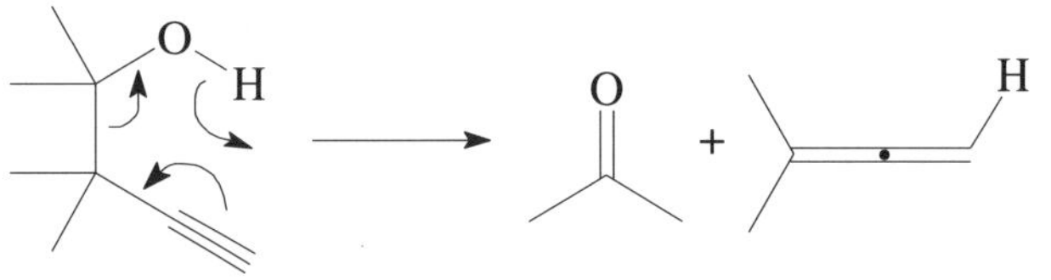

13   **14**

The reaction conditions for the ene reaction of simple starting materials are, for
example, 220 °C for 20 h in an aromatic solvent like trichlorobenzene. Lewis
acid-catalyzed intramolecular reactions have been described, e.g. with $FeCl_3$ in
dichloromethane at −78 °C.[4] Yields strongly depend on substrate structure.

1. K. Alder, F. Pascher, A. Schmitz, *Ber. Dtsch. Chem. Ges.* **1943**, *76*, 27–53.
2. H. M. R. Hoffmann, *Angew. Chem.* **1969**, *90*, 597–618;
   Angew. Chem. Int. Ed. Engl. **1969**, *8*, 556.
3. W. Oppolzer, V. Snieckus, *Angew. Chem.* **1978**, *90*, 506–516;
   Angew. Chem. Int. Ed. Engl. **1978**, *17*, 476.

4. L. F. Tietze, U. Beifuß, *Synthesis* **1988**, 359–362.
5. L. F. Tietze, U. Beifuß, *Angew. Chem.* **1985**, *97*, 1067–1069; *Angew. Chem. Int. Ed. Engl.* **1985**, *24*, 1042.
6. W. Oppolzer, A. Nakao, *Tetrahedron Lett.* **1986**, *27*, 5471–5474.
7. A. Viola, J. J. Collins, N. Filipp, *Tetrahedron* **1981**, *37*, 3772–3811.
8. H. Hopf, R. Kirsch, *Tetrahedron Lett.* **1985**, *26*, 3327–3330.

Ester Pyrolysis

Alkenes by pyrolysis of carboxylic esters

Carboxylic esters **1**, having an O-alkyl group with a β-hydrogen, can be cleaved thermally into the corresponding carboxylic acid **2** and an alkene **3**.[1,2] This reaction often is carried out in the gas phase; generally the use of a solvent is not necessary.

The reaction proceeds by an E_i-mechanism. The β-hydrogen and the carboxylate are cleaved synchronously from the substrate molecule, while forming a new bond. This elimination reaction belongs to the class of *syn*-eliminations; in the case of the ester pyrolysis, the substrate molecule passes through a six-membered cyclic transition state **4**:

For the formation of the new double bond, the general rules for eliminations do apply. Following *Bredt's rule*, no double bond to a bridgehead carbon atom will be formed. If the elimination can lead to a conjugated system of unsaturated groups, this pathway will be favored. Otherwise the Hofmann rule will be followed, which favors an elimination towards the less substituted carbon center.

With cyclic substrates, the formation of the new double bond depends on the availability of a *cis*-β-hydrogen, which is required for the *syn*-elimination mechanism. In the case of six-membered ring substrates this means that the β-hydrogen has to be equatorial for elimination with an axial acetate group (OAc), and axial for elimination with an equatorial acetate. The six-membered cyclic transition state **4** does not have to be completely coplanar. For example the pyrolysis of the cyclohexane derivative **5**, bearing an axial acetate group, yields the alkene **6** only,[3] by elimination of the axial acetate and the equatorial β-hydrogen:

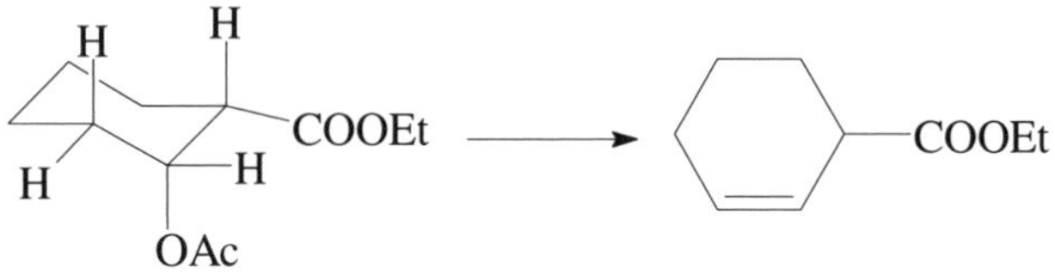

In the case of the cyclohexane derivative **7** however, that bears an equatorial acetate group, two axial cis-β-hydrogens are available, and elimination in both directions is possible. The pyrolysis of **7** yields the two elimination products **8** and **6**. Formation of product **8** is strongly favored, because the new double bond is in conjugation to the ester carbonyl group.[3]

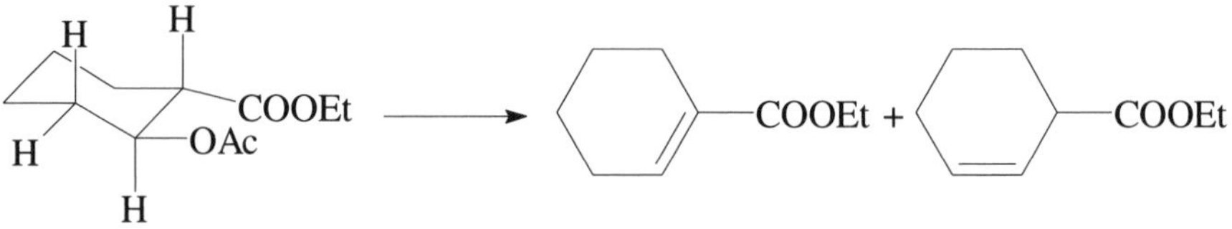

Rearrangements and other side-reactions are rare. The ester pyrolysis is therefore of some synthetic value, and is used instead of the dehydration of the corresponding alcohol. The experimental procedure is simple, and yields are generally high. Numerous alkenes have been prepared by this route for the first time. For the preparation of higher alkenes ($> C_{10}$), the pyrolysis of the corresponding alcohol in the presence of acetic anhydride may be the preferable method.[4] The pyrolysis of lactones **9** leads to unsaturated carboxylic acids **10**:[5]

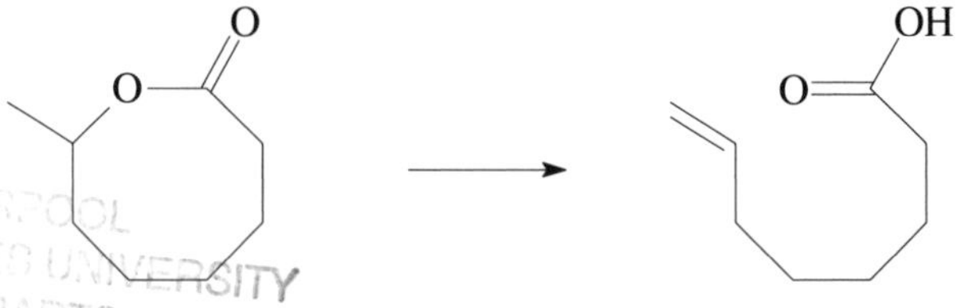

Since the *syn*-elimination mechanism requires formation of a six-membered cyclic transition state, this reaction is not possible for five- or six-membered lactones, but may be applied to higher homologs.

1.   C. H. DePuy, R. W. King, *Chem. Rev.* **1960**, *60*, 431–457.
2.   A. Maccoll in *The Chemistry of Alkenes* (Ed.: S. Patai), Wiley, New York, **1964**, p. 217–221.
3.   W. A. Bailey, R. A. Baylouny, *J. Am. Chem. Soc.* **1959**, *81*, 2126–2129.
4.   D. W. Aubrey, A. Barnatt, W. Gerrard, *Chem. Ind. (London)* **1965**, 681.
5.   W. J. Bailey, C. N. Bird, *J. Org. Chem.* **1977**, *42*, 3895–3899.

F

Favorskii Rearrangement

Carboxylic esters from α-haloketones

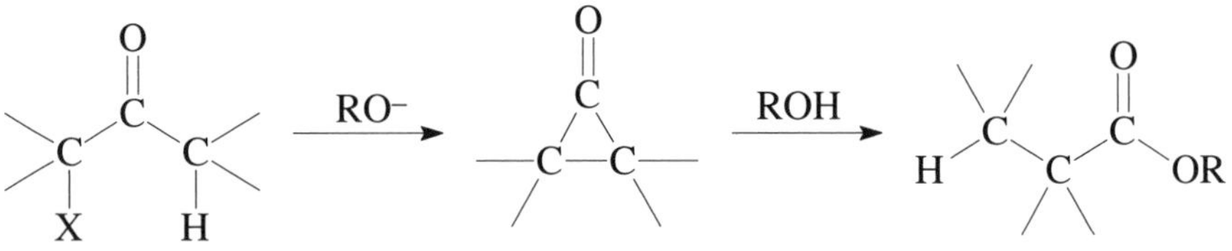

α-Halo ketones **1**, when treated with a base, can undergo a rearrangement reaction to give a carboxylic acid or a carboxylic acid derivative **3**, e.g. an ester or amide, depending on the base used. This reaction is called the *Favorskii rearrangement*.[1,2] As base a hydroxide, alkoxide or amine is used; the halogen substituent can be a chlorine, bromine or iodine atom.

In the initial step[3,4] the α-halo ketone **1** is deprotonated by the base at the α'-carbon to give the carbanion **4**, which then undergoes a ring-closure reaction by an intramolecular substitution to give the cyclopropanone derivative **2**. The halogen substituent functions as the leaving group:

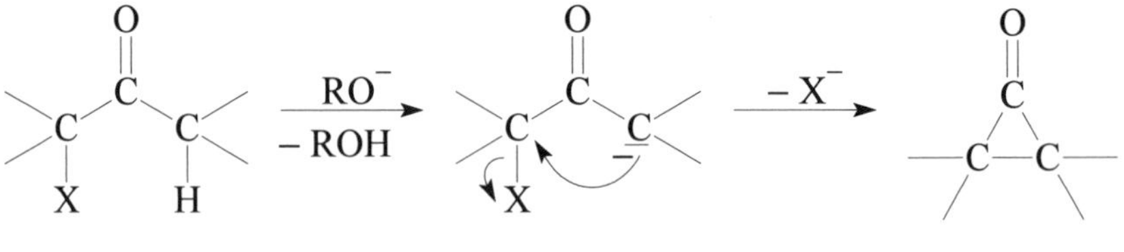

Nucleophilic addition of the base to the intermediate **2** leads to ring opening. With a symmetrically substituted cyclopropanone, cleavage of either C_α–CO bond leads to the same product. With unsymmetrical cyclopropanones, that bond is broken preferentially that leads to the more stable carbanion **5**:

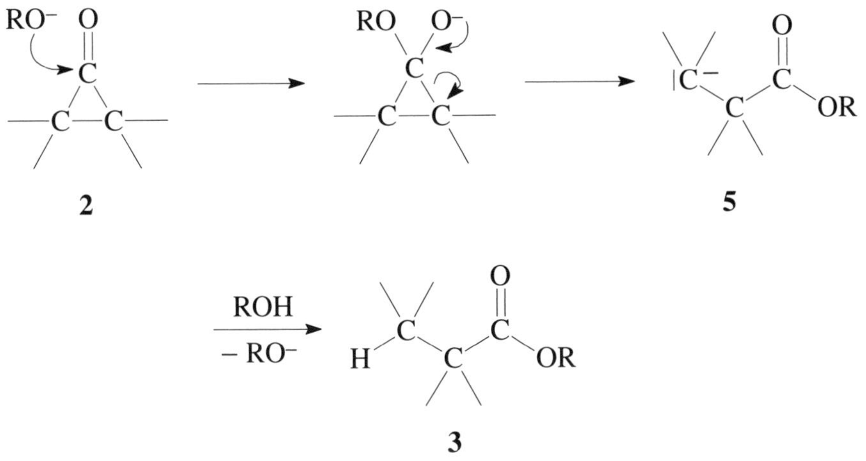

The carbanionic species thus formed is protonated to give the final product **3**. The use of an alkoxide as base leads to formation of a carboxylic ester as rearrangement product; use of a hydroxide will lead to formation of a carboxylic acid salt:

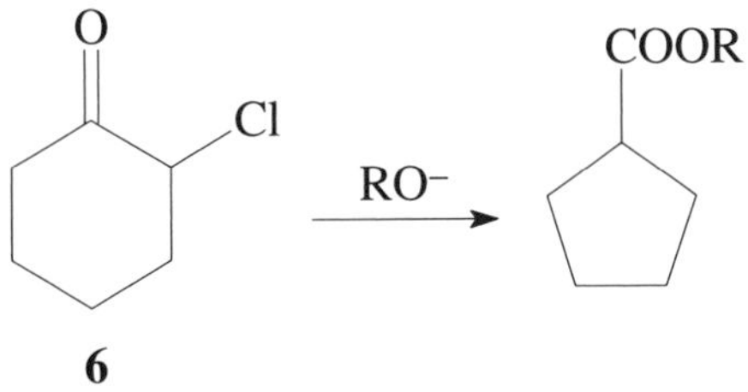

With cyclic α-halo ketones, e.g. 2-chloro cyclohexanone **6**, the Favorskii rearrangement leads to a ring contraction by one carbon atom. This type of reaction has for example found application as a key step in the synthesis of cubane by *Eaton* and *Cole*[5] for the construction of the cubic carbon skeleton:

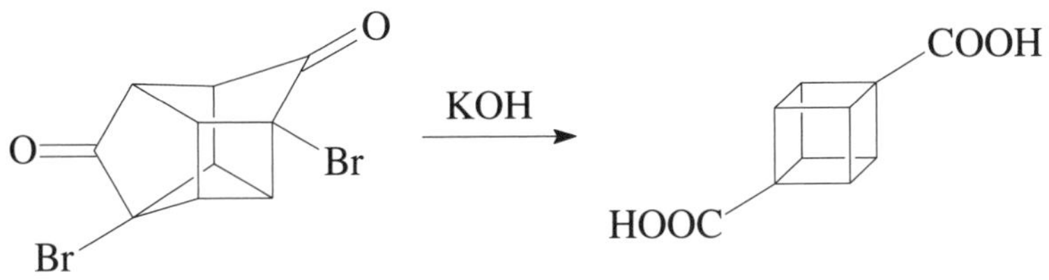

Under Favorskii conditions α,α-dihalo ketones **7**, as well as α,α'-dihalo ketones, bearing one α'-hydrogen, rearrange to give α,β-unsaturated esters **8**:[6]

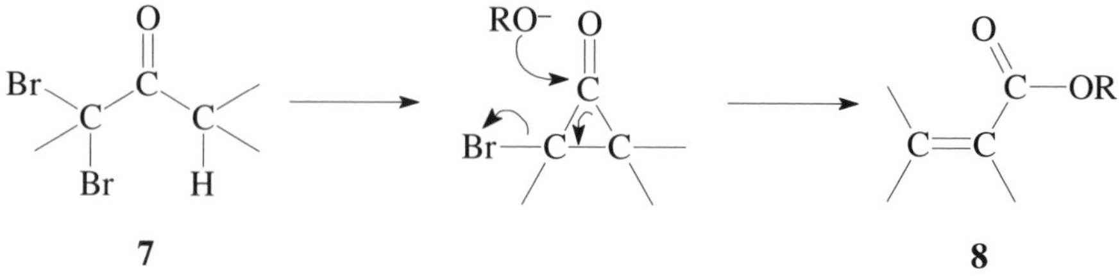

7   8

The rearrangement with ring contraction probably is the most important synthetic application of the Favorskii reaction; it is for example used in the synthesis of steroids. Yields can vary from good to moderate. As solvents diethyl ether or alcohols are often used. With acyclic α-halo ketones bearing voluminous substituents in α'-position, yields can be low; a *tert*-butyl substituent will prevent the rearrangement.

1. A. Favorskii, *J. Prakt. Chem.* **1895**, *51*, 533–563.
2. A. S. Kende, *Org. React.* **1960**, *11*, 261–316.
3. C. Rappe in *The Chemistry of the Carbon-Halogen Bond* (Ed.: S. Patai), Wiley, New York, **1973**, *Vol. 2*, p. 1084–1101.
4. H. H. Wasserman, G. M. Clark, P. C. Turley, *Top. Curr. Chem.* **1974**, *47*, 73–156.
5. P. E. Eaton, T. W. Cole, Jr., *J. Am. Chem. Soc.* **1964**, *86*, 962.
6. A. Abad, M. Arnó, J. R. Pedro, E. Seone, *Tetrahedron Lett.* **1981**, 1733–1736.

Finkelstein Reaction

Exchange of the halogen in alkyl halides

$$R-X \underset{-X^-}{\overset{+Y^-}{\rightleftharpoons}} R-Y$$

The synthesis of alkyl halides from alkyl halides is called the *Finkelstein reaction*.[1–3]

For preparative use it is necessary to shift the equilibrium in favor of the desired product. This may for example be achieved by taking advantage of different solubilities of the reactants.

With primary alkyl halides **1** the Finkelstein reaction proceeds by a S_N2-mechanism. An alkali halide is used to deliver the nucleophilic halide anion:[3]

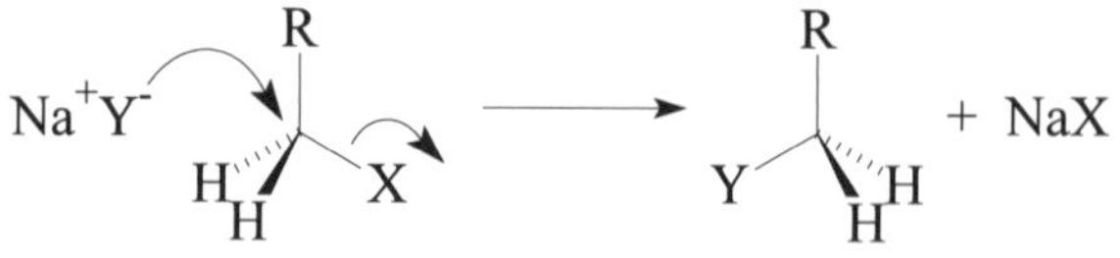

1

Of preparative importance is the substitution of chloride or bromide or iodide, since the more reactive alkyl iodides are better substrates for further transformations. Alkyl iodides often are difficult to prepare directly, which is why the conversion of readily accessible chlorides or bromides *via* a Finkelstein reaction is often preferred.

Differences in solubility of the reactants may for example be utilized as follows. Sodium iodide is much more soluble in acetone than are sodium chloride or sodium bromide. Upon treatment of an alkyl chloride or bromide with sodium iodide in acetone, the newly formed sodium chloride or bromide precipitates from the solution and is thus removed from equilibrium. Alkyl iodides can be conveniently prepared in good yields by this route. Alkyl bromides are more reactive as the corresponding chlorides. Of high reactivity are α-halogen ketones, α-halogen carboxylic acids and their derivatives, as well as allyl and benzyl halides.

Secondary or tertiary alkyl halides are much less reactive. For example an alkyl dichloride with a primary and a secondary chloride substituent reacts selectively by exchange of the primary chloride. The reactivity with respect to the Finkelstein reaction is thus opposite to the reactivity for the hydrolysis of alkyl chlorides. For the Finkelstein reaction on secondary and tertiary substrates Lewis acids may be used,[4] e.g. $ZnCl_2$, $FeCl_3$ or Me_3Al.

Alkyl fluorides can be prepared by the Finkelstein reaction.[5,6] The fluoride anion is a bad leaving group; the reverse reaction thus does not take place easily, and the equilibrium lies far to the right. As reagents potassium fluoride, silver fluoride or gaseous hydrogen fluoride may be used.

1. W. H. Perkin, B. F. Duppa, *Justus Liebigs Ann. Chem.* **1859**, *112*, 125–127.
2. H. Finkelstein, *Ber. Dtsch. Chem. Ges.* **1910**, *43*, 1528–1535.
3. A. Roedig, *Methoden Org. Chem. (Houben-Weyl)* **1960**, Vol. 5/4 p. 595–605.
4. J. A. Miller, M. J. Nunn, *J. Chem. Soc., Perkin Trans. 1*, **1976**, 416–420.
5. A. L. Henne, *Org. React.* **1944**, *2*, 49–93.
6. S. Rozen, R. Filler, *Tetrahedron* **1985**, *41*, 1111–1153.

Fischer Indole Synthesis

Indoles from aryl hydrazones

Heating of an aryl hydrazone **1** in the presence of a catalyst leads to elimination of ammonia and formation of an indole **2**. This reaction is known as the *Fischer indole synthesis*,[1-7] and is somewhat related to the *Benzidine rearrangement*.

A mechanism, that has been proposed by *G. M. Robinson* and *R. Robinson*,[8] consists of three steps. Initially the phenyl hydrazone **1** undergoes a reversible rearrangement to give the reactive ene-hydrazine **3**:

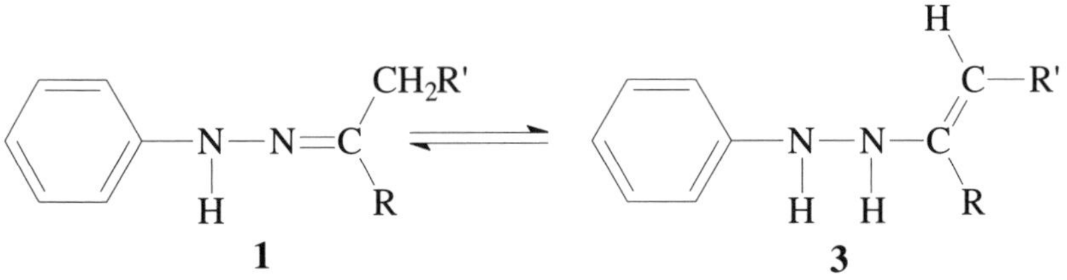

A [3.3]-sigmatropic rearrangement of **3** leads to formation of a new carbon–carbon bond and the cationic species **4**:

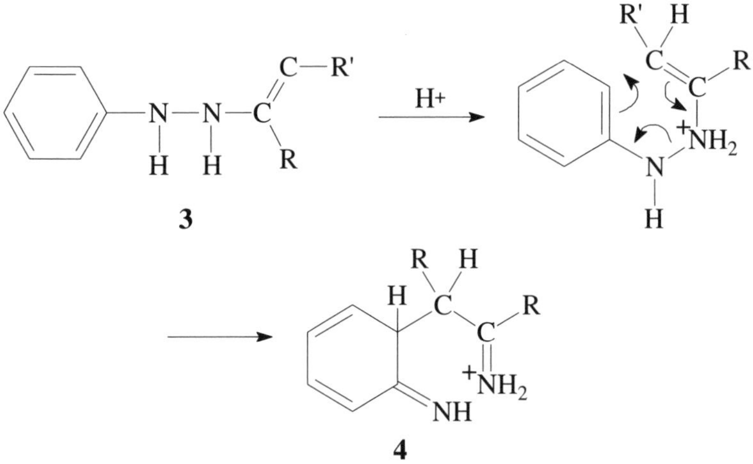

That electrocyclic reaction is related to the *Claisen rearrangement* of phenyl vinyl ether. In a final step a cyclization takes place with subsequent elimination of ammonia to yield the indole **2**:

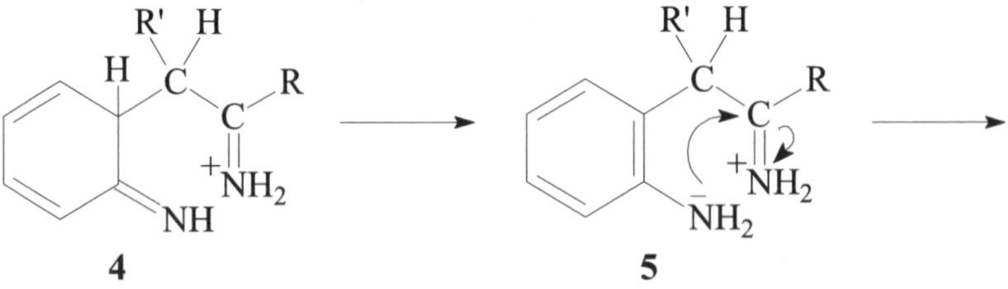

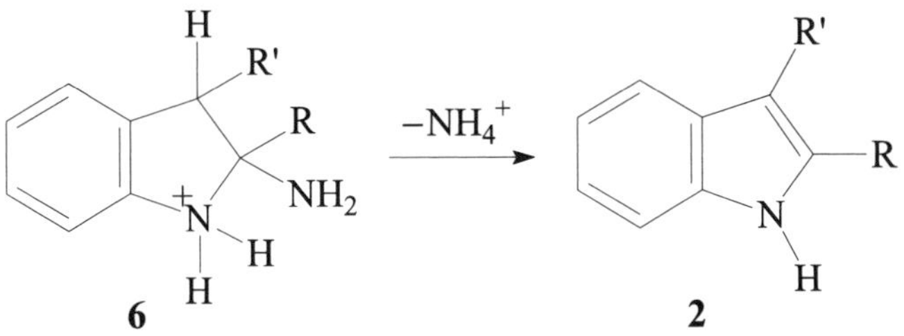

The mechanism outlined above is supported by experimental findings. An intermediate **5** has been isolated,[9–11] and has been identified by ^{13}C- and ^{15}N-nuclear magnetic resonance spectroscopy.[12] Side-products have been isolated, which are likely to be formed from intermediate **4**.[3] ^{15}N-isotope labeling experiments have shown that only the nitrogen remote from the phenyl group is eliminated as ammonia.[13]

Metal halides like zinc chloride are used as Lewis-acid catalysts. Other Lewis-acids or protic acids, as well as transition metals, have found application also. The major function of the catalyst seems to be the acceleration of the second step—the formation of the new carbon–carbon bond.

The hydrazones **1** used as starting materials are easily prepared by reaction of an aldehyde or ketone **8** with an aryl hydrazine **7**:

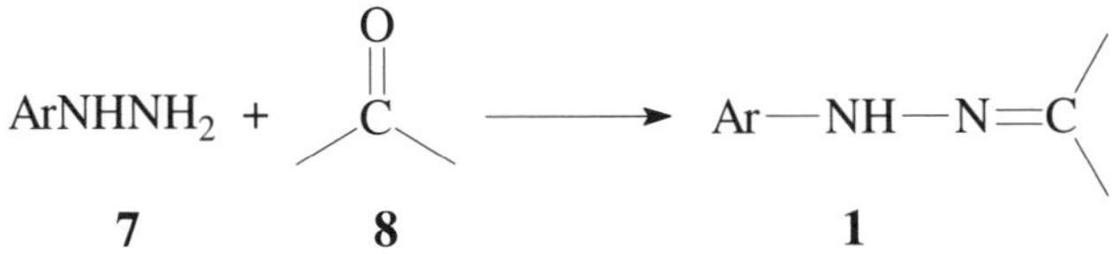

In order to allow further transformation to an indole, the carbonyl compound **8** must contain an α-methylene group. The hydrazone **1** needs not to be isolated. An equimolar mixture of arylhydrazine **7** and aldehyde or ketone **8** may be treated directly under the reaction conditions for the Fischer indole synthesis.[3]

Another route to suitable arylhydrazones is offered by the *Japp–Klingemann reaction*.

The Fischer indole synthesis is of wide scope, and can be used for the preparation of substituted indoles and related systems. For example reaction of the phenylhydrazone **9**, derived from cyclohexanone, yields the tetrahydrocarbazole **10**:[5,6,7]

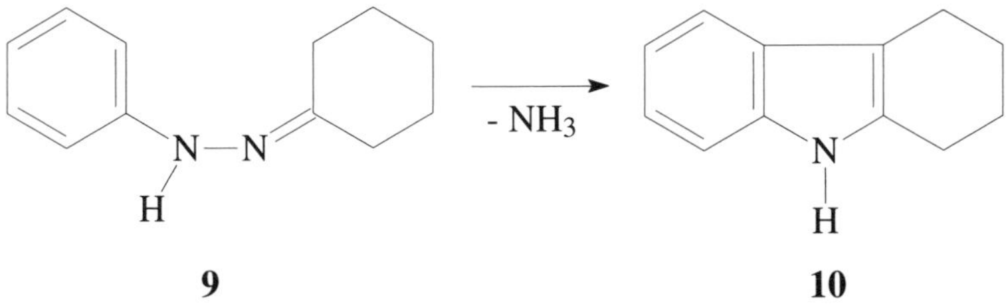

1. E. Fischer, F. Jourdan, *Ber. Dtsch. Chem. Ges.* **1883**, *16*, 2241–2245.
2. B. Robinson, *Chem. Rev.* **1969**, *69*, 227–250.
3. B. Robinson, *Chem. Rev.* **1963**, *63*, 373–401.
4. I. I. Grandberg, V. I. Sorodkin, *Russ. Chem. Rev.* **1974**, *43*, 115–128.
5. H. J. Shine, *Aromatic Rearrangements*, American Elsevier, New York, **1969**, p. 190–207.
6. R. J. Sundberg, *The Chemistry of Indoles*, Academic Press, New York, **1970**, p. 142–163.
7. B. Robinson, *The Fischer Indole Synthesis*, Wiley, New York, **1982**.
8. G. M. Robinson, R. Robinson, *J. Chem. Soc.* **1918**, *113*, 639–643.
9. P. L. Southwick, B. McGrew, R. R. Enge, G. E. Milliman, R. J. Owellen, *J. Org. Chem.* **1963**, *28*, 3058–3065.
10. P. L. Southwick, J. A. Vida, B. M. Fitzgerald, S. K. Lee, *J. Org. Chem.*, **1968**, *33*, 2051–2056.
11. T. P. Forrest, F. M. F. Chen, *J . Chem. Soc., Chem. Commun.* **1972**, 1067.
12. A. W. Douglas, *J. Am. Chem. Soc.* **1978**, *100*, 6463–6469; *J. Am. Chem. Soc.* **1979**, *101*, 5676–5678.
13. K. Clusius, H. R. Weisser, *Helv. Chim. Acta* **1952**, *35*, 400–406.

Friedel–Crafts Acylation

Acylation of aromatic compounds

The most important method for the synthesis of aromatic ketones **3** is the *Friedel–Crafts acylation*.[1–4] An aromatic substrate **1** is treated with an acyl chloride **2** in the presence of a Lewis-acid catalyst, to yield an acylated aromatic compound. Closely related reactions are methods for the formylation, as well as an alkylation procedure for aromatic compounds, which is also named after *Friedel* and *Crafts*.

The reaction is initiated by formation of a donor–acceptor complex **4** from acyl chloride **2**, which is thereby activated, and the Lewis acid, e.g. aluminum trichloride. Complex **4** can dissociate into the acylium ion **5** and the aluminum tetrachloride anion; **4** as well as **5** can act as an electrophile in a reaction with the aromatic substrate:

$$\longrightarrow \quad R-\overset{+}{C}=O + AlCl_4^-$$

5

Depending on the specific reaction conditions, complex **4** as well as acylium ion **5** have been identified as intermediates; with a sterically demanding substituent R, and in polar solvents the acylium ion species **5** is formed preferentially.[5] The electrophilic agent **5** reacts with the aromatic substrate, e.g. benzene **1**, to give an intermediate σ-complex—the cyclohexadienyl cation **6**. By loss of a proton from intermediate **6** the aromatic system is restored, and an arylketone is formed that is coordinated with the carbonyl oxygen to the Lewis acid. Since a Lewis-acid molecule that is coordinated to a product molecule is no longer available to catalyze the acylation reaction, the catalyst has to be employed in equimolar quantity. The product-Lewis acid complex **7** has to be cleaved by a hydrolytic workup in order to isolate the pure aryl ketone **3**.

Product complex **7** as well as the free product **3** are much less reactive towards further electrophilic substitution as is the starting material; thus the formation of polyacylated products is not observed. If the starting material bears one or more non-deactivating substituents, the direction of acylation can be predicted by the general rules for aromatic substitution.

Drawbacks as known from the *Friedel–Crafts alkylation* are not found for the Friedel–Crafts acylation. In some cases a decarbonylation may be observed as a side-reaction, e.g. if loss of CO from the acylium ion will lead to a stable carbenium species **8**. The reaction product of the attempted acylation will then be rather an alkylated aromatic compound **9**:

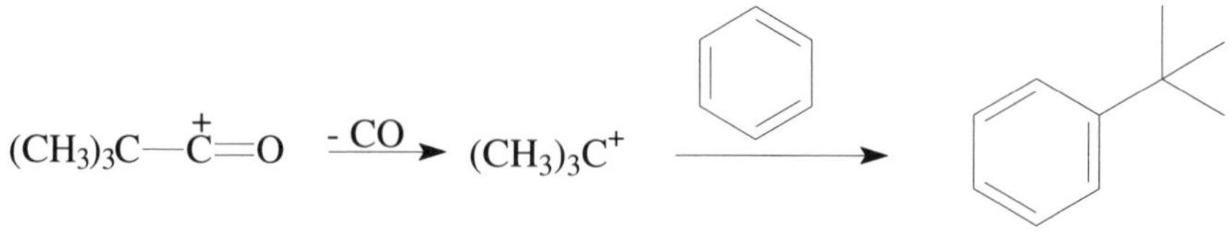

$$(CH_3)_3C\overset{+}{\underset{}{-}}\overset{}{C}=O \xrightarrow{\text{- CO}} (CH_3)_3C^+ \longrightarrow$$

$$\mathbf{8} \qquad\qquad\qquad\qquad \mathbf{9}$$

An important application of the Friedel–Crafts acylation is the intramolecular reaction, leading to ring closure. This variant is especially useful for the closure of six-membered rings, but five-membered ring products and larger rings are also accessible:

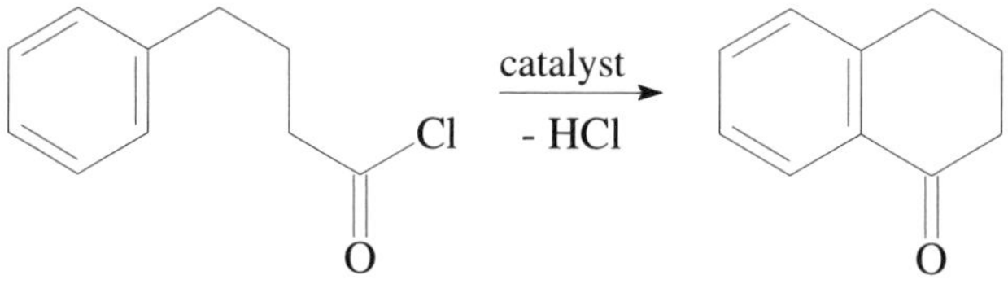

$$\xrightarrow[\text{- HCl}]{\text{catalyst}}$$

As acylating agent, a carboxylic anhydride may be used instead of the acyl halide. The reaction then yields the arylketone together with a carboxylic acid, each of which forms a complex with the Lewis acid used. The catalyst therefore has to be employed in at least twofold excess:

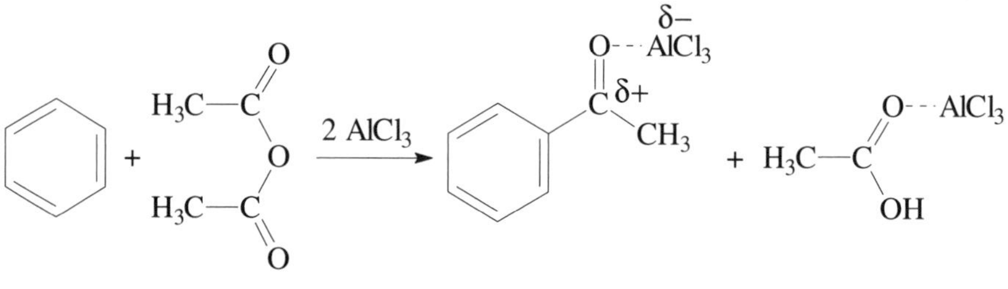

$$\xrightarrow{\text{2 AlCl}_3}$$

With a mixed anhydride two different arylketones may be formed. Reaction of a cyclic anhydride of a dicarboxylic acid, e.g. succinic anhydride, leads to formation of an arylketo acid.[2]

A carboxylic acid may also be employed directly as acylating agent, without being first converted into an acyl halide; in that case a protic acid is used as catalyst.

As an illustrating example for the application of the Friedel–Crafts acylation in the synthesis of complex molecules, its use in the synthesis of [2.2.2]cyclophane **13** by *Cram and Truesdale*[6] shall be outlined. The reaction of [2.2]paracyclophane **10** with acetyl chloride gives the acetyl-[2.2]paracyclophane **11**, which is converted into the pseudo-geminal disubstituted phane **12** by a *Blanc reaction*, and further to the triple bridge hydrocarbon **13**:

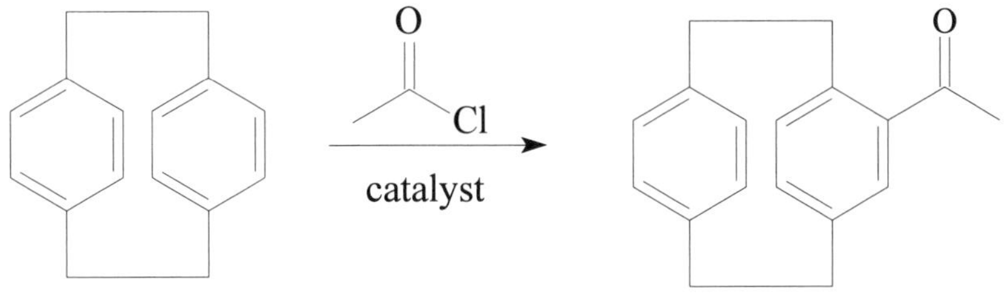

10   **11**

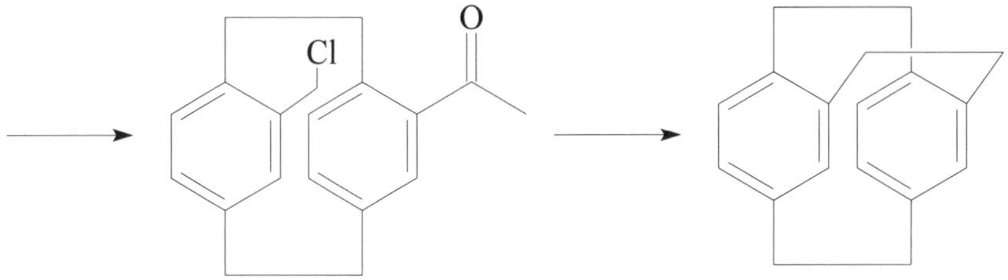

12   **13**

The Friedel–Crafts reaction is one of the most important reactions in organic chemistry. Nitrobenzene does not react; it may even be used as solvent. Phenols are acylated at oxygen; the phenyl ester thus obtained, can be converted into an *o*- or *p*-acylphenol by the *Fries reaction*. Many aromatic heterocycles do react; however pyridine as well as quinoline are unreactive. As catalyst a Lewis acid, e.g. $AlCl_3$, $ZnCl_2$, BF_3, SbF_5,[7] or a protic acid such as H_2SO_4, H_3PO_4 and $HClO_4$ is used. The necessity of large amounts of catalyst has been outlined above. In some cases, a Friedel–Crafts acylation can be carried out with small amounts or even without catalyst; however the application of higher temperatures is then generally required.[8]

1. G. A. Olah, *Friedel-Crafts and Related Reactions*, Wiley, New York, **1963–1964**, Vol. 1 and 2.
2. E. Berliner, *Org. React.* **1949**, *5*, 229–289.
3. R. Taylor, *Electrophilic Aromatic Substitution*, Wiley, New York, **1990**, p. 222–238.
4. B. Chevrier, R. Weiss, *Angew. Chem.* **1974**, *86*, 12–21;
    Angew. Chem. Int. Ed. Engl. **1974**, *13*, 1.
5. D. Cassimatis, J. P. Bonnin, T. Theophanides, *Can. J. Chem.* **1970**, *48*, 3860–3871.
6. D. J. Cram, E. A. Truesdale, *J. Am. Chem. Soc.* **1973**, *95*, 5825–5827.
7. D. E. Pearson, C. A. Buehler, *Synthesis* **1972**, 533–542.
8. G. G. Yakobson, G. G. Furin, *Synthesis* **1980**, 345–364.

Friedel–Crafts Alkylation

Alkylation of aromatic compounds

$$\mathbf{1} \qquad \mathbf{2} \qquad \mathbf{3}$$

The synthesis of an alkylated aromatic compound **3** by reaction of an aromatic substrate **1** with an alkyl halide **2**, catalyzed by a Lewis acid, is called the *Friedel–Crafts alkylation*.[1–4] This method is closely related to the *Friedel–Crafts acylation*. Instead of the alkyl halide, an alcohol or alkene can be used as reactant for the aromatic substrate under Friedel–Crafts conditions. The general principle is the intermediate formation of a carbenium ion species, which is capable of reacting as the electrophile in an electrophilic aromatic substitution reaction.

The initial step is the coordination of the alkyl halide **2** to the Lewis acid to give a complex **4**. The polar complex **4** can react as electrophilic agent. In cases where the group R can form a stable carbenium ion, e.g. a *tert*-butyl cation, this may then act as the electrophile instead. The extent of polarization or even cleavage of the R–X bond depends on the structure of R as well as the Lewis acid used. The addition of carbenium ion species to the aromatic reactant, e.g. benzene **1**, leads to formation of a σ-complex, e.g. the cyclohexadienyl cation **6**, from which the aromatic system is reconstituted by loss of a proton:

$$\mathbf{2} \qquad\qquad \mathbf{4} \qquad\qquad \mathbf{5}$$

$$\mathbf{1} \qquad\qquad\qquad \mathbf{6}$$

$$\mathbf{3}$$

That mechanism is supported by the detection of such σ-complexes at low temperatures.[5,6] An analogous mechanism can be formulated with a polarized species **4** instead of the free carbenium ion **5**.

If the alkyl halide contains more than one, equally reactive C-halogen centers, these will generally react each with one aromatic substrate molecule. For example dichloromethane reacts with benzene to give diphenylmethane, and chloroform will give triphenylmethane. The reaction of tetrachloromethane with benzene however stops with the formation of triphenyl chloromethane **7** (trityl chloride), because further reaction is sterically hindered:

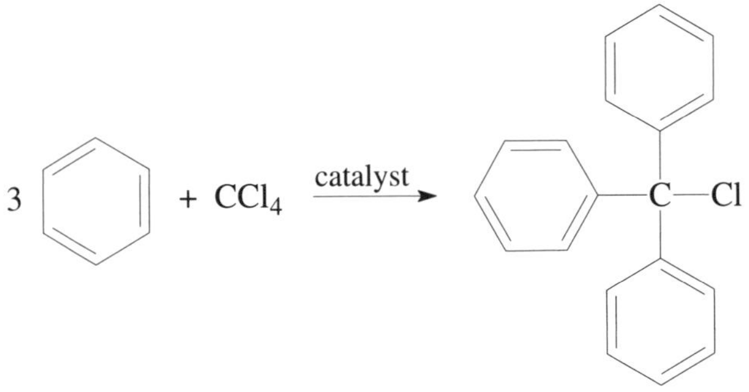

7

The intramolecular variant[3] of the Friedel–Crafts alkylation is also synthetically useful, especially for the closure of six-membered rings, e.g. the synthesis of tetraline **8**; but five- and seven-membered ring products are also accessible:

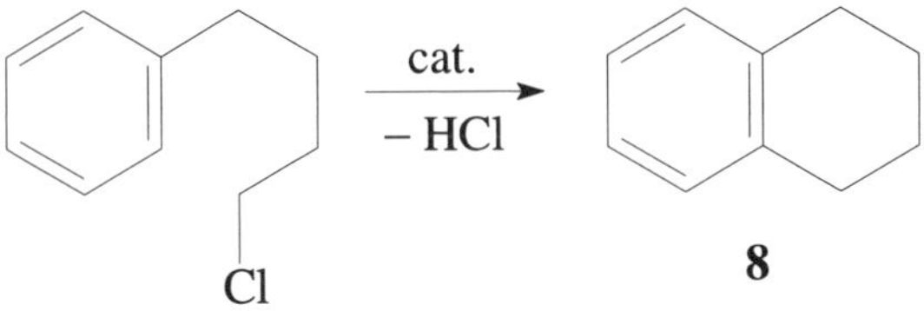

8

The alkylation with alkenes can be catalyzed by protons. The carbon–carbon double bond of the alkene is protonated according to *Markownikoff's rule*, to give a carbenium ion **10**, which then reacts by the above mechanism to yield the alkylated aromatic product **11**:

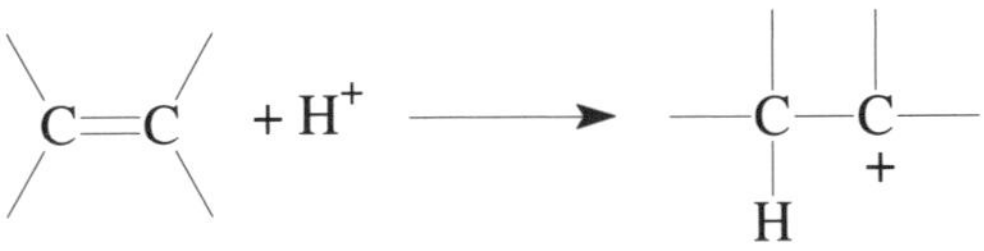

8   **10**

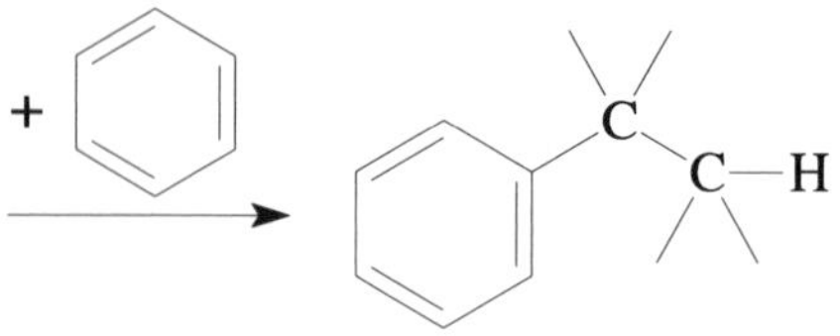

11

Alcohols can be converted into reactive species by reaction with a Lewis acid, e.g. $AlCl_3$, or by protonation and subsequent loss of H_2O to give a carbenium ion **12**.

$$ROH + AlCl_3 \longrightarrow ROAlCl_2 \longrightarrow R^+ + OAlCl_2^-$$
12

$$ROH + H^+ \longrightarrow ROH_2^+ \longrightarrow R^+ + H_2O$$
12

In contrast to the Friedel–Crafts acylation, the alkylation is a reversible reaction. That feature can be used for the regioselective synthesis of substituted aromatic derivatives.[6] The *t*-butyl group can be used as a bulky protecting group, that can be removed later. In the following example, an *ortho* substituted phenol is to be synthesized: without the *t*-butyl-group *para* to the substituent R, the usual reactivity with respect to an incoming second substituent would lead to a mixture or *ortho*- and *para*-substituted product. With the *t*-butyl group blocking the *para*-position, the sulfonation occurs *ortho* to R only. After conversion of the sulfonic acid to the phenol, and removal of the *t*-butyl group, the desired *ortho*-substituted phenol **13** is obtained:

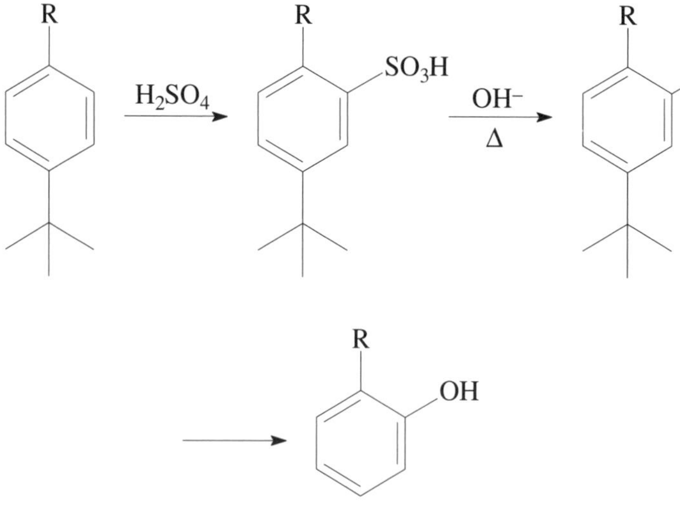

13

The applicability of the Friedel–Crafts alkylation reaction in organic synthesis is somewhat limited for the following reasons. Due to the activating effect of an alkyl group connected to an aromatic ring, the monoalkylated reaction product is more reactive towards electrophilic substitution than the original starting material. This effect favors the formation of di- or even poly-substituted products. The scope of the reaction is limited by the reactivity of certain starting materials. Naphthalenes and related polycyclic aromatic substrates may undergo side reactions because of their high reactivity towards the catalyst, and give low yields of monoalkylated product. Many aromatic heterocycles are not suitable substrates for a Friedel–Crafts alkylation. Functional groups like $-OH$, $-NH_2$, $-OR$, that coordinate to the Lewis acid also should not be present on the aromatic ring. Another problem is the formation of rearranged products, either from reaction of rearranged carbenium ions or migration of the alkyl substituent at the aromatic ring. When benzene **1** is treated with 1-bromopropane under Friedel–Crafts conditions, the rearranged product i-propylbenzene (cumene) **15** is obtained as the major product, together with the expected n-propylbenzene **14**:

| **1** | **14** | **15** |

Since the alkylation reaction is reversible, a rearrangement of the initial alkylation product can take place, resulting in a migration of the alkyl group on the aromatic ring. This can be used for the deliberate isomerization of alkylated products. Because of these complications it can be more effective to prepare an alkylated aromatic derivative by first conducting a Friedel–Crafts acylation, and then reduce the keto group to a methylene group in order to get the alkyl side chain. This route has one additional step, but avoids the drawbacks mentioned above.

As catalysts Lewis acids such as $AlCl_3$, $TiCl_4$, SbF_5, BF_3, $ZnCl_2$ or $FeCl_3$ are used. Protic acids such as H_2SO_4 or HF are also used, especially for reaction with alkenes or alcohols. A recent development is the use of acidic polymer-resins, e.g. Nafion-H, as catalysts for Friedel–Crafts alkylations.[8]

1. C. Friedel, J. M. Crafts, *J. Chem. Soc.* **1877**, *32*, 725.
2. C. C. Price, *Org. React.* **1946**, *3*, 1–82.
3. G. A. Olah, *Friedel–Crafts and Related Reactions*, Wiley, New York, **1963–1964**, Vol. 1 and 2.
4. R. Taylor, *Electrophilic Aromatic Substitution*, Wiley, New York, **1990**, p. 187–203.
5. G. A. Olah, S. J. Kuhn, *J. Am. Chem. Soc.* **1958**, *80*, 6541–6545.
6. F. Effenberger, *Chem. Unserer Zeit* **1979**, *13*, 87–94.
7. G. G. Yakobson, G. G. Furin, *Synthesis* **1980**, 345–364.
8. G. A. Olah, P. S. Iyer, G. K. S. Prakash, *Synthesis* **1986**, 513–531.

Friedländer Quinoline Synthesis

Condensation of *o*-aminobenzaldehydes with α-methylene carbonyl compounds

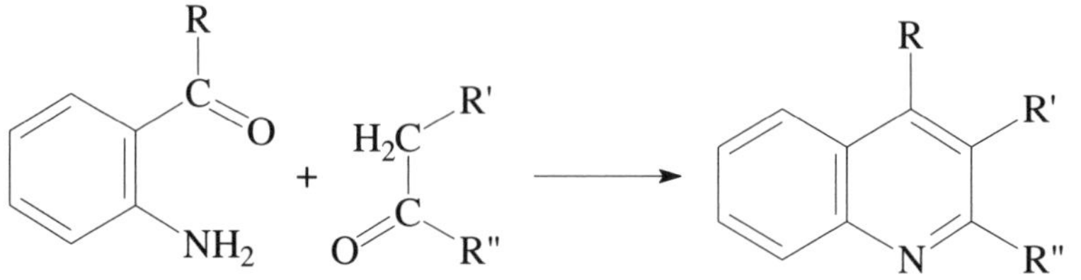

Quinolines **3** can be obtained from reaction of *ortho*-aminobenzaldehydes or *o*-aminoarylketones **1** with α-methylene carbonyl compounds.[1-3] Various modified procedures are known; a related reaction is the *Skraup quinoline synthesis*.

The mechanistic pathway of the ordinary Friedländer synthesis is not rigorously known. Two steps are formulated. In a first step a condensation reaction, catalyzed by acid or base, takes place, that can lead to formation of two different types of products: (a) an imine (Schiff base) **4**, or (b) an α,β-unsaturated carbonyl compound **5**:

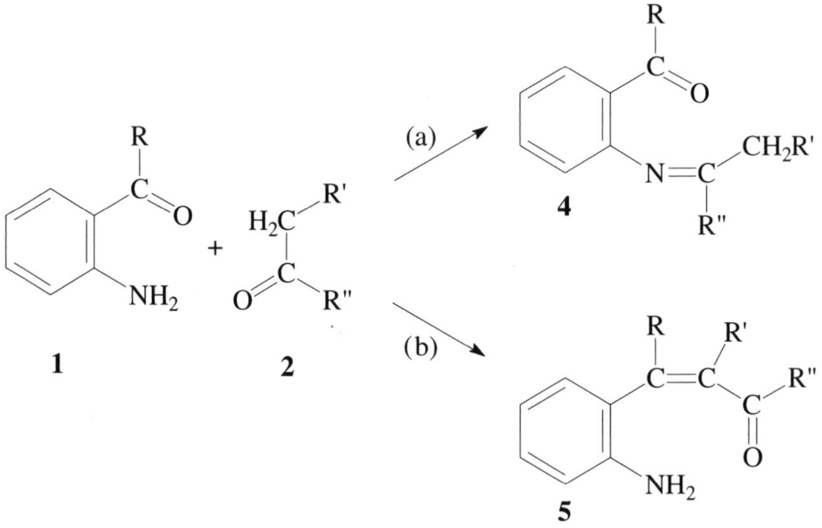

Although that reaction has been known for more than one hundred years, it is not clear whether the reaction proceeds *via* pathway (a) or (b) or both. Since the reaction works with a large number of different substrates and under various reaction conditions, e.g. catalyzed by acid or base, or without a catalyst, it is likely that the actual mechanistic pathway varies with substrate and reaction conditions.[3]

The next step in both cases is a dehydrative cyclization to yield the quinoline **3**:

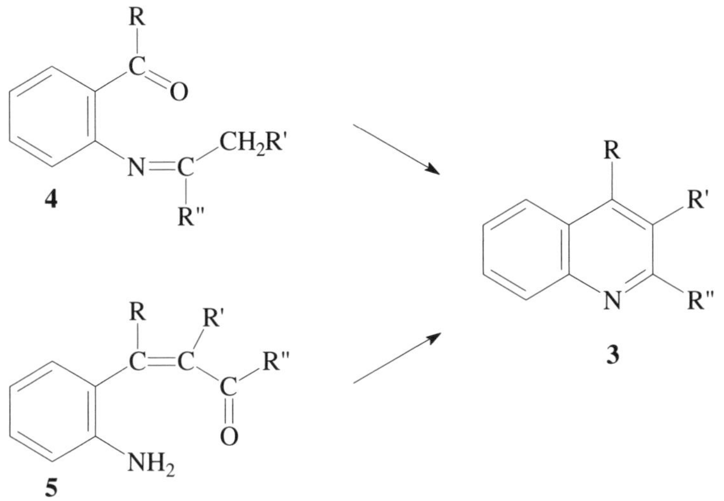

Since various substituents are tolerated, the Friedländer reaction is of preparative value for the synthesis of a large variety of quinoline derivatives. The benzene ring may bear for example alkyl, alkoxy, nitro or halogen substituents. Substituents R, R' and R'' also are variable.[3] The reaction can be carried out with various carbonyl compounds, that contain an enolizable α-methylene group. The reactivity of that group is an important factor for a successful reaction.

Usually the reaction is carried out in the presence of a basic catalyst, or simply by heating the reactants without solvent and catalyst.

As basic catalysts KOH, NaOH or piperidine are used. As acidic catalysts are used HCl, H_2SO_4, polyphosphoric acid or *p*-toluenesulfonic acid.

Although the uncatalyzed Friedländer reaction requires more drastic conditions, i.e. temperatures of 150–200 °C, it often gives better yields of quinolines.[3]

Certain quinolines can be prepared by heating a single suitable compound. For example acetanilide **6** rearranges upon heating in the presence of zinc chloride as catalyst, to give a mixture of *o*- and *p*-acetylaniline **7** and **8**. These two reactants then do undergo the condensation reaction to yield flavaniline **9** that has found application as a dyestuff:[4]

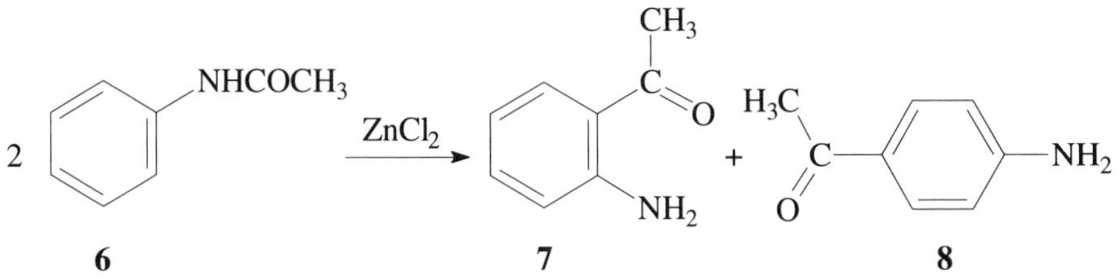

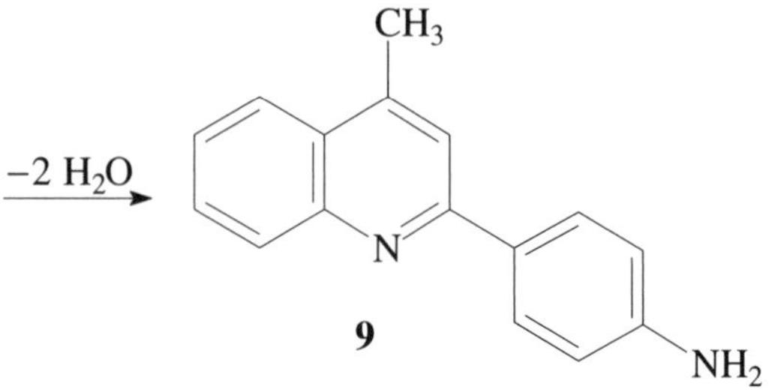

$$\xrightarrow{-2\ H_2O}$$

9

The Friedländer quinoline synthesis is particular useful for the preparation of 3-substituted quinolines, which are less accessible by other routes. A drawback however is the fact that the required *o*-aminobenzaldehydes or *o*-aminoarylketones are not as easy to prepare as, e.g., the anilines that are required for the Skraup synthesis.

1. P. Friedländer, *Ber. Dtsch. Chem. Ges.* **1883**, *16*, 1833–1839.
2. G. Jones, *Chem. Heterocycl. Compd.* **1977**, *32(1)*, 181–207.
3. C. Cheng, S. Yan, *Org. React.* **1982**, *28*, 37–201.
4. E. Besthorn, O. Fischer, *Ber. Dtsch. Chem. Ges.* **1883**, *16*, 68–75.

Fries Rearrangement

Acylphenols from phenyl esters

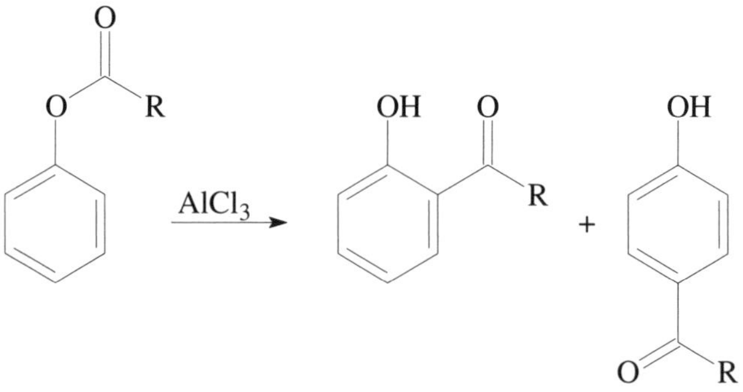

$$\xrightarrow{AlCl_3}$$

1   **2**   **3**

Phenolic esters (**1**) of aliphatic and aromatic carboxylic acids, when treated with a Lewis acid as catalyst, do undergo a rearrangement reaction to yield *ortho*- and *para*-acylphenols **2** and **4** respectively. This *Fries rearrangement* reaction[1,2] is an important method for the synthesis of hydroxyaryl ketones.

The reaction mechanism[3–5] is not rigorously known. Evidence for an intramolecular pathway as well as an intermolecular pathway has been found;

however crossover experiments did not lead to clear distinction. Results from extensive studies[3] suggest that a phenyl ester can rearrange by both pathways in the same reaction. The actual reactivity depends on substrate structure, reaction temperature, the solvent used, and the kind and concentration of Lewis acid used. Usually at least equimolar amounts of Lewis acid are employed.

The Lewis acid can coordinate to the substrate at either one of the oxygen centers, or even both when used in excess:[3]

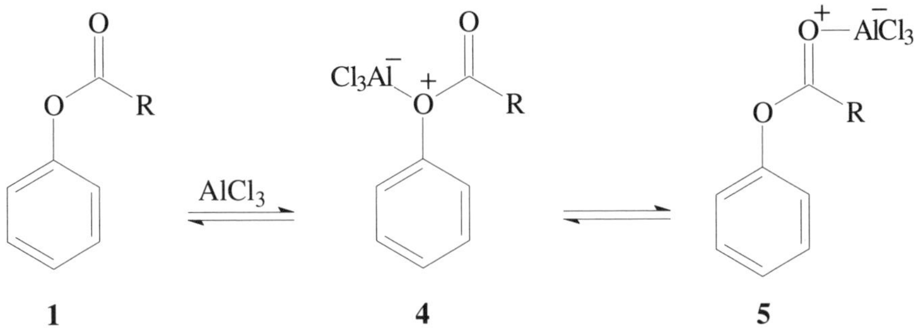

The Lewis acid complex **4** can cleave into an ion-pair that is held together by the solvent cage, and that consists of an acylium ion and a Lewis acid-bound phenolate. A π-complex **6** is then formed, which further reacts *via* electrophilic aromatic substitution in the *ortho-* or *para*-position:

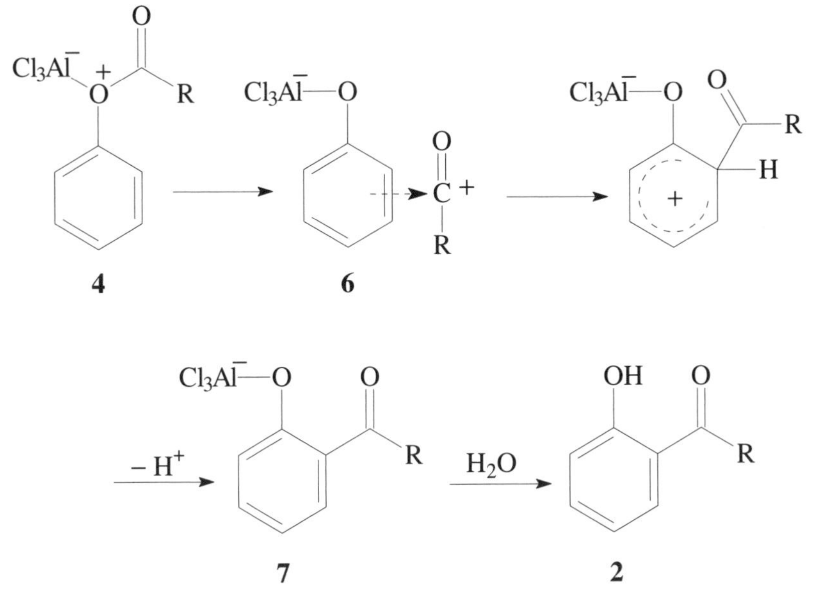

The mechanism for that step is closely related to that of the *Friedel–Crafts acylation*. Upon subsequent hydrolysis the *o*-substituted Lewis acid-coordinated phenolate **7** is converted to the free *o*-acylphenol **2**. By an analogous route, involving an electrophilic aromatic substitution *para* to the phenolate oxygen, the corresponding *para*-acylphenol is formed.

Since the Fries rearrangement is a equilibrium reaction, the reverse reaction may be used preparatively under appropriate experimental conditions.[2,6] An instructive example,[2] which shows how the regioselectivity depends on the reaction temperature, is the rearrangement of *m*-cresyl acetate **8**. At high temperatures the *ortho*-product **9** is formed, while below 100 °C the *para*-derivative **10** is formed:

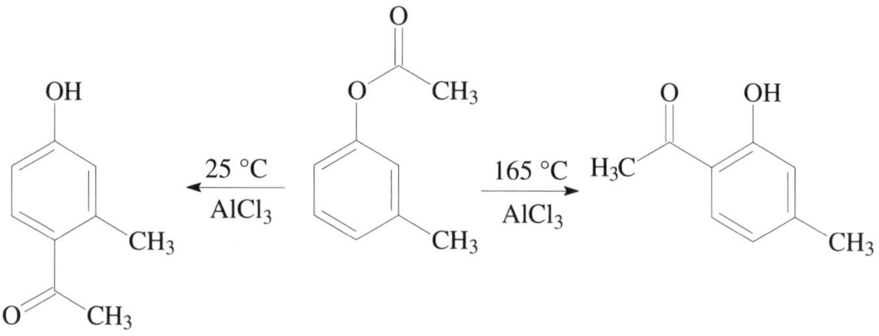

A photochemical variant, the so-called *photo-Fries rearrangement*,[7] proceeds *via* intermediate formation of radical species. Upon irradiation the phenyl ester molecules (**1**) are promoted into an excited state **11**. By homolytic bond cleavage the radical-pair **12** is formed that reacts to the semiquinone **13**, which in turn tautomerizes to the *p*-acylphenol **3**. The corresponding *ortho*-derivative is formed in an analogous way:

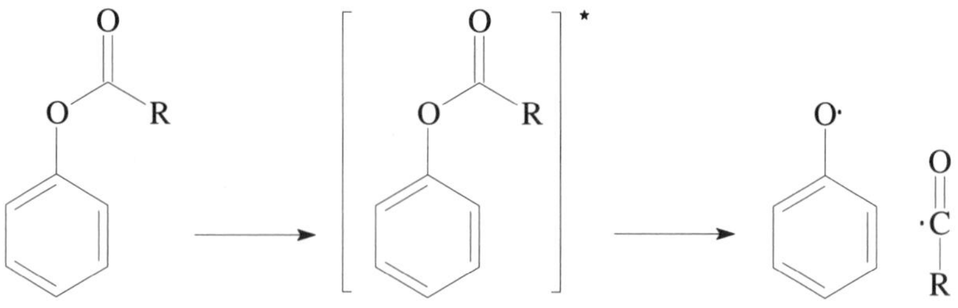

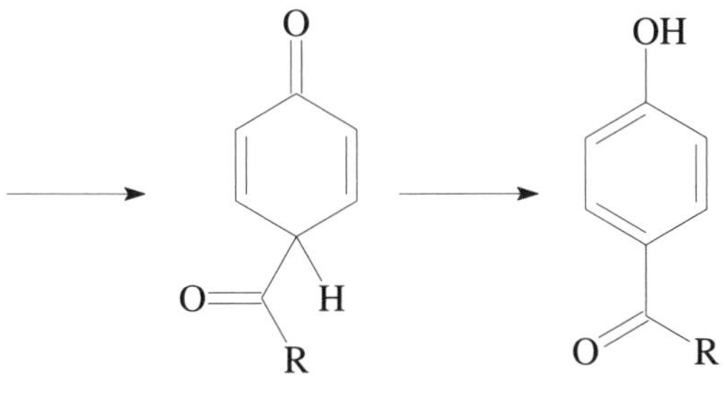

| 13 | 3 |

As catalysts for the Fries rearrangement reaction are for example used: aluminum halides,[3] zinc chloride, titanium tetrachloride,[8] boron trifluoride and trifluoromethanesulfonic acid.[7]

1. K. Fries, G. Finck, *Ber. Dtsch. Chem. Ges.* **1908**, *41*, 4271–4284.
2. A. H. Blatt, *Org. React.* **1942**, *1*, 342–369.
3. M. J. S. Dewar, L. S. Hart, *Tetrahedron* **1970**, *26*, 973–1000.
4. Y. Ogata, H. Tabuchi, *Tetrahedron* **1964**, *20*, 1661–1666.
5. A. Warshawsky, R. Kalir, A. Patchornik, *J. Am. Chem. Soc.* **1978**, *100*, 4544–4550.
6. F. Effenberger, R. Gutmann, *Chem. Ber.* **1982**, *115*, 1089–1102.
7. D. Bellus, P. Hrdlovic, *Chem. Rev.* **1967**, *67*, 599–609.
8. R. Martin, P. Demerseman, *Synthesis* **1989**, 25–28.

G

Gabriel Synthesis

Primary amines from *N*-substituted phthalimides

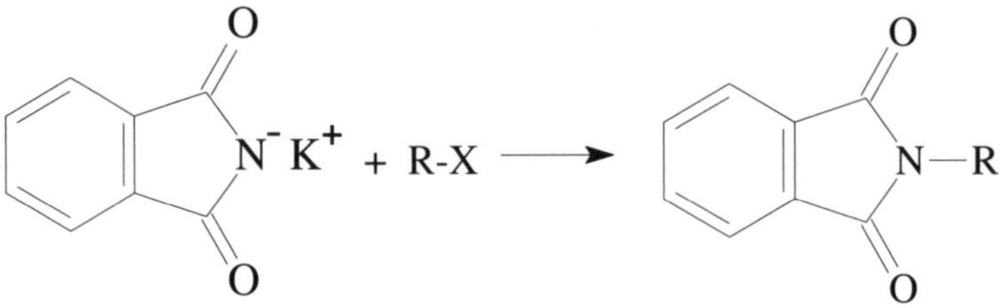

The reaction of potassium phthalimide **1** with an alkyl halide **2** leads to formation of a *N*-alkyl phthalimide **3**,[1,2] which can be cleaved hydrolytically or by reaction with hydrazine (*Ing–Manske* variant)[3] to yield a primary amine **5**. This route owes its importance as a synthetic method to the fact that primary amines are prepared selectively, not contaminated with secondary or tertiary amines.

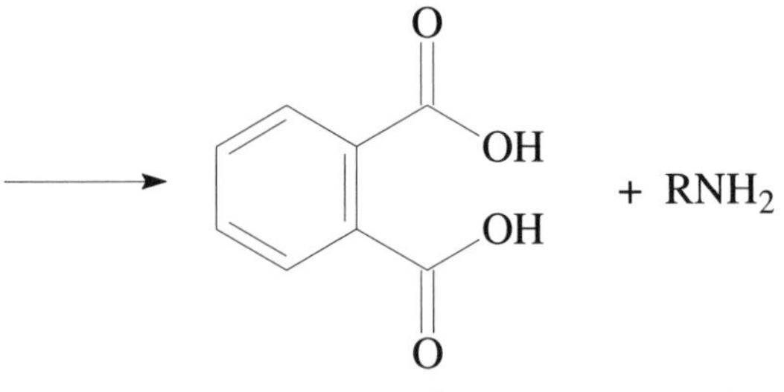

The two-step procedure includes formation of a *N*-substituted phthalimide **3**, and its subsequent cleavage to the primary amine **5**. Phthalimide (which can be obtained from reaction of phthalic acid with ammonia) shows NH-acidity, since the negative charge of the phthalimide anion (the conjugated base) is stabilized by resonance. Phthalimide is even more acidic than related 1,3-diketones, since the nitrogen center is more electronegative than carbon. The phthalimide anion acts as nucleophile in reaction with an alkyl halide; the substitution reaction is likely to proceed by a S_N2-mechanism:

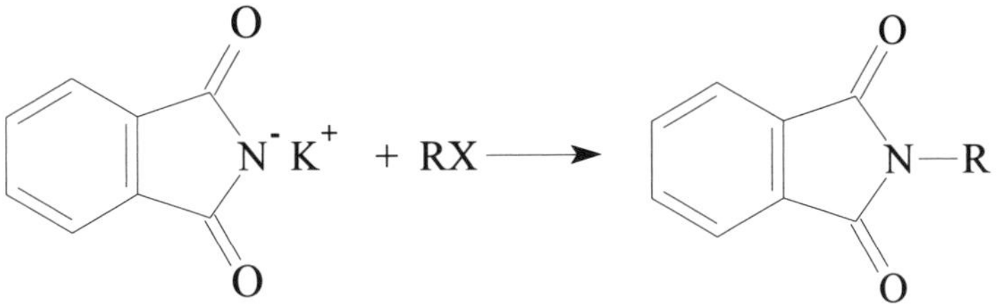

$$\text{1} \quad + \quad \text{RX} \quad \xrightarrow{\hspace{1cm}} \quad \text{3}$$

A further alkylation of the nitrogen is not possible. In a second step the N-substituted phthalimide **3** is hydrolyzed to give the desired amine **5** and phthalic acid **4**:

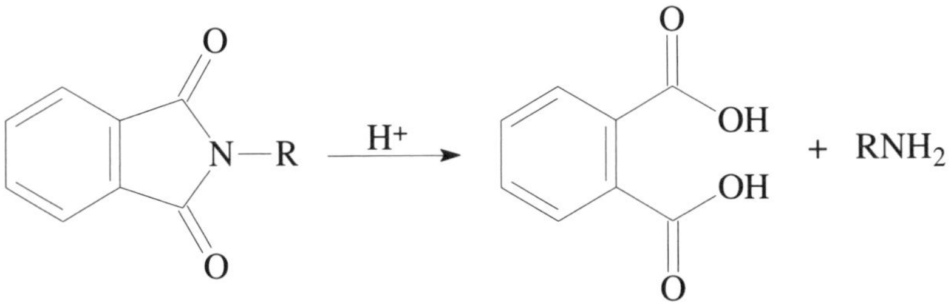

$$\text{3} \quad \xrightarrow{\text{H}^+} \quad \text{4} \quad + \quad \text{RNH}_2 \quad \text{5}$$

The hydrolytic cleavage is usually slow, and requires drastic reaction conditions. A more elegant method is presented by the *Ing–Manske* procedure,[3] where the N-alkylated imide is treated with hydrazine under milder conditions. In addition to the desired amine **5**, the cyclic phthalic hydrazide **6** is then formed:

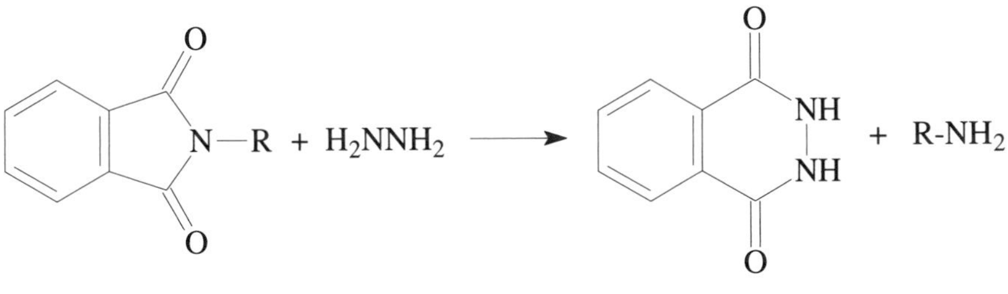

$$\text{3} \quad + \quad \text{H}_2\text{NNH}_2 \quad \xrightarrow{\hspace{1cm}} \quad \text{6} \quad + \quad \text{R-NH}_2 \quad \text{5}$$

The Gabriel synthesis is often carried out by heating the starting materials without a solvent for several hours at a temperature of 150 °C or higher. The use of solvents like dimethylformamide can lead to better results. In a number of solvents—e.g. toluene—the phthalimide is insoluble; the reaction can however be conducted in the presence of a phase transfer catalyst.[4]

The hydrazinolysis is usually conducted in refluxing ethanol, and is a fast process in many cases. Functional groups, that would be affected under hydrolytic conditions, may be stable under hydrazinolysis conditions. The primary amine is often obtained in high yield. The Gabriel synthesis is for example recommended for the synthesis of isotopically labeled amines and amino acids.[2] α-Amino acids 9 can be prepared by the Gabriel route, if a halomalonic ester—e.g. diethyl bromomalonate 7—is employed as the starting material instead of the alkyl halide:

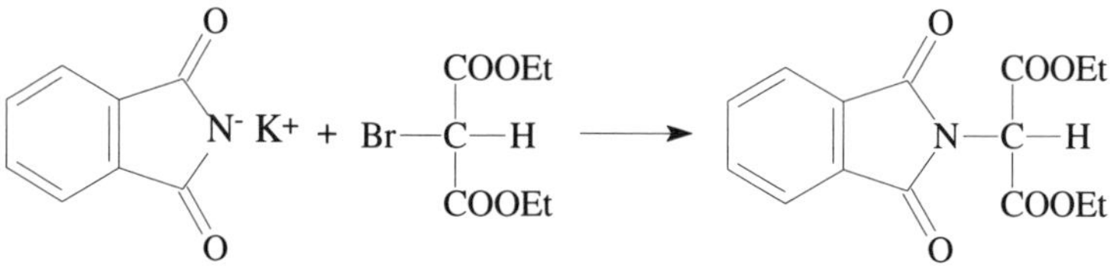

The N-phthalimidomalonic ester 8 can be further alkylated at the malonic carbon center with most alkyl halides, or with an α,β-unsaturated carbonyl compound; thus offering a general route to α-amino acids 9.

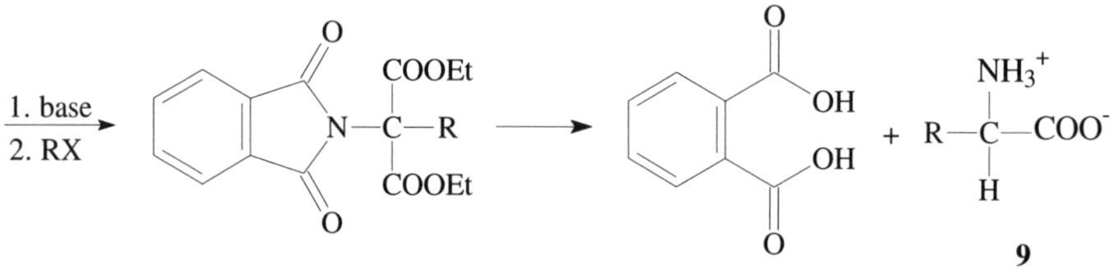

1. S. Gabriel, *Ber. Dtsch. Chem. Ges.* **1887**, *20*, 2224–2236.
2. M. S. Gibson, R. W. Bradshaw, *Angew. Chem.* **1968**, *80*, 986–996; *Angew. Chem. Int. Ed. Engl.* **1968**, *7*, 919.
3. H. R. Ing, R. H. F. Manske, *J. Chem. Soc.* **1926**, 2348–2351.
4. D. Landini, F. Rolla, *Synthesis* **1976**, 389–391.

Gattermann Synthesis

Formylation of aromatic compounds

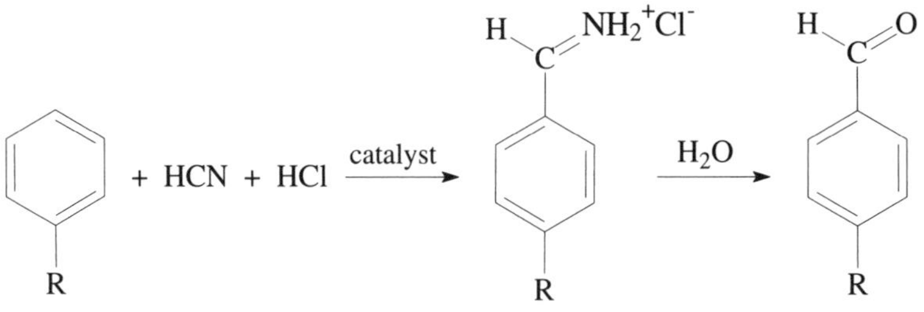

The preparation of a formyl-substituted aromatic derivative **3** from an aromatic substrate **1** by reaction with hydrogen cyanide and gaseous hydrogen chloride in the presence of a catalyst is called the *Gattermann synthesis*.[1,2] This reaction can be viewed as a special variant of the *Friedel–Crafts acylation* reaction.

Mechanistically it is an electrophilic aromatic substitution reaction. The electrophilic species (**4**—its exact structure is not known) is generated in a reaction of hydrogen cyanide and hydrogen chloride (gas) and a Lewis acid:

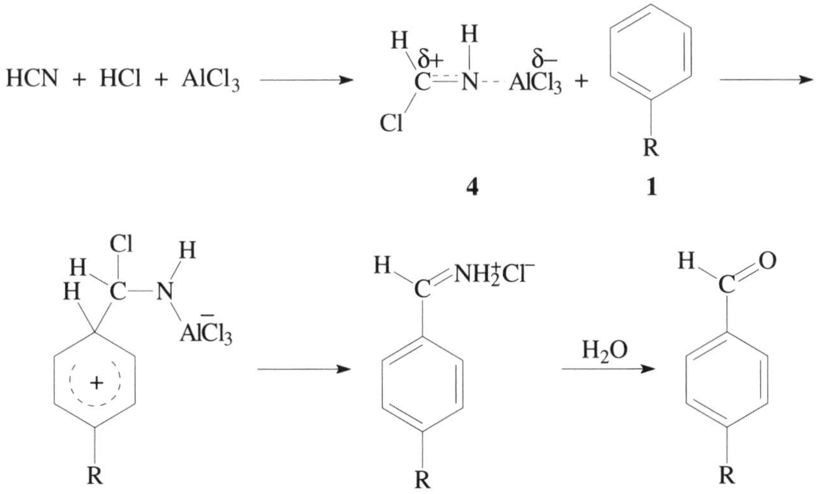

The electrophile **4** adds to the aromatic ring to give a cationic intermediate **5**. Loss of a proton from **5** and concomitant rearomatization completes the substitution step. Subsequent hydrolysis of the iminium species **2** yields the formylated aromatic product **3**. Instead of the highly toxic hydrogen cyanide, zinc cyanide can be used.[3] The hydrogen cyanide is then generated *in situ* upon reaction with the hydrogen chloride. The zinc chloride, which is thereby formed, then acts as Lewis acid catalyst.

The applicability of the Gattermann synthesis is limited to electron-rich aromatic substrates, such as phenols and phenolic ethers. The introduction of the formyl group occurs preferentially *para* to the activating substituent (compare *Friedel–Crafts acylation*). If the *para*-position is already substituted, then the *ortho*-derivative will be formed.

An analogous reaction is the *Houben–Hoesch reaction*,[4,5] (sometimes called the *Hoesch reaction*) using nitriles **7** to give aryl ketones **8**. This reaction also is catalyzed by Lewis acids; often zinc chloride or aluminum chloride is used. The Houben–Hoesch reaction is limited to phenols—e.g. resorcinol **6**—phenolic ethers and certain electron-rich aromatic heterocycles:[6,7]

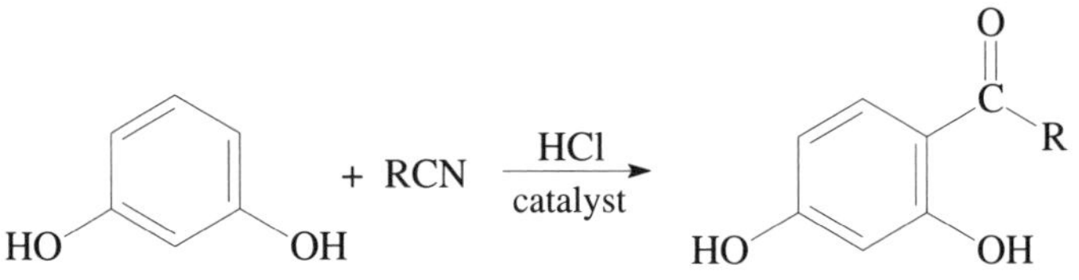

6	**7**	**8**

The synthetic importance of the Houben–Hoesch reaction is even more limited by the fact that aryl ketones are also available by application of the Friedel–Crafts acylation reaction.

Another formylation reaction, which is named after Gattermann, is the *Gattermann–Koch reaction*.[8] This is the reaction of an aromatic substrate with carbon monoxide and hydrogen chloride (gas) in the presence of a Lewis acid catalyst. Similar to the Gattermann reaction, the electrophilic agent **9** is generated, which then reacts with the aromatic substrate in an electrophilic aromatic substitution reaction to yield the formylated aromatic compound **10**:

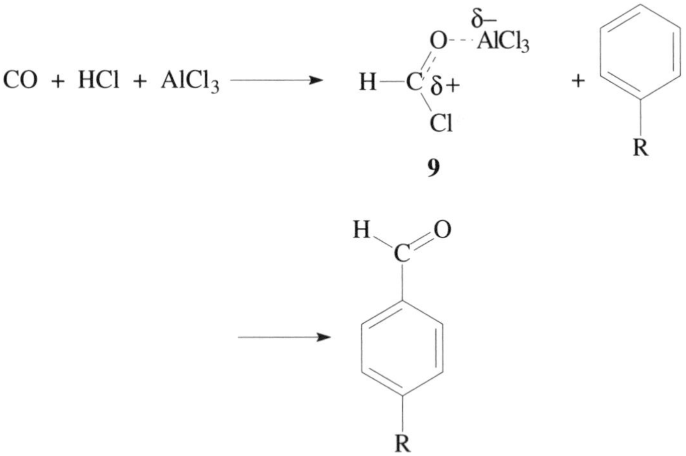

9

10

In order to achieve high yields, the reaction usually is conducted by application of high pressure. For laboratory use, the need for high-pressure equipment, together with the toxicity of carbon monoxide, makes that reaction less practicable. The scope of that reaction is limited to benzene, alkyl substituted and certain other electron-rich aromatic compounds.[9] With mono-substituted benzenes, the *para*-formylated product is formed preferentially. Super-acidic catalysts have been developed[3], for example generated from trifluoromethanesulfonic acid, hydrogen fluoride and boron trifluoride; the application of elevated pressure is then not necessary.

While the Friedel–Crafts acylation is a general method for the preparation of aryl ketones, and of wide scope, there is no equivalently versatile reaction for the preparation of aryl aldehydes. There are various formylation procedures known, each of limited scope. In addition to the reactions outlined above, there is the *Vilsmeier reaction*, the *Reimer–Tiemann reaction*, and the *Rieche formylation reaction*.[10–12] The latter is the reaction of aromatic compounds with 1,1-dichloromethyl ether as formylating agent in the presence of a Lewis acid catalyst. This procedure has recently gained much importance.

1. L. Gattermann, *Ber. Dtsch. Chem. Ges.* **1898**, *31*, 1149–1152.
2. W. E. Truce, *Org. React.* **1957**, *9*, 37–72.
3. G. A. Olah, L. O. Hannesian, M. Arvanaghi, *Chem. Rev.* **1987**, *87*, 671–686.
4. K. Hoesch, *Ber. Dtsch. Chem. Ges.* **1915**, *48*, 1122–1133.
5. J. Houben, *Ber. Dtsch. Chem. Ges.* **1926**, *59*, 2878–2891.
6. P. S. Spoerri, A. S. DuBois, *Org. React.* **1949**, *5*, 387–412.
7. E. A. Jeffery, D. P. N. Satchell, *J. Chem. Soc. B*, **1966**, 579–586.
8. L. Gattermann, J. A. Koch, *Ber. Dtsch. Chem. Ges.* **1897**, *30*, 1622–1624.
9. N. N. Cronnse, *Org. React.* **1949**, *5*, 290–300.
10. A. Rieche, H. Gross, E. Höft, *Chem. Ber.* **1960**, *93*, 88–94.
11. F. P. DeHaan, G. L. Delker, W. D. Covey, A. F. Bellomo, J. A. Brown, D. M. Ferrara, R. H. Haubrich, E. B. Lander, C. J. MacArthur, R. W. Meinhold, D. Neddenriep, D. M. Schubert, R. G. Stewart, *J. Org. Chem.* **1984**, *49*, 3963–3966.
12. G. Simchen, *Methoden Org. Chem. (Houben-Weyl)* **1983**, Vol. E3, p. 19–27.

Glaser Coupling Reaction

Coupling of terminal alkynes

$$2\ R-C\!\equiv\!C-H \xrightarrow[\text{base}]{\text{catalyst}} R-C\!\equiv\!C-C\!\equiv\!C-R$$

$$\mathbf{1} \qquad\qquad\qquad\qquad \mathbf{2}$$

The *Glaser reaction*[1,2] is an oxidative coupling of terminal alkynes **1** to yield a symmetrical *bis*-acetylene **2**; the coupling step is catalyzed by a copper salt. Closely related is the *Eglinton reaction*,[3] which differs from the Glaser reaction mainly by the use of stoichiometric amounts of copper salt as oxidizing agent.

Acetylene and terminal alkynes are CH-acidic compounds; the proton at the carbon–carbon triple bond can be abstracted by a suitable base. Such a deprotonation is the initial step of the Glaser reaction as well as the Eglinton reaction. Both reactions proceed by very similar mechanisms; therefore only one is outlined in the following:[4,5]

$$R—C{\equiv}C—H \xrightarrow{\text{base}} R—C{\equiv}CI^{-}$$

1   **3**

The acetylide anion **3** is likely to form an alkynyl-copper complex by reaction with the cupric salt. By electron transfer the copper-II ion is reduced, while the acetylenic ligands dimerize to yield the *bis*-acetylene **2**:

$$2\ R—C{\equiv}CI^{-} \xrightarrow{Cu^{2+}} R—C{\equiv}C—C{\equiv}C—R$$

3   **2**

The Glaser coupling reaction is carried out in aqueous ammonia or an alcohol/ammonia solution in the presence of catalytic amounts of a copper-I salt. The required copper-II species for reaction with the acetylide anion $R-C{\equiv}C^{-}$ are generated by reaction with an oxidant—usually molecular oxygen. For the Eglinton procedure, equimolar amounts of a copper-II salt are used in the presence of pyridine as base.

The Glaser reaction and the Eglinton reaction are well suited for the preparation of cyclic oligo-ynes.[6] This has been used by Sondheimer and coworkers in the synthesis of annulenes.[7] For example the 1,5-hexadiyne **5** under Glaser conditions gave a trimeric coupling product—the cyclic *hexa*-yne **6**—together with other oligomers. Product **6** was then converted to the [18]annulene **7**:

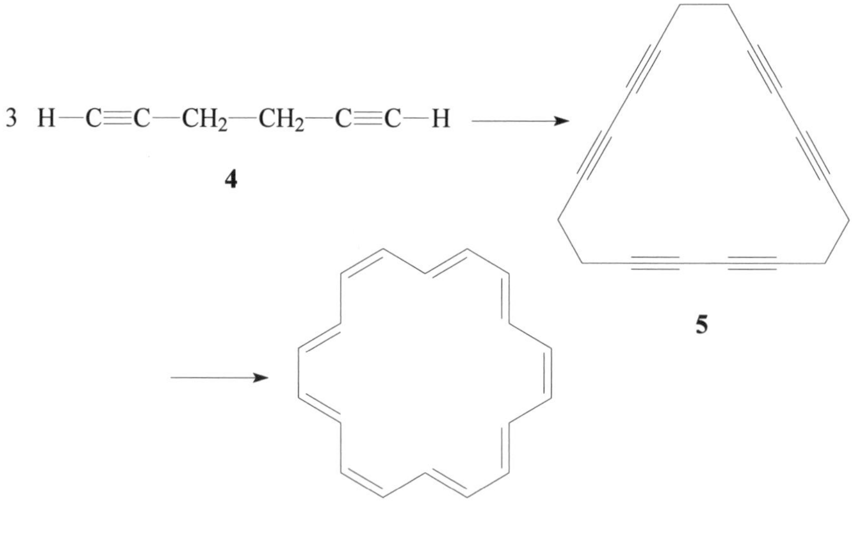

$$3\ H—C{\equiv}C—CH_2—CH_2—C{\equiv}C—H \longrightarrow$$

4

5

6

The two reactions described above can be applied for the synthesis of symmetrical *bis*-acetylenes only. Unsymmetrical bis-acetylenes can be prepared by using the *Cadiot–Chodkiewicz reaction*.[8,9] For that method a terminal alkyne **1** is reacted with a bromoalkyne **8** in the presence of a copper catalyst, to yield an unsymmetrical coupling product **9**:

$$R-C\equiv C-H \;+\; Br-C\equiv C-R' \quad\xrightarrow{\;Cu^+\;}\quad R-C\equiv C-C\equiv C-R'$$

$$\textbf{1} \qquad\qquad \textbf{7} \qquad\qquad\qquad\qquad \textbf{8}$$

All three coupling procedures are suitable to give high yields under mild reaction conditions. Many functional groups do not interfere. For the application in organic synthesis the Eglinton variant may be more convenient than the Glaser method; a drawback however is the need for stoichiometric amounts of copper salt.

1. C. Glaser, *Ber. Dtsch. Chem. Ges.* **1869**, *2*, 422–424.
2. L. I. Simandi in *The Chemistry of Triple-Bonded Functional Groups, Supp. C* (Ed.: S. Patai, Z. Rappoport), Wiley, New York, **1983**, Vol. 1, p. 529–534.
3. L. G. Fedenok, V. M. Berdnikov, M. S. Shvartsberg, *J. Org. Chem. USSR* **1973**, *9*, 1806–1809.
4. G. Eglinton, A. R. Galbraith, *Chem. Ind. (London)* **1956**, 737–738.
5. A. A. Clifford, W. A. Waters, *J. Chem. Soc.* **1963**, 3056–3062.
6. N. Nakagawa in *The Chemistry of the Carbon-Carbon Triple Bond* (Ed.: S. Patai), Wiley, New York, **1978**, Vol. 2, p. 654–656.
7. F. Sondheimer, R. Wolovsky, *J. Am. Chem. Soc.* **1962**, *84*, 260–269.
8. W. Chodkiewicz, *Ann. Chim. (Paris)* **1957**, *13/2*, 819–869.
9. N. Ghose, D. R. M. Walton, *Synthesis* **1974**, 890–891.

Glycol Cleavage

Oxidative cleavage of vicinal diols

The oxidative cleavage of the central carbon–carbon bond in a vicinal diol **1**, by reaction with lead tetraacetate or periodic acid, yields two carbonyl compounds **2** and **3** as products.

Lead tetraacetate $Pb(OAc)_4$ reacts with one of the hydroxyl groups to give an intermediate lead compound **4** and acetic acid. The rate determining step is the formation of a five-membered ring product **5** by further loss of acetic acid. Ring opening by carbon–carbon bond cleavage affords the carbonyl compounds **2** and **3**, together with lead-II-acetate:

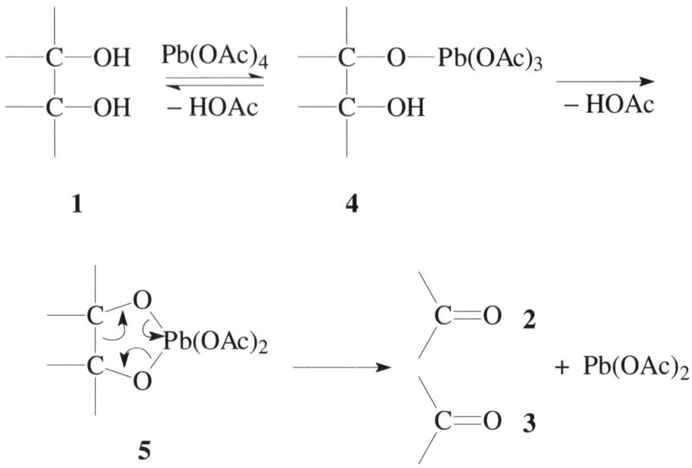

This mechanism applies to cis-1,2-diols and to open-chain 1,2-diols that can arrange in cisoid conformation. *Trans*-1,2-diols also do undergo the cleavage reaction, but at considerably slower rate, and by a different mechanism. For the latter case an acid-catalyzed decomposition of intermediate **4** to give the carbonyl products **2** and **3** is assumed, without going through a cyclic intermediate:

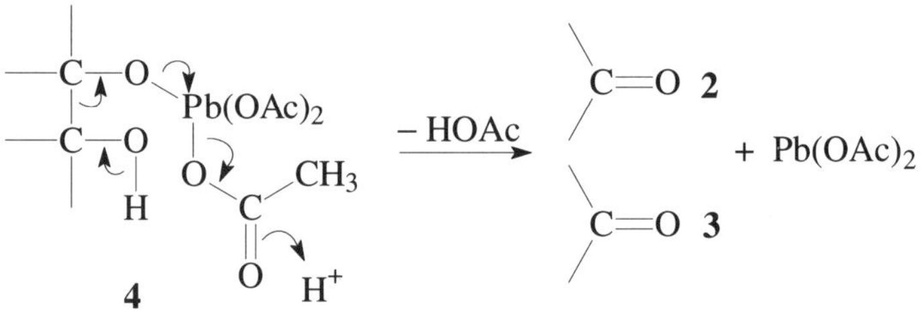

Under basic conditions, the cleavage would be initiated by deprotonation of a free hydroxyl group, as shown in **4**.

The cleavage of 1,2-diols **1** by periodic acid is associated with the name of the French chemist *Malaprade*.[4] The reaction mechanism[5,6] is related to that outlined above, and is likely to involve a five-membered ring periodate ester intermediate **7**:

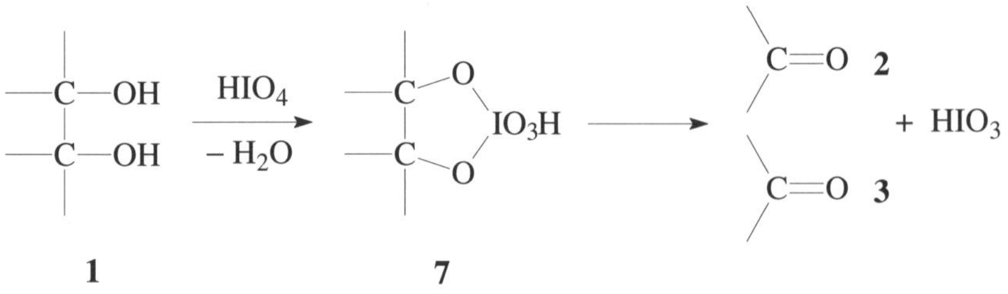

Lead tetraacetate and periodic acid complement one another in their applicability as reagents for glycol cleavage. The water sensitive lead tetraacetate is used in organic solvents, while periodic acid can be used for cleavage of water-soluble diols in aqueous medium.

The carbon–carbon double bond of an alkene **8** can be cleaved oxidatively, by a dihydroxylation reaction-glycol cleavage sequence:

This two-step sequence is a valuable alternative to the direct double bond cleavage by *ozonolysis*.

1. R. Criegee, L. Kraft, B. Rank, *Justus Liebigs Ann. Chem.* **1933**, *507*, 159–197.
2. R. Criegee, E. Höger, G. Huber, P. Kruck, F. Marktscheffel, H. Schellenberger, *Justus Liebigs Ann. Chem.* **1956**, *599*, 81–125.
3. G. M. Rubottom in *Oxidation in Organic Chemistry* (Ed.: W. S. Trahanovsky), Academic Press, New York, **1982**, p. 27–37.
4. M. L. Malaprade, *Bull. Soc. Chim. France* **1928**, *43*, 683–696.
5. E. J. Jackson, *Org. React.* **1944**, *2*, 341–375.
6. B. Sklarz, *Q. Rev. Chem. Soc.* **1967**, *21*, 3–28.

Gomberg–Bachmann Reaction

Synthesis of biaryls

Arenediazonium species **1** can be reacted with another aromatic substrate **2**, by the *Gomberg–Bachmann* procedure,[1,2] to yield biaryl compounds **3**. The intramolecular variant is called the *Pschorr reaction*.[3]

An arenediazonium ion **1** in aqueous alkaline solution is in equilibrium with the corresponding diazohydroxide **4**.[4] The latter can upon deprotonation react with diazonium ion **1**, to give the so-called 'anhydride' **5**. An intermediate product **5** can decompose to a phenyl radical **6** and the phenyldiazoxy radical **7**, and molecular nitrogen. Evidence for an intermediate diazoanhydride **5** came from crossover experiments:[4]

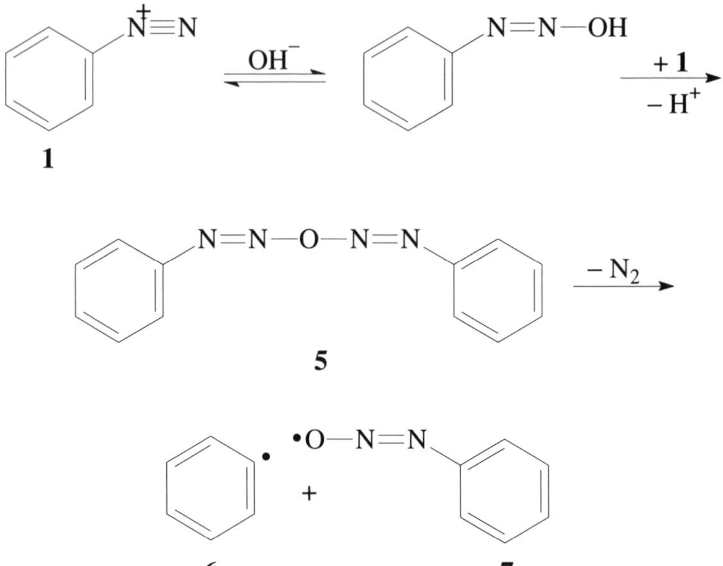

1

5

6   **7**

The very reactive phenyl radical reacts with the aromatic substrate **2**, present in the reaction mixture. Subsequent loss of a hydrogen radical, which then combines with **7** to give **4**, yields a biaryl coupling product; e.g. the unsymmetrical biphenyl derivative **3**:

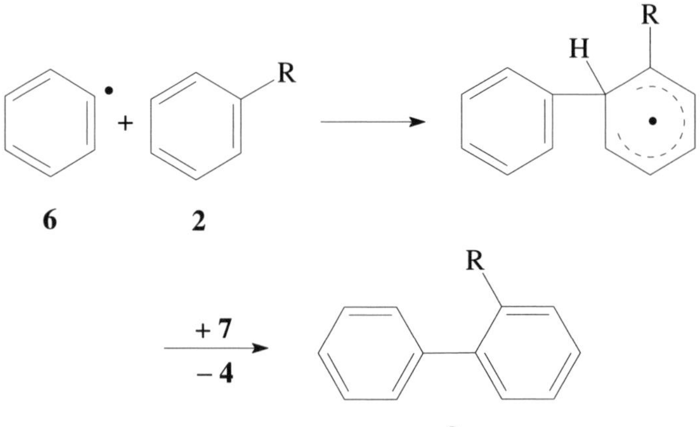

6   **2**

3

With a substituted aromatic ring compound **2**, mixtures of isomeric coupling products may be formed; the *ortho*-product usually predominates. The rules for regiochemical preferences as known from electrophilic aromatic substitution reactions (see for example *Friedel–Crafts acylation*), do not apply here.

Symmetrical biphenyls are also accessible by that procedure, but can often be prepared more conveniently by other routes.

In contrast to the Gomberg–Bachmann reaction, the intramolecular variant, the Pschorr reaction,[5] is carried out in strongly acidic solution, and in the presence of copper powder. Diazonium biphenyl ethers are converted to dibenzofurans, e.g. **8 → 9**:[3]

8   **9**

The Pschorr reaction also works with substrates containing a bridge other than oxygen. Thus various tricyclic products containing a biaryl subunit are accessible, e.g. carbazoles and fluorenes.

When the arenediazonium compound **1** is treated with sodium acetate, instead of alkali hydroxide, the reaction proceeds *via* an intermediate nitroso compound, and is called the *Hey reaction*.[6,7]

The Gomberg–Bachmann reaction is usually conducted in a two-phase system, an aqueous alkaline solution, that also contains the arenediazonium salt, and an organic layer containing the other aromatic reactant. Yields can be improved by use of a phase transfer catalyst.[8] Otherwise yields often are below 40%, due to various side reactions taking place. The Pschorr reaction generally gives better yields.

1. M. Gomberg, W. E. Bachmann, *J. Am. Chem. Soc.* **1924**, *46*, 2339–2343.
2. R. Bolton, G. Williams, *Chem. Soc. Rev.* **1986**, *15*, 261–289.
3. D. F. DeTar, *Org. React.* **1957**, *9*, 409–462.
4. C. Rüchardt, E. Merz, *Tetrahedron Lett.* **1964**, 2431–2436.
5. R. Pschorr, *Ber. Dtsch. Chem. Ges.* **1896**, *29*, 496–501.
6. J. Elks, J. W. Haworth, P. H. Hey, *J.Chem. Soc.* **1940**, 1284–1286.
7. D. R. Augood, G. H. Williams, *Chem. Rev.* **1957**, *57*, 123–190.
8. J. R. Beadle, S. H. Korzeniowski, D. E. Rosenberg, B. J. Garcia-Slanga, G. W. Gokel, *J. Org. Chem.* **1984**, *49*, 1594–1603.

Grignard Reaction

Addition of organomagnesium compounds to polarized multiple bonds

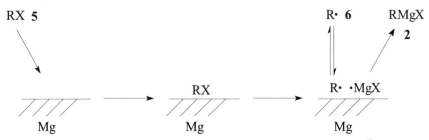

1	**2**	**3**	**4**

Organomagnesium compounds of the general structure RMgX (**2**) can be prepared by reaction of an alkyl or aryl halide RX with magnesium metal, and are called Grignard reagents. Such a reagent can add to a polarized double or triple bond in the so-called *Grignard reaction*.[1,2] Suitable reactants for this versatile reaction are for example aldehydes, ketones, esters, nitriles, carbon dioxide and other substrates containing polar functional groups such as $C=N-$, $C=S$, $S=O$, $N=O$. Most common and of synthetic importance is the reaction of a carbonyl compound **1**, to give a magnesium alkoxide **3**, which yields an alcohol **4** upon hydrolytic workup.

Grignard reagents are a very important class of organometallic compounds. For their preparation an alkyl halide or aryl halide **5** is reacted with magnesium metal. The formation of the organometallic species takes place at the metal surface; by transfer of an electron from magnesium to a halide molecule, an alkyl or aryl radical species **6** respectively is formed. Whether the intermediate radical species stays adsorbed at the metal surface (the *A-model*)[3], or desorbs into solution (the *D-model*)[4], still is in debate:

RX **5**   R· **6**   RMgX

                                                               2

//// ——→ //// ——→ ////
        RX   R· ·MgX

Mg   Mg   Mg

At the metal surface, the radical species R· and ·MgX combine to form the Grignard reagent **2**, which subsequently desorbs from the surface into solution. Macroscopically, the overall process is observed as a continuous decrease of the amount of magnesium metal.

In the alkyl halide **5**, the carbon–halogen bond is polarized, because of the higher electronegativity of the halogen. The carbon center next to the halogen has a formal $\delta+$ charge, and is electrophilic in its reactivity. In the Grignard reagent **2** however, the polarization is reversed; the carbon center next to the magnesium is carbanionic in nature; it has a formal $\delta-$ charge, and reacts as a nucleophile. This reversal of reactivity is called *Umpolung*.[12]

Since the formation of the Grignard compound takes place at the metal surface, a metal oxide layer deactivates the metal, and prevents the reaction from starting. Such an unreactive metal surface can be activated for instance by the addition of small amounts of iodine or bromine.

The solvent used plays an important role, since it can stabilize the organomagnesium species through complexation. Nucleophilic solvents such as ethers—e.g. diethyl ether or tetrahydrofuran—are especially useful. The magnesium center gets coordinated by two ether molecules as ligands.

$$
\begin{array}{c}
Et\diagdown\quad\diagup Et \\
O \\
\downarrow \\
R-Mg-X \\
\uparrow \\
O \\
Et\diagup\quad\diagdown Et
\end{array}
$$

In a nucleophilic solvent, the organomagnesium species not only exists as RMgX, but is rather described by the *Schlenk equilibrium*:

$$2\,RMgX \; \rightleftharpoons \; R_2Mg + MgX_2$$
$$\textbf{2}$$

In addition dimeric species are formed, being in equilibrium with the monomeric RMgX. The Schlenk equilibrium is influenced by substrate structure, the nature of the solvent, concentration and temperature.

The many diverse Grignard reactions, resulting from possible substrates and reaction conditions, cannot be described by a uniform reaction mechanism. The reaction of a ketone **1** with a Grignard reagent can be rationalized by a polar mechanism, as well as a *radical* mechanism.[5,6] The polar mechanism is formulated as the transfer of the group R together with the binding electron pair onto the carbonyl carbon center, and the formation of a magnesium–oxygen bond. The overall result is the formation of the magnesium alkoxide **3** of a tertiary alcohol:

$$
\begin{array}{ccc}
\diagdown \;\; \delta+\; \delta- & & \Big| \\
C = O & \longrightarrow & R-C-OMgX \\
\diagup \;\; & & \Big| \\
R-MgX & & \textbf{3} \\
\delta- \;\; \delta+ & &
\end{array}
$$

An alternative radical mechanism is formulated as the transfer of a single electron from the Grignard reagent **2** onto the carbonyl group (single electron transfer mechanism —*SET mechanism*). The intermediate pair of radicals **7** then combines to form product **3**:

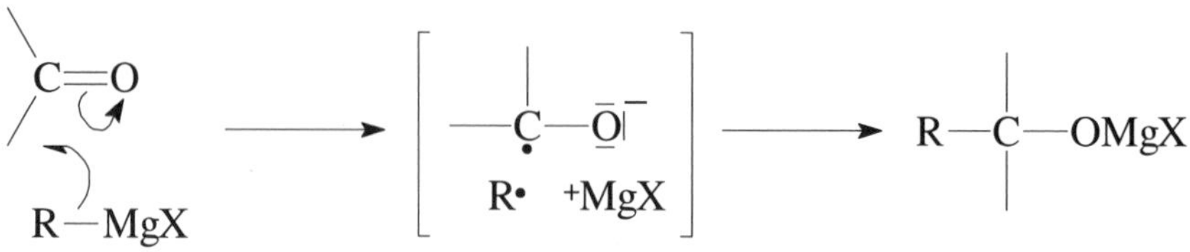

$$7 \qquad\qquad 3$$

The actual mechanism by which a particular reaction proceeds strongly depends on the nature of the organomagnesium reagent. For instance benzophenone reacts with methylmagnesium bromide by a polar mechanism, while the reaction with *t*-butylmagnesium chloride proceeds for steric reasons by a SET-mechanism.

The carbonyl carbon of an unsymmetrical ketone is a prochiral center; reaction with a Grignard reagent **2** (R ≠ R′, R″) can take place on either face of the carbonyl group with equal chance. The products **8a** and **8b** are consequently formed in equal amounts as racemic mixture, as long as no asymmetric induction becomes effective:[7]

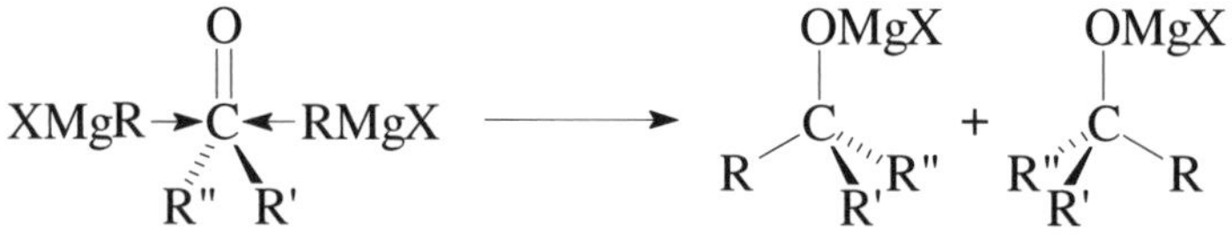

$$\textbf{1} \qquad\qquad \textbf{8a} \qquad \textbf{8b}$$

By treatment of a racemic mixture of an aldehyde or ketone that contains a chiral center—e.g. 2-phenylpropanal **9**—with an achiral Grignard reagent, four stereoisomeric products can be obtained; the diastereomers **10** and **11** and the respective enantiomer of each.

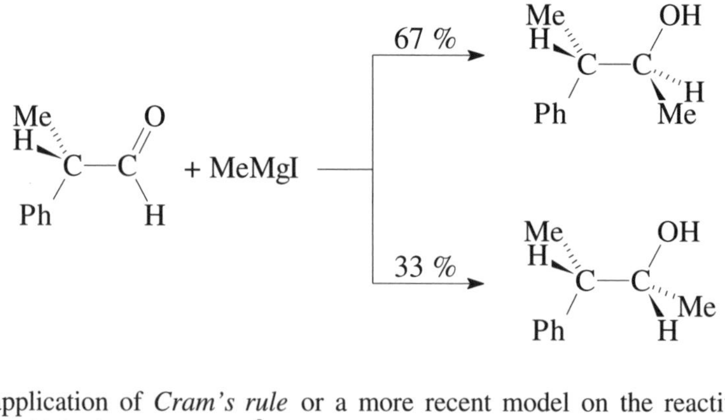

By application of *Cram's rule* or a more recent model on the reactivity of α-chiral aldehydes or ketones,[8] a prediction can be made, which stereoisomer will be formed predominantly, if the reaction generates an additional chiral center.

With compounds that contain acidic hydrogens—e.g. water, alcohols, phenols, enols, carboxylic acids, primary or secondary amines—the Grignard reagent RMgX reacts to give the corresponding hydrocarbon RH. For that reason, a Grignard reaction must be run in dry solvent, and by strictly excluding moisture. Grignard reagents also react with molecular oxygen present in the reaction mixture, resulting in a lower yield of the desired product. Solvolysis with D_2O can be used to introduce a deuterium atom selectively at a particular carbon center.

$$RMgX \xrightarrow{\ D_2O\ } RD$$

The acidity of certain reactants can be used for an exchange of the group R in a Grignard reagent RMgX. For example the alkyne **12** reacts with a Grignard compound **2** to give the alkynylmagnesium derivative **13** and the less acidic hydrocarbon **14**:

$$R'-C\equiv C-H + RMgX \longrightarrow R'-C\equiv C-MgX + RH$$

$$\quad\text{**12**}\qquad\quad\text{**2**}\qquad\qquad\qquad\text{**13**}\qquad\qquad\text{**14**}$$

Grignard reagents that contain a β-hydrogen—e.g. **15**—can reduce a carbonyl substrate by transfer of that hydrogen as a side-reaction. The so-called *Grignard reduction* is likely to proceed *via* a six-membered cyclic transition state **16**; the alkyl group of alkylmagnesium compound **15** is thereby converted into an alkene **17**.

$$\text{**1**}\qquad\text{**15**}\qquad\qquad\text{**16**}$$

Another side-reaction can be observed with sterically hindered ketones that contain an α-hydrogen—e.g. **18**. By transfer of that hydrogen onto the group R of RMgX **2**, the ketone **18** is converted into the corresponding magnesium enolate **19**, and the hydrocarbon RH **14** is liberated:

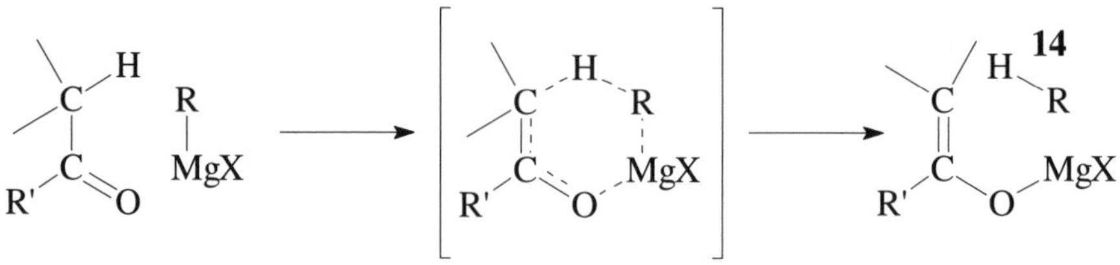

18   **2**   **19**

Tertiary magnesium alkoxides **20**, bearing a β-hydrogen, may undergo a dehydration reaction upon acidic workup, and thus yield an alkene **21**:

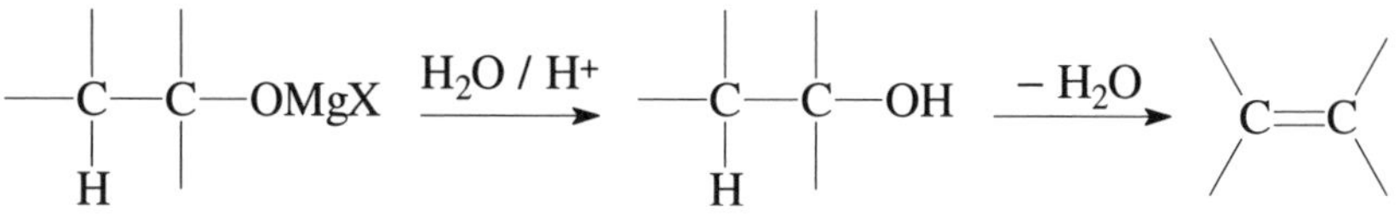

20   **21**

Grignard reagents can react as nucleophiles with a large variety of carbonyl substrates; in the following scheme the products obtained after a hydrolytic workup are shown. The scheme gives an impression of the versatility of the Grignard reaction:

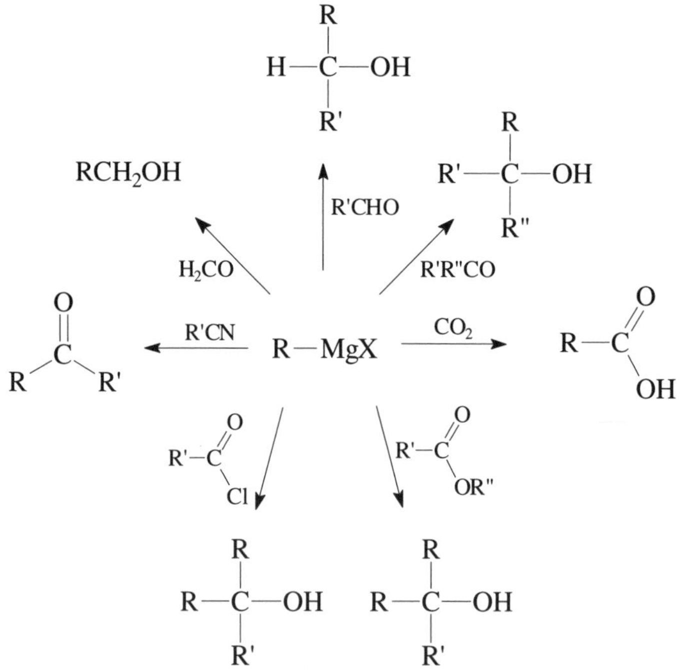

The conversion of a nitrile R'—CN into a ketone R'—CO—R demonstrates that polarized multiple bonds other than C=O also react with Grignard reagents, and that such reactions are synthetically useful. Esters **22** and acid chlorides can react subsequently with two equivalents of RMgX: the initially formed tetravalent product from the first addition reaction can decompose to a ketone that is still reactive, and reacts with a second RMgX. The final product **23** then contains two substituents R, coming from the Grignard reagent:

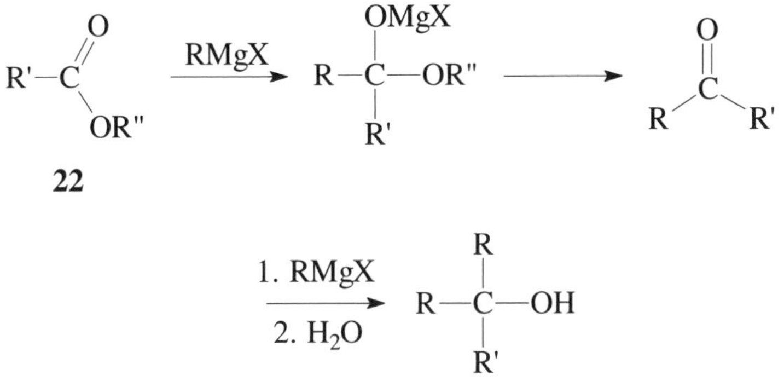

22

23

Another type of Grignard reaction is the alkylation with alkyl halides.[9] Upon treatment of a Grignard reagent RMgX with an alkyl halide **5**, a *Wurtz*-like coupling reaction takes place.

$$RMgX + R'X \longrightarrow R—R' + MgX_2$$

2   **5**

This reaction can be used for the synthesis of hydrocarbons; but it may also take place as a side-reaction during generation of a Grignard reagent from an alkyl halide and magnesium, then leading to formation of undesired side-products.

Grignard reagents do react with epoxides **24** by an S_N2-mechanism, resulting in a ring-opening reaction. An epoxide carbon bearing no additional substituent—i.e. a methylene group—is more reactive towards nucleophilic attack than a substituted one:

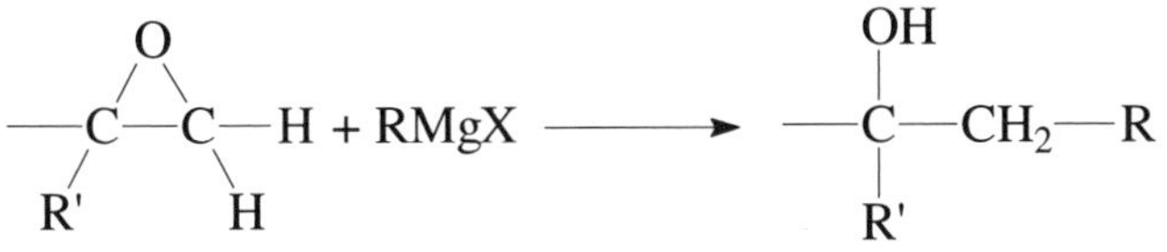

24

The Grignard reaction is one of the most important reactions in organic chemistry, because of its versatility in carbon–carbon bond formation. For some of the Grignard reactions mentioned above, alternative procedures using organolithium compounds have been developed, and may give better results. The formation of Grignard reagents from slow reacting alkyl or aryl halides has been made possible by recently developed modified procedures like the application of ultrasound,[10] or variants using activated magnesium.[11]

1.   V. Grignard, *C. R. Acad. Sci.* **1900**, *130*, 1322–1324.
2.   K. Nützel, H. Gilman, G. F. Wright, *Methoden Org. Chem. (Houben-Weyl)* **1973**, Vol. 13/2a, p. 49–527.
3.   H. M. Walborsky, *Acc. Chem. Res.* **1990**, *23*, 286–293.
4.   J. F. Garst, *Acc. Chem. Res.* **1991**, *24*, 95–97.
5.   E. C. Ashby, *Pure Appl. Chem.* **1980**, *52*, 545–569.
6.   M. Orchin, *J. Chem. Educ.* **1989**, *66*, 586–588.
7.   E. C. Ashby, J. T. Laemmle, *Chem. Rev.* **1975**, *75*, 521–546.
8.   M. Nogradi, *Stereoselective Synthesis*, VCH, Weinheim, **1986**, p. 131–140.
9.   J. C. Stowell, *Chem. Rev.* **1984**, *84*, 409–435.
10.   C. J. Einhorn, J. Einhorn, J.-L. Luche, *Synthesis* **1989**, 787–813.
11.   A. Fürstner, *Angew. Chem.* **1993**, *105*, 171–197;
      Angew. Chem. Int. Ed. Engl. **1993**, *32*, 164.
12.   D. Seebach, *Angew, Chem. Int. Ed. Engl.* **1979**, *18*, 239.

H

Haloform Reaction

Oxidative cleavage of methyl ketones

$$\underset{1}{\overset{O}{\underset{R}{\parallel}}\underset{R}{\overset{C}{\diagdown}}CH_3} \quad \xrightarrow[OH^-]{X_2} \quad \underset{2}{RCOO^-} + \underset{3}{HCX_3}$$

Methyl ketones **1**, as well as acetaldehyde, are cleaved into a carboxylate anion **2** and a trihalomethane **3** (a haloform) by the *Haloform reaction*.[1,2] The respective halogen can be chlorine, bromine or iodine.

The methyl group of a methyl ketone is converted into an α,α,α-trihalomethyl group by three subsequent analogous halogenation steps, that involve formation of an intermediate enolate anion (**4–6**) by deprotonation in alkaline solution, and introduction of one halogen atom in each step by reaction with the halogen. A halogen substituent α to the carbonyl group makes an adjacent hydrogen more acidic, and further halogenation will take place at the same carbon center:

$$\underset{1}{\overset{O}{\underset{R}{\parallel}}\overset{C}{\diagdown}CH_3} \quad \xrightarrow{OH^-} \quad \left[\underset{4}{\overset{O}{\underset{R}{\parallel}}\overset{C}{\diagdown}\bar{C}H_2} \longleftrightarrow \overset{\bar{|O|}}{\underset{R}{\diagdown}}\overset{C}{\diagup}CH_2 \right] \xrightarrow{X_2}$$

$$\overset{O}{\underset{R}{\parallel}}\overset{C}{\diagdown}CH_2X \;+\; X^- \xrightarrow{OH^-} \overset{O}{\underset{R}{\parallel}}\overset{C}{\diagdown}\bar{C}HX \xrightarrow{X_2} \overset{O}{\underset{R}{\parallel}}\overset{C}{\diagdown}CHX_2$$

5

$$\xrightarrow{\text{OH}^-} \quad \underset{6}{R-\overset{\overset{\displaystyle O}{\|}}{C}-\underline{C}\bar{X}_2} \quad \xrightarrow{X_2} \quad \underset{7}{R-\overset{\overset{\displaystyle O}{\|}}{C}-CX_3}$$

The α,α,α-trihaloketone **7** can further react with the hydroxide present in the reaction mixture. The hydroxide anion adds as a nucleophile to the carbonyl carbon; the tetravalent intermediate suffers a carbon–carbon bond cleavage:[3]

$$\underset{7}{R-\overset{\overset{\displaystyle O}{\|}}{C}-CX_3} + OH^- \longrightarrow R-\underset{\underset{\displaystyle OH}{|}}{\overset{\overset{\displaystyle |\overset{..}{O}|^-}{|}}{C}}-CX_3 \longrightarrow$$

$$R-\overset{\overset{\displaystyle O}{\diagup\!\!\diagdown}}{\underset{\displaystyle OH}{C}} + CX_3^- \longrightarrow \underset{2}{RCOO^-} + \underset{3}{HCX_3}$$

The reaction also works with primary and secondary methyl carbinols **8**. Those starting materials are first oxidized under the reaction conditions to the corresponding carbonyl compound **1**:

$$X\!-\!X + \underset{\underset{\displaystyle 8}{}}{R-\underset{\underset{\displaystyle H}{|}}{\overset{\overset{\displaystyle OH}{|}}{C}}-CH_3} + OH^- \longrightarrow X^- + HX + \underset{1}{R-\overset{\overset{\displaystyle O}{\|}}{C}-CH_3} + H_2O$$

With ketones bearing α'-hydrogens, a halogenation at that position is a possible side-reaction, and may lead to cleavage of the substrate.[2]

Fluorine cannot be used, although trifluoroketones can be cleaved into carboxylate and trifluoromethane. The haloform reaction can be conducted under mild conditions—at temperatures ranging from 0–10 °C—in good yields; even a sensitive starting material like methylvinylketone can be converted into acrylic acid in good yield.

Besides its synthetic importance, the haloform reaction is also used to test for the presence of a methylketone function or a methylcarbinol function in a molecule. Such compounds will upon treatment with iodine and an alkali hydroxide lead to formation of iodoform (*iodoform test*). The iodoform is easily identified by its yellow colour, its characteristic odour and the melting point.

1. A. Lieben, *Justus Liebigs Ann. Chem.* **1870** *Supp. 7*, 218–236.
2. S. K. Chakrabartty in *Oxidation in Organic Chemistry, Part C* (Ed.: W. S. Trahanovsky), Academic Press, New York, **1978**, p. 343–370.
3. C. Zucco, C. F. Lima, M. C. Rezende, J. F. Vianna, F. Nome, *J. Org. Chem.* **1987**, *52*, 5356–5359.

Hantzsch Pyridine Synthesis

1,4-Dihydropyridines from condensation of β-ketoesters with aldehydes and ammonia

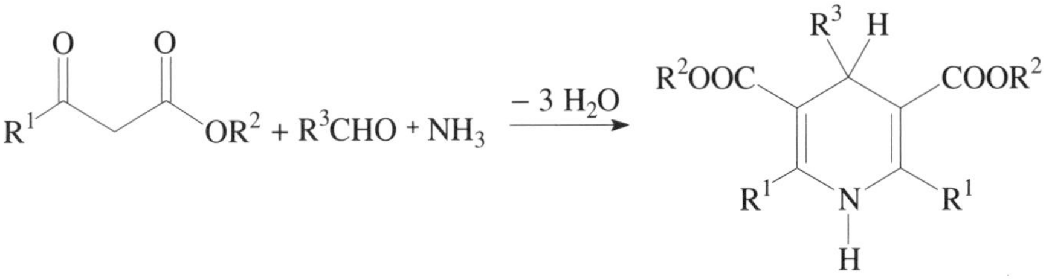

A general method for the construction of a pyridine ring is the *Hantzsch synthesis*.[1–4] A condensation reaction of two equivalents of a β-ketoester **1** with an aldehyde **2** and ammonia leads to a 1,4-dihydropyridine **3**, which can be oxidized to the corresponding pyridine **4**—for example by nitric acid:

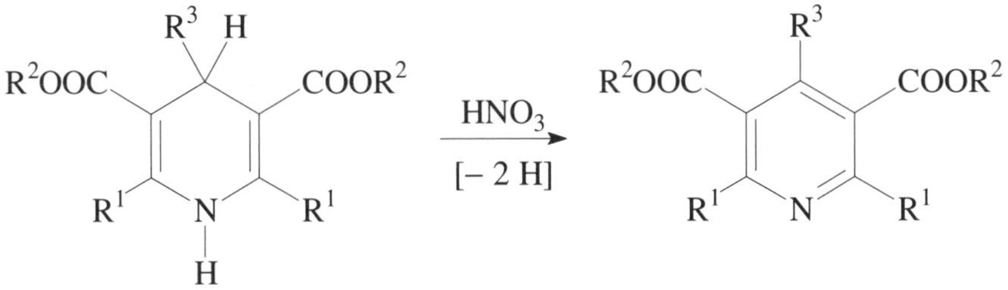

In general the oxidation does not affect the substituent R^3 at C-4; however if R^3 is a benzyl group $PhCh_2-$, this will be cleaved from C-4, and a hydrogen is retained in that position (unusual oxidation to yield pyridine).

The classical synthesis started from acetoacetic ester (**1**, $R^1 = CH_3$, $R^2 = C_2H_5$) and acetaldehyde (**2**, $R^3 = CH_3$). By subsequent cleavage of the substituents from C-3 and C-5, the collidine **5** was obtained ($R^1 = R^3 = CH_3$):[1]

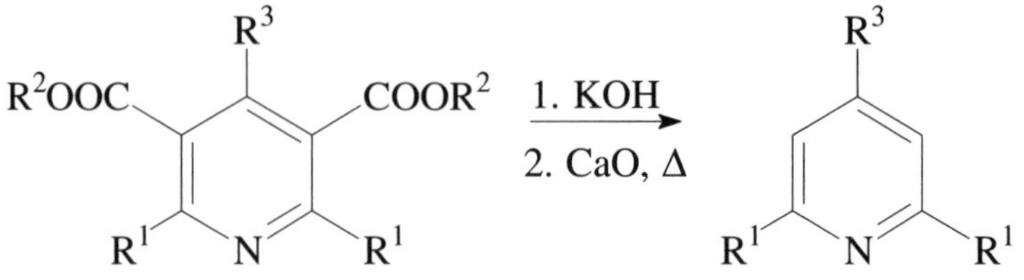

$$R^2OOC \underset{R^1 \quad N \quad R^1}{\overset{R^3}{\diagup \diagdown}} COOR^2 \quad \xrightarrow[2.\ CaO,\ \Delta]{1.\ KOH} \quad R^1 \underset{N}{\overset{R^3}{\diagup \diagdown}} R^1$$

$$\textbf{4} \qquad\qquad\qquad\qquad\qquad \textbf{5}$$

The initial step of the Hantzsch synthesis is likely to be a *Knoevenagel condensation* reaction of aldehyde **2** and β-ketoester **1** to give the α,β-unsaturated ketoester **6**:

$$\underset{H}{\overset{R^3}{\diagdown}} C{=}O + H_2C \underset{\underset{O}{\overset{\parallel}{C}-OR^2}}{\overset{\overset{O}{\overset{\parallel}{C}-R^1}}{\diagup}} \quad \xrightarrow{-\ H_2O} \quad \underset{H}{\overset{R^3}{\diagdown}} C \underset{\underset{O}{\overset{\parallel}{C}-OR^2}}{\overset{\overset{O}{\overset{\parallel}{C}-R^1}}{=}}$$

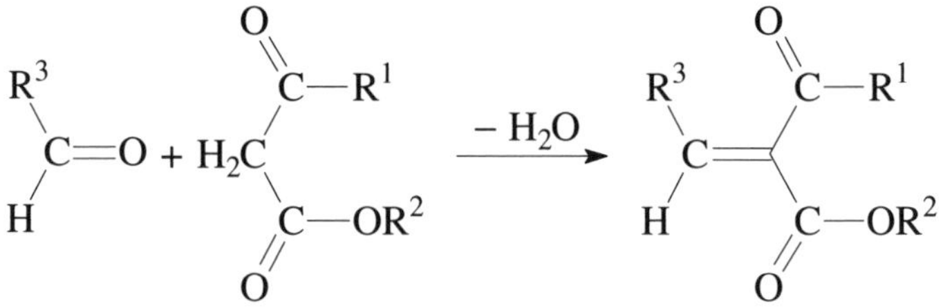

$$\textbf{2} \qquad\qquad \textbf{1} \qquad\qquad\qquad\qquad \textbf{6}$$

From β-ketoester **1** and ammonia the enamine **7** and water is formed:

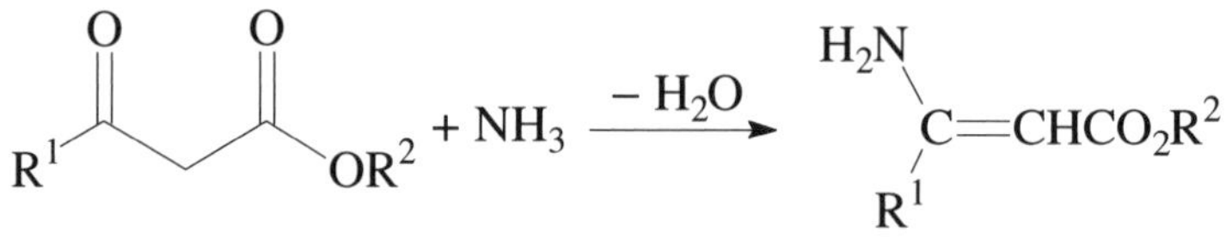

$$R^1 \underset{}{\overset{O \quad\ O}{\diagdown\diagup\diagdown}} OR^2 \; + NH_3 \quad \xrightarrow{-\ H_2O} \quad \underset{R^1}{\overset{H_2N}{\diagdown}} C{=}CHCO_2R^2$$

$$\textbf{1} \qquad\qquad\qquad\qquad\qquad \textbf{7}$$

The ring synthesis then proceeds in subsequent steps by condensation of the unsaturated ketoester **6** and enamine **7** to yield a 1,4-dihydropyridine **3**:

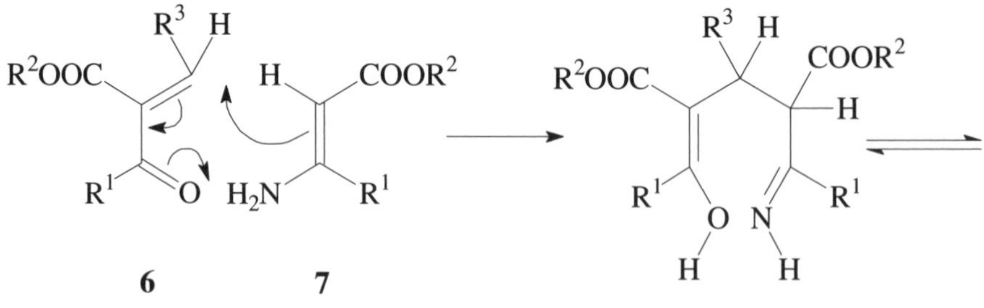

$$\qquad\qquad \textbf{6} \qquad\qquad \textbf{7}$$

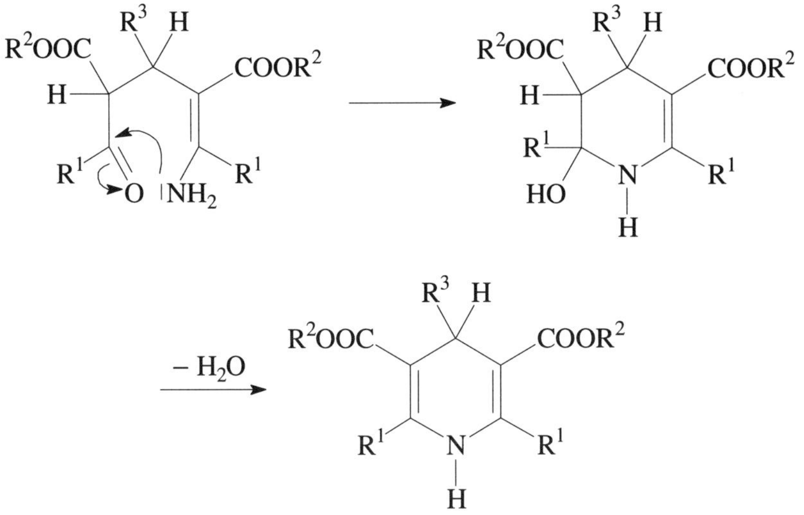

1,4-Dihydropyridines not only are intermediates for the synthesis of pyridines, but also are themselves an important class of N-heterocycles;[5,6] an example is the coenzyme NADH. Studies on the function of NADH led to increased interest in the synthesis of dihydropyridines as model compounds. Aryl-substituted dihydropyridines have been shown to be physiologically active as calcium antagonists. Some derivatives have found application in the therapy of high blood pressure and angina pectoris.[7] For that reason the synthesis of 1,4-dihydropyridines has been the subject of intensive research and industrial use. The Hantzsch synthesis has thus become an important reaction.

Many dihydropyridines that are of therapeutic interest are unsymmetrically substituted at C-3 and C-5. The synthesis of such compounds is possible from separately prepared Knoevenagel condensation products 6, as is outlined in the following scheme for nitrendipine 8, which is used in the medical treatment of high blood pressure.[4]

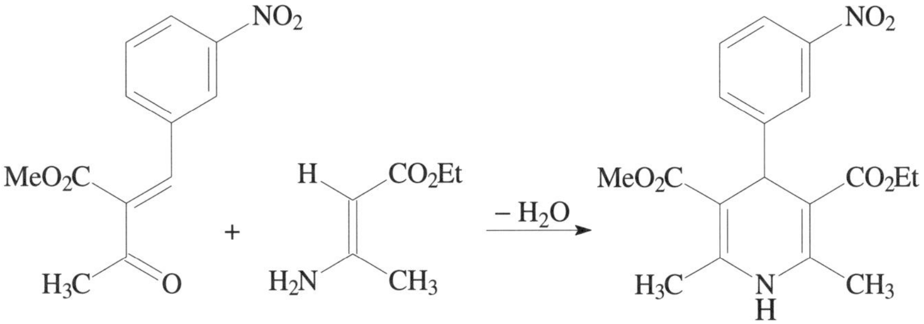

The reaction is of wide scope. Instead of ester groups as substituents at C-3 and C-5, other acceptor substituents—e.g. oxo, cyano, sulfonyl or nitro groups—can be employed in order to stabilize the 1,4-dihydropyridine system.

1.　A. Hantzsch, *Justus Liebigs Ann. Chem.* **1882**, *215*, 1–82.
2.　F. Brody, P. R. Ruby in *The Chemistry of Heterocyclic Compounds, Pyridine and its Derivatives, Vol. 14, Part 1*, (Ed.: E. Klingsberg), Wiley, New York, **1960**, p. 500–503.
3.　R. E. Lyle in *The Chemistry of Hetercyclic Compounds, Pyridine and its Derivatives, Vol. 14, Suppl. Part 1*, (Ed.: R. A. Abramovitch), Wiley, New York, **1974**, p. 139–143.
4.　F. Bossert, H. Meyer, E. Wehinger, *Angew. Chem.* **1981**, *93*, 755–763; *Angew. Chem. Int. Ed. Engl.* **1981**, *20*, 762.
5.　U. Eisner, J. Kuthan, *Chem. Rev.* **1972**, *72*, 1–42.
6.　D. M. Stout, A. I. Meyers, *Chem. Rev.* **1982**, *82*, 223–243.
7.　S. Goldmann, J. Stoltefuß, *Angew. Chem.* **1991**, *103*, 1587–1605; *Angew. Chem. Int. Ed. Engl.* **1991**, *30*, 1559.

Heck Reaction

Arylation or vinylation of alkenes

1

A modern and very important reaction in organic chemistry is the *Heck reaction*,[1-4] i.e. the palladium-catalyzed carbon–carbon bond coupling of an alkyl, aryl or vinyl group to an alkene **1**. The actual reactive coupling species is a palladium complex, generated from a halide RX (X = Br, I), that adds to the olefinic substrate.

In the following, the reaction mechanism is formulated for aryl halides; analogous mechanisms can be written for vinyl and alkyl halides. Arylpalladium complexes can be prepared by various methods. Aryl halides, arylmercury compounds or other aryl derivatives can be reacted with various palladium compounds. As a stabilizing ligand triphenylphosphine is often employed. A typical reagent for the coupling of an aryl derivative ArX to an alkene, consists of palladium acetate **2** (1%), triphenylphosphine (2%) and a stoichiometric amount of triethylamine. The latter is necessary for the regeneration of the catalyst during reaction.

A catalytic amount of reactive palladium-(0)-complex **3** is likely to be formed when the palladium-(II)-acetate **2** oxidizes a small amount of the alkene:[2]

$$Pd(OAc)_2 + 2\,PPh_3 + \underset{\textbf{1}}{\overset{\displaystyle \overset{H}{\underset{}{C=C}}}{}} \longrightarrow Pd(PPh_3)_2 + \underset{\textbf{3}}{\overset{\displaystyle \overset{OAc}{\underset{}{C=C}}}{}} + HOAc$$

$$\underset{\textbf{2}}{} \qquad\qquad \underset{\textbf{1}}{} \qquad\qquad \underset{\textbf{3}}{}$$

The catalytic cycle of the Heck reaction can be formulated with four reactions:[2,5,6]

(a) Formation of an arylpalladium complex **4** from the palladium-(0) complex **3** and the aryl derivative **5** by oxidative addition.
(b) Addition of complex **4** to the alkene (olefin insertion)
(c) A β-elimination reaction from complex **6** releases the substituted alkene **7**.
(d) Regeneration of the palladium-(0) complex **3** by reaction with a base, e.g. triethylamine:

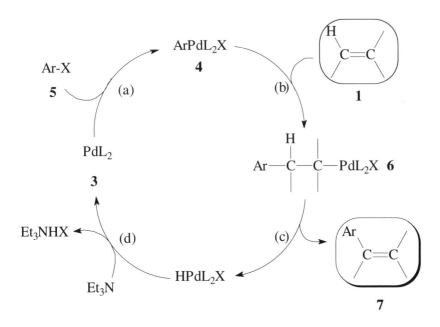

The regioselectivity of the addition of complex **4** to a substituted alkene is mainly influenced by steric factors. The substitution of hydrogen occurs preferentially at the carbon center that has the larger number of hydrogens. The Heck reaction of hex-1-ene **8** with bromobenzene leads to a 4 : 1 mixture of phenylhexenes **9** and **10**, where **9** is obtained as a mixture of E/Z-isomers:

$$Ph\diagdown\diagup\diagup\diagdown \quad \textbf{9}$$

The coupling of bromo- or iodobenzene to styrene yields regioselectively a mixture of *E*- and *Z*-stilbene. An electron-withdrawing substituent at the olefinic double bond often improves the regioselectivity, while an electron-donor substituted alkene gives rise to formation of regioisomers.

An alkylation *via* Heck reaction is limited to alkylpalladium complexes where the alkyl group does not bear β-hydrogens. For example methyl, benzyl and neopentyl groups can be coupled to alkenes.

With respect to the olefinic substrate, various functional groups are tolerated, e.g. an ester, ether, carboxy or cyano group. Primary and secondary allylic alcohols, e.g. **11**, react with concomitant migration of the double bond, to give an enol derivative that tautomerizes to the corresponding aldehyde (e.g. **12**) or ketone.

The Heck reaction is considered to be the best method for carbon–carbon bond formation by substitution of an olefinic proton. In general yields are good to very good. Sterically demanding substituents however may reduce the reactivity of the alkene. Polar solvents such as methanol, acetonitrile, *N,N*-dimethyl formamide or hexamethylphosphoric triamide are often used. Reaction temperatures are ranging from 50–160 °C. There are various other important palladium-catalyzed reactions known, where organo-palladium complexes are employed; however these reactions must not be confused with the Heck reaction.

1.   R. F. Heck, H. A. Dieck, *J. Am. Chem. Soc.* **1974**, *96*, 1133–1136.
2.   R. F. Heck, *Org. React.* **1982**, *27*, 345–390.
3.   A. D. Ryabov, *Synthesis* **1985**, 233–252.
4.   R. F. Heck, *Palladium Reagents in Organic Syntheses*, Academic Press, New York, **1985**;
     A. de Meijere, F. E. Meyer, *Angew. Chem.* **1994**, *106*, 2473–2506; *Angew. Chem. Int. Ed. Engl.* **1994**, *33*, 2379.
     W. A. Herrmann et al., *Angew. Chem.* **1995**, *107*, 1989–1992; *Angew. Chem. Int Ed. Engl.* **1995**, *34*, 18.

W. Cabri, I. Candiani, *Acc. Chem. Res.* **1995**, *28*, 2–7.
M. Beller, K. Kühlein, *Synlett* **1995**, 441–442.
5. L. G. Volkova, I. Y. Levitin, M. E. Volpin, *Russ. Chem. Rev.* **1975**, *44*, 552–560.
6. R. F. Heck, *Acc. Chem. Res.* **1979**, *12*, 146–151.

Hell–Volhard–Zelinskii Reaction

α-Halogenation of carboxylic acids

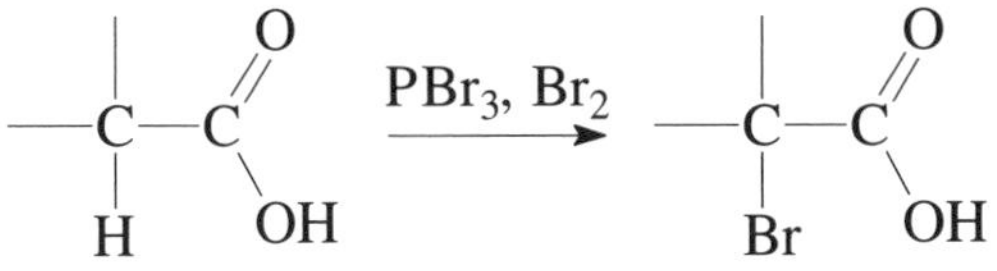

By application of the *Hell–Volhard–Zelinskii reaction*,[1,2] an α-hydrogen of a carboxylic acid **1** can be replaced by bromine or chlorine to give an α-bromo- or α-chlorocarboxylic acid **2** respectively.

In the following the reaction is outlined for an α-bromination. The reaction mechanism involves formation of the corresponding acyl bromide **3** by reaction of carboxylic acid **1** with phosphorus tribromide PBr_3. The acyl bromide **3** is in equilibrium with the enol derivative **4**, which further reacts with bromine to give the α-bromoacyl bromide **5**:

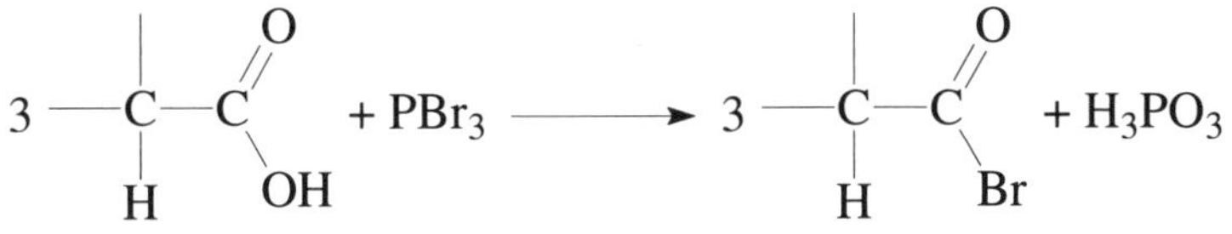

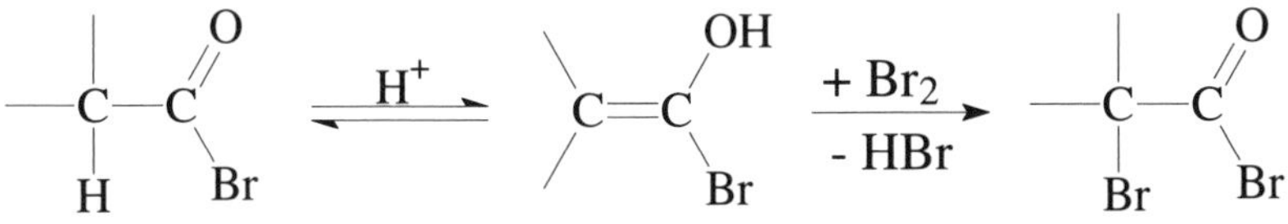

The α-bromoacyl bromide **5** converts unreacted carboxylic acid **1** by an exchange reaction into the more reactive acyl bromide **3**, which subsequently becomes α-brominated as formulated above:

Instead of phosphorus tribromide, red phosphorus can be used as catalyst. The phosphorus tribromide is then formed *in situ*. Carboxylic acids that enolize easily will also react without a catalyst present.

The formulated mechanism is supported by the finding that no halogen from the phosphorus trihalide is transfered to the α-carbon of the carboxylic acid. For instance, the reaction of a carboxylic acid with phosphorus tribromide and chlorine yields exclusively an α-chlorinated carboxylic acid. In addition, carboxylic acid derivatives that enolize easily—e.g. acyl halides and anhydrides—do react without a catalyst present.

A second α-hydrogen may also be replaced by halogen. In some cases it may be difficult to obtain the *mono*-halogenated product. α-Fluorinated or α-iodinated carboxylic acids cannot be prepared by this method.

The α-bromo or α-chloro carboxylic acids **2** are versatile intermediates for further synthetic transformations. For example they can be converted into α-hydroxy carboxylic acids by reaction with water; by reaction with cyanide α-cyanocarboxylic acids **7** are obtained, which can be further converted to 1,3-dicarboxylic acids **8** by hydrolysis of the cyano group. Reaction of an α-halocarboxylic acid **2** with ammonia leads to formation of an α-amino acid **9**:

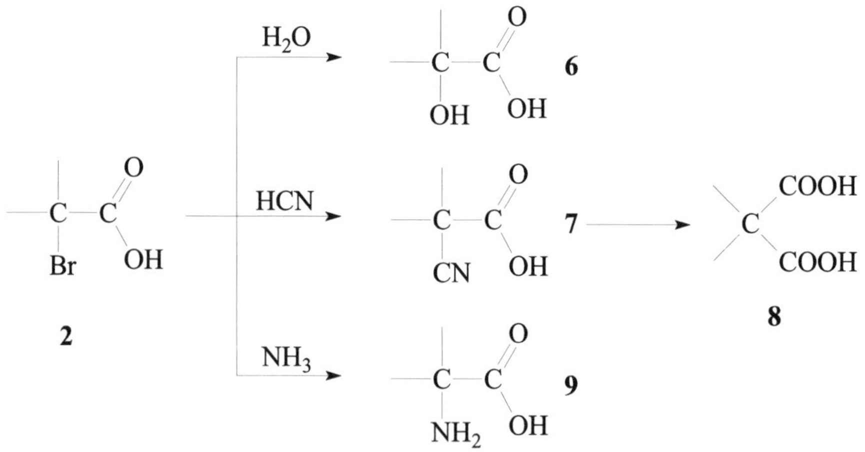

The preparation of α-iodocarboxylic acids is of particular interest, since iodide is a better leaving group as is chloride or bromide. A similar α-iodination with a phosphorus trihalide as catalyst is not known. However the iodination can be achieved in the presence of chlorosulfonic acid; mechanistically the intermediate formation of a ketene **10** by dehydration of the carboxylic acid is assumed:

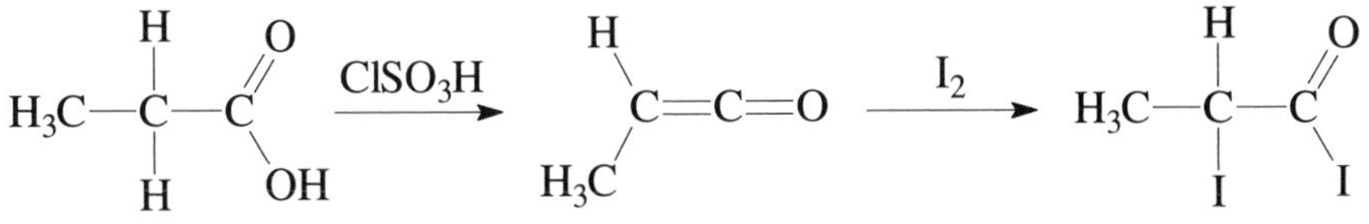

10

The ketene as a more reactive species is iodinated by reaction with iodine. Bromine or chlorine as substituents may also be introduced by this method.[3]

1. C. Hell, *Ber. Dtsch. Chem. Ges.* **1881**, *14*, 891–893.
2. H. J. Harwood, *Chem. Rev.* **1962**, *62*, 99–154.
3. Y. Ogata, K. Tomizawa, *J. Org. Chem.* **1979**, *44*, 2768–2770.

Hofmann Elimination Reaction

Alkenes from amines

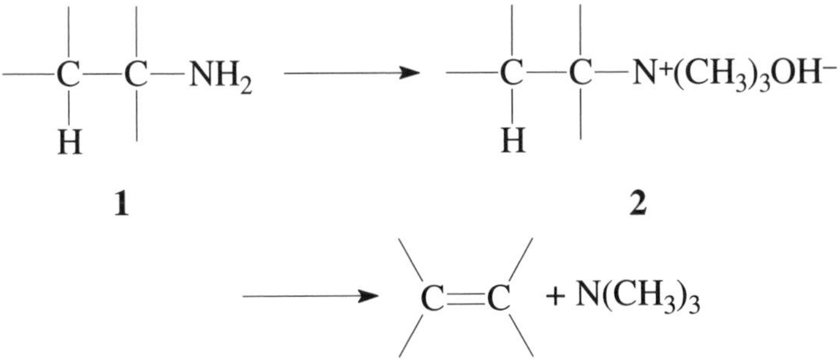

1   **2**

3   **4**

The preparation of an alkene **3** from an amine **1** by application of a β-elimination reaction is an important method in organic chemistry. A common procedure is the *Hofmann elimination*,[1,2] where the amine is first converted into a quaternary ammonium salt by exhaustive methylation. Another route for the conversion of amines to alkenes is offered by the *Cope elimination*.

　　Primary, secondary and tertiary amines can serve as starting materials. The amine, e.g. **1**, is first treated with excess methyl iodide, to generate the quaternary ammonium iodide **5**. Subsequent treatment with silver oxide in water gives the corresponding ammonium hydroxide **2**:

$$\overset{\displaystyle |}{\underset{\displaystyle H}{C}} - \overset{\displaystyle |}{\underset{\displaystyle |}{C}} - NH_2 \quad \xrightarrow{CH_3I} \quad \overset{\displaystyle |}{\underset{\displaystyle H}{C}} - \overset{\displaystyle |}{\underset{\displaystyle |}{C}} - N^+(CH_3)_3I^- \quad \xrightarrow{Ag_2O}$$

$$\quad\quad\quad\quad 1 \quad\quad\quad\quad\quad\quad\quad\quad\quad\quad 5$$

$$\overset{\displaystyle |}{\underset{\displaystyle H}{C}} - \overset{\displaystyle |}{\underset{\displaystyle |}{C}} - N^+(CH_3)_3OH^-$$

$$2$$

The elimination reaction takes place upon heating of the ammonium hydroxide to a temperature of 100–200 °C, often under reduced pressure:

$$\overset{\displaystyle |}{\underset{\displaystyle \underset{HO^-}{H}}{C}} - \overset{\displaystyle |}{\underset{\displaystyle |}{C}} - N^+(CH_3)_3 \quad \xrightarrow{\Delta} \quad \overset{\diagdown}{\underset{\diagup}{C}} = \overset{\diagup}{\underset{\diagdown}{C}} + N(CH_3)_3 + H_2O$$

$$\quad\quad 2 \quad\quad\quad\quad\quad\quad\quad\quad\quad\quad 3 \quad\quad\quad 4$$

In general the β-elimination proceeds by a E2-mechanism. It involves cleavage of trimethylamine and a β-hydrogen from the original substrate alkyl group; see scheme above—**2** → **3**. In some cases—depending on substrate structure and reaction conditions—evidence for a E1cB-mechanism has been found:

$$\overset{\displaystyle |}{\underset{\displaystyle \underset{HO^-}{H}}{C}} - \overset{\displaystyle |}{\underset{\displaystyle |}{C}} - N^+(CH_3)_3 \quad \underset{\xrightarrow{\hspace{1cm}}}{\overset{-H^+}{\rightleftharpoons}} \quad \overset{\displaystyle |}{\underset{\displaystyle |}{\underset{..}{C}}} - \overset{\displaystyle |}{\underset{\displaystyle |}{C}} - N^+(CH_3)_3$$

$$\quad\quad 2$$

$$\xrightarrow{\hspace{2cm}} \quad \overset{\diagdown}{\underset{\diagup}{C}} = \overset{\diagup}{\underset{\diagdown}{C}} + N(CH_3)_3$$

$$\quad\quad\quad\quad\quad 3 \quad\quad\quad 4$$

For the elimination of trimethylamine and water from the *erythro-* and *threo-*isomer of trimethyl-1,2-diphenylpropylammonium iodide **6** and **7** respectively, by treatment with sodium ethoxide, a stereospecific *trans*-elimination has been found to take place; thus supporting a E2-mechanism. From the *erythro*-isomer **6** the Z-alkene **8** was obtained, while the *threo*-isomer **7** yielded the E-alkene **9**:

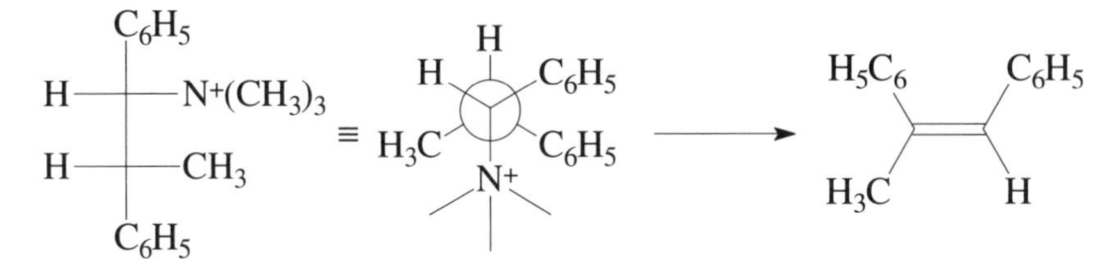

6
erythro

8

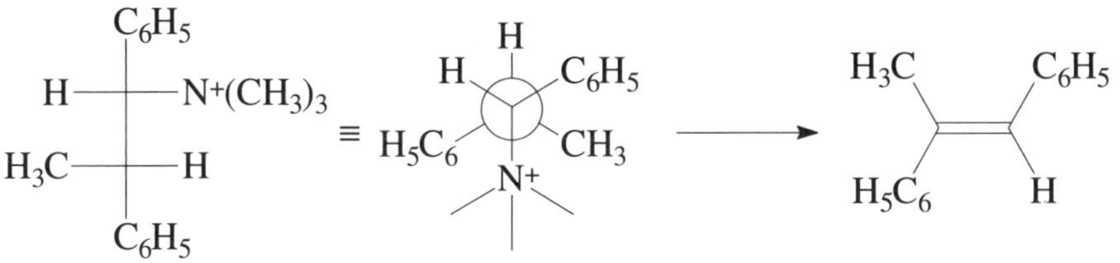

7
threo

9

If however a t-butoxide was used as base, only the thermodynamically favored E-alkene **9** was formed, suggesting a E1cB-mechanism in that case. It has been shown that a $Z \rightarrow E$-isomerization does not occur under these reaction conditions.

When the nitrogen is part of a ring, as for example in N-methylpyrolidine **10**, the olefinic product resulting from one elimination step still contains the nitrogen as a tertiary amino group. A second quaternization/elimination sequence is then necessary to eliminate the nitrogen function from the molecule; as final product a diene is then obtained:

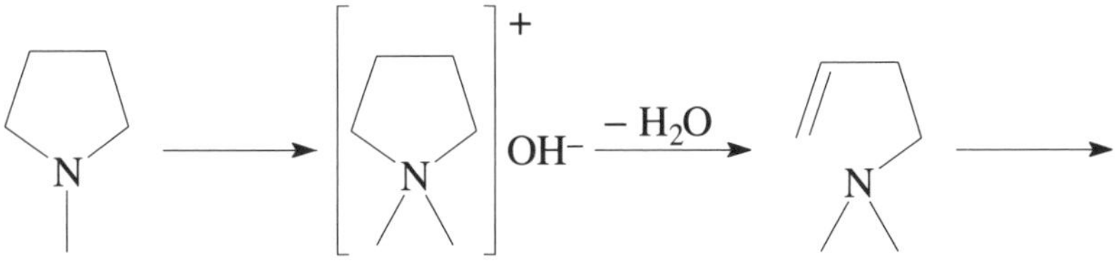

10

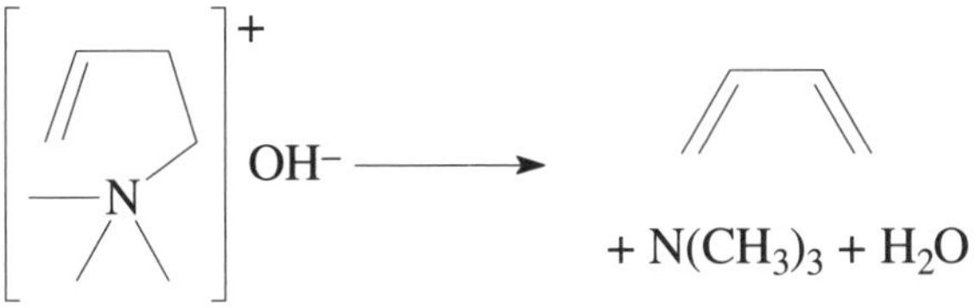

$+ N(CH_3)_3 + H_2O$

With starting materials containing a bridgehead-nitrogen, e.g. quinolizidine **11**, a third quaternization/elimination sequence is necessary for complete elimination of the nitrogen; as final product a triene is then obtained:

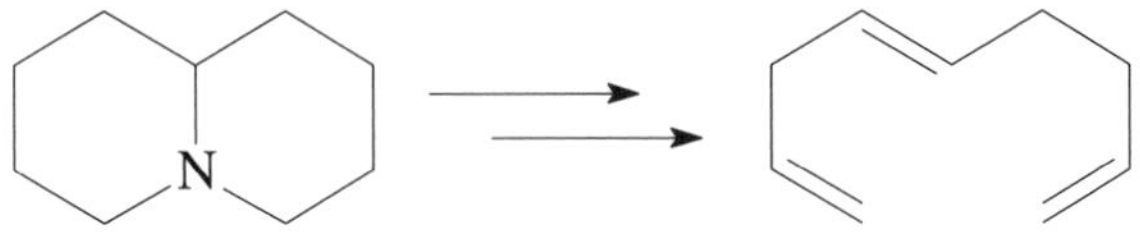

11

The number of reaction sequences required for liberation of trimethylamine **4** indicates the degree of incorporation of a particular nitrogen into the molecular skeleton. Because of that feature, the Hofmann elimination has been used for the structural analysis of natural products, e.g. alkaloids.

As a side-reaction, a nucleophilic substitution to give an alcohol **12** is often observed:

$$R-N^+(CH_3)_3 + OH^- \longrightarrow ROH + N(CH_3)_3$$
12

With substrates, where a β-hydrogen as well as a β'-hydrogen is available for elimination, product formation follows the so-called *Hofmann rule*, which states that the less substituted alkene will be formed preferentially. For example from 2-aminobutane **13**, but-1-ene **14** is formed preferentially, while but-2-ene **15** is formed in minor amounts only:

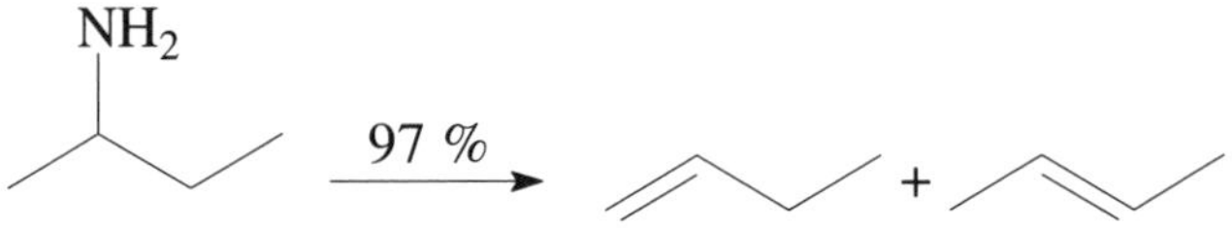

13   **14**   **15**

                            95 %   5 %

In addition to its application for structural analysis, the Hofmann elimination also is of synthetic importance. For instance the method has been used to prepare *E*-cyclooctene **16**, as well as higher homologs:[3]

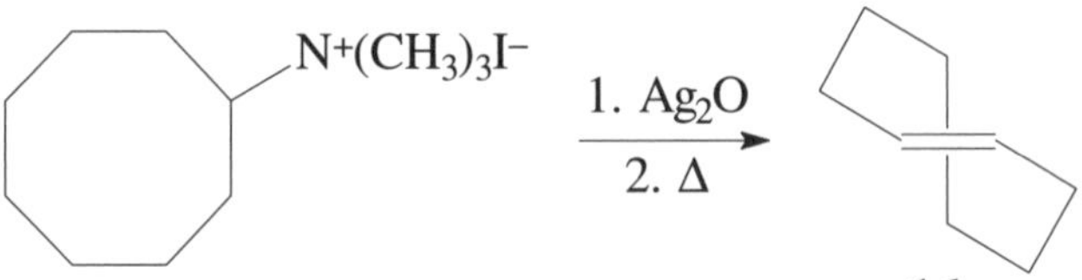

16

Such reactions usually lead to formation of *E*- and *Z*-isomers, with the more strained *E*-isomer predominating.

A variant, the *1,6-Hofmann elimination*, has become a standard method for the synthesis of [2.2]paracyclophanes **17**; although it often gives low yields.

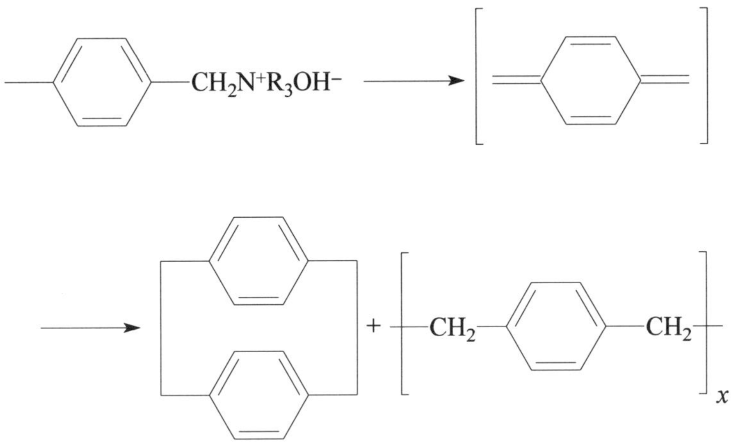

17

This method can also be used to synthesize multilayer phanes.[4]

1. A. W. Hofmann, *Justus Liebigs Ann. Chem.* **1851**, *78*, 253–286.
2. A. C. Cope, E. R. Trumbull, *Org. React.* **1960**, *11*, 317–493.
3. K. Ziegler, H. Wilms, *Justus Liebigs Ann. Chem.* **1950**, *567*, 1–43.
4. F. Vögtle, P. Neumann, *Synthesis* **1973**, 85–103.

Hofmann Rearrangement

Primary amines from carboxylic amides

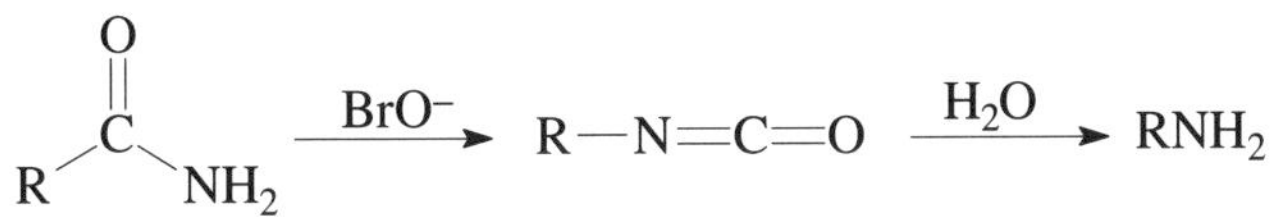

By application of the *Hofmann rearrangement* reaction,[1,2] a carboxylic amide is converted into an amine, with concomitant chain degradation by one carbon (*Hofmann degradation*). Formally this reaction can be considered as a removal of the carbonyl group from a carboxylic amide. The reaction is closely related to the *Curtius rearrangement* and the *Lossen rearrrangement*. In each case the rearrangement is initiated by generating an electron-sextet configuration at nitrogen.

For the Hofmann rearrangement reaction, a carboxylic amide **1** is treated with hypobromite in aqueous alkaline solution. Initially an N-bromoamide **4** is formed. With two electron-withdrawing substituents at nitrogen the N-bromoamide shows NH-acidity, and can be deprotonated by hydroxide to give the anionic species **5**.

The next step involves cleavage of bromide from the nitrogen center and migration of the group R, to give an isocyanate **2**. Here the question arises, whether the N−Br bond is cleaved first and the migration of R takes place afterwards, or both steps proceed in a concerted way? So far most findings suggest a concerted process.[3] In general the rearrangement of a chiral group R takes place without racemization. The isocyanate **2** under these reaction conditions reacts with water, to form the carbaminic acid **6**. This addition product is unstable, and decomposes to yield the amine **3** and carbon dioxide or rather carbonate in alkaline solution:

The N-bromoamide, its anion as well as the isocyanate have been identified as intermediates; thus supporting the reaction mechanism as formulated above.

Generally yields are good. R can be alkyl or aryl. Modern variants of the Hofmann rearrangement use lead tetraacetate[4] or iodosobenzene[5] instead of hypobromite.

1. A. W. Hofmann, *Ber. Dtsch. Chem. Ges.* **1881**, *14*, 2725–2736.
2. E. S. Wallis, J. F. Lane, *Org. React.* **1946**, *3*, 267–306.
3. T. Imamoto, Y. Tsuno, Y. Yukawa, *Bull. Chem. Soc. Jpn.* **1971**, *44*, 1632–1638.
4. H. E. Baumgarten, H. L. Smith, A. Staklis, *J. Org. Chem.* **1975**, *40*, 3554–3561.
5. A. S. Radhakrishna, C. G. Rao, R. K. Varma, B. B. Singh, S. P. Batnager, *Synthesis* **1983**, 538.

Hunsdiecker Reaction

Alkyl bromides from carboxylates

$$
\underset{1}{\overset{\displaystyle \overset{O}{\underset{\|}{\;}}}{R-C-O-Ag}} + Br_2 \longrightarrow \underset{2}{RBr + AgBr + CO_2}
$$

Silver carboxylates **1** can be decarboxylated by treatment with bromine, to yield alkyl bromides **2** in the so-called *Hunsdiecker reaction*.[1,2]

The reaction is likely to proceed by a radical-chain mechanism, involving intermediate formation of carboxyl radicals, as in the related *Kolbe electrolytic synthesis*. Initially the bromine reacts with the silver carboxylate **1** to give an acyl hypobromite species **3** together with insoluble silver bromide, which precipitates from the reaction mixture. The unstable acyl hypobromite decomposes by homolytic cleavage of the O—Br bond, to give a bromo radical and the carboxyl radical **4**. The latter decomposes further to carbon dioxide and the alkyl radical **5**, which subsequently reacts with hypobromite **3** to yield the alkyl bromide **2** and the new carboxyl radical **4**:[3]

By trapping experiments, the intermediate radical species have been identified, thus supporting the mechanism as formulated above.

Suitable substrates for the Hunsdiecker reaction are first of all aliphatic carboxylates. Aromatic carboxylates do not react uniformly. Silver benzoates with electron-withdrawing substituents react to the corresponding bromobenzenes, while electron-donating substituents can give rise to formation of products where an aromatic hydrogen is replaced by bromine. For example the silver *p*-methoxybenzoate **6** is converted to 3-bromo-4-methoxybenzoic acid **7** in good yield:

The silver carboxylate employed as starting material can be prepared from the corresponding carboxylic acid and silver oxide. In order to achieve the conversion to an alkyl bromide through decarboxylation in good yield, the silver carboxylate must be sufficiently pure. Bromine is often used as reagent in the Hunsdiecker reaction, though chlorine or iodine may also be employed. As solvent, carbon tetrachloride is most often used. In general yields are moderate to good.

In a modified procedure the free carboxylic acid is treated with a mixture of mercuric oxide and bromine in carbon tetrachloride; the otherwise necessary purification of the silver salt is thereby avoided. This procedure has been used in the first synthesis of [1.1.1]propellane 10. Bicyclo[1.1.1]pentane-1,3-dicarboxylic acid 8 has been converted to the dibromide 9 by the modified Hunsdiecker reaction. Treatment of 9 with t-butyllithium then resulted in a debromination and formation of the central carbon–carbon bond; thus generating the propellane 10.[4]

A complementary method is the *Kochi reaction*.[5,6] This reaction is especially useful for the generation of secondary and tertiary alkyl chlorides through decarboxylation of carboxylic acids, where the classical method may not work. As reagents, lead tetraacetate and lithium chloride are then employed:

$$RCOOH + Pb(OAc)_4 + LiCl \longrightarrow RCl + CO_2 + LiPb(OAc)_3 + HOAc$$

Another decarboxylation reaction that employs lead tetraacetate under milder conditions, has been introduced by Grob et al.[7] In that case N-chlorosuccinimide is used as chlorinating agent and a mixture of N,N-dimethylformamide and acetic acid as solvent.

1. H. Hunsdiecker, C. Hunsdiecker, *Ber. Dtsch. Chem. Ges.* **1942**, *75*, 291–297.
2. C. V. Wilson, *Org. React.* **1957**, *9*, 332–387.
3. R. G. Johnson, R. K. Ingham, *Chem. Rev.* **1956**, *56*, 219–269.
4. K. B. Wiberg, *Acc. Chem. Res.* **1984**, *17*, 379–386.
5. J. K. Kochi, *J. Org. Chem.* **1965**, *30*, 3265–3271.

6. R. A. Sheldon, J. K. Kochi, *Org. React.* **1972**, *19*, 279–421.
7. K. B. Becher, M. Geisel, C. A. Grob, F. Kuhnen, *Synthesis* **1973**, 493–495.

Hydroboration

Addition of boranes to alkenes

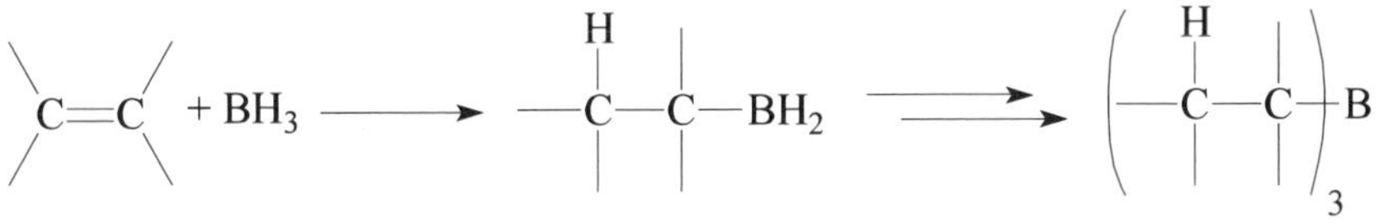

$$\underset{\textbf{1}}{\diagdown_{/}C{=}C\diagup^{\diagdown}} + \underset{\textbf{2}}{BH_3} \longrightarrow \underset{\textbf{3}}{-\overset{H}{\underset{|}{C}}-\overset{|}{\underset{|}{C}}-BH_2} \Longrightarrow \left(-\overset{H}{\underset{|}{C}}-\overset{|}{\underset{|}{C}}-\right)_3 B$$

Borane **2** adds to carbon–carbon double bonds without the need of catalytic activation. This reaction has been discovered and thoroughly investigated by *H. C. Brown*,[1] and is called *hydroboration*.[2–5] It permits a regioselective and stereospecific conversion of alkenes to a variety of functionalized products.

Free borane (**2**) exists as gaseous dimer—the diborane B_2H_6. In addition Lewis acid/Lewis base-complexes, as for example formed in an ethereal solvent, e.g. **4**, are commercially available:

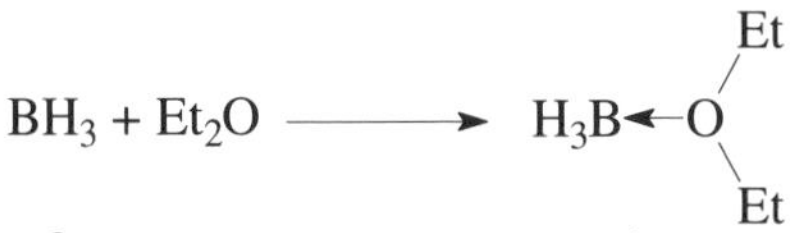

$$\underset{\textbf{2}}{BH_3} + Et_2O \longrightarrow \underset{\textbf{4}}{H_3B{\leftarrow}O\overset{\diagup Et}{\diagdown_{Et}}}$$

In reaction with an alkene, initially a three-membered ring Lewis acid/Lewis base-complex **5** is formed, where the carbon–carbon double bond donates π-electron density into the empty p-orbital of the boron center. This step resembles the formation of a bromonium ion in the electrophilic addition of bromine to an alkene:

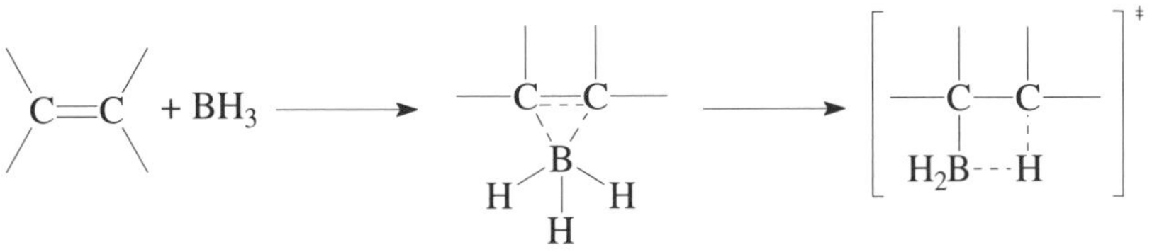

$$\underset{\textbf{3}}{-\overset{|}{\underset{H_2B}{C}}-\overset{|}{\underset{H}{C}}-}$$

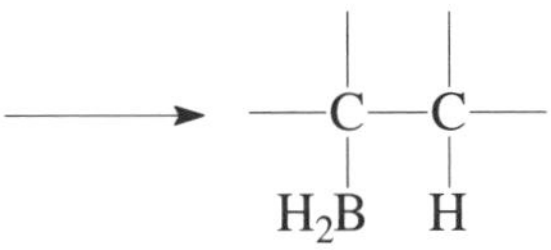

In the next step, one of the borane-hydrogens is transfered to a sp^2-carbon center of the alkene and a carbon–boron bond is formed, *via* a four-membered cyclic transition state **6**. A *mono*-alkylborane $R-BH_2$ molecule thus formed can react the same way with two other alkene molecules, to yield a trialkylborane R_3B. In case of *tri*- and *tetra*-substituted alkenes—e.g. 2-methylbut-2-ene **7** and 2,3-dimethylbut-2-ene **9**—which lead to sterically demanding alkyl-substituents at the boron center, borane will react with only two or even only one equivalent of alkene, to yield a *di*alkylborane or *mono*alkylborane respectively:

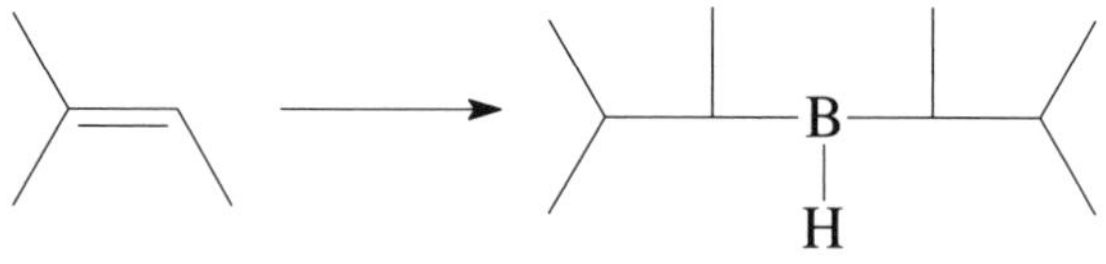

7   **8**

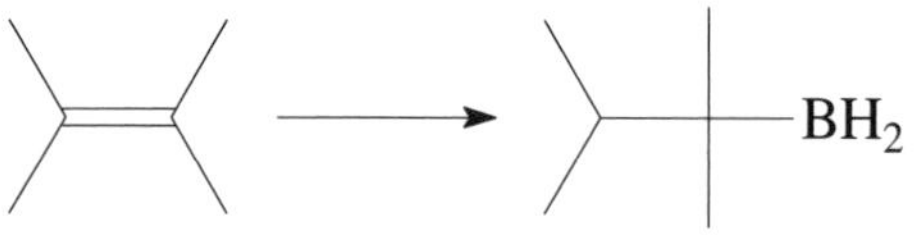

9   **10**

The hydroboration is a *syn*-stereospecific reaction. For example reaction with 1-methylcyclopentene **11** yields the 1,2-*trans*-disubstituted product **12** only:

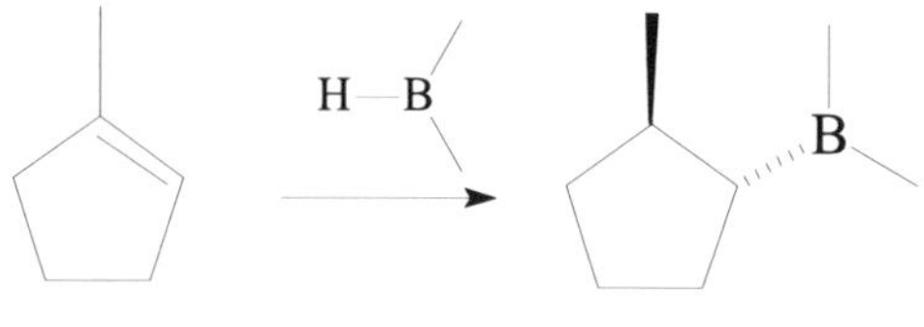

11   **12**

The hydroboration is a regioselective reaction. In general the addition will lead to a product, where the boron is connected to the less substituted or less sterically hindered carbon center. If the olefinic carbons do not differ much in reactivity or their sterical environment, the regioselectivity may be low. It can be enhanced by use of a less reactive alkylborane—e.g. disiamylborane **8**:

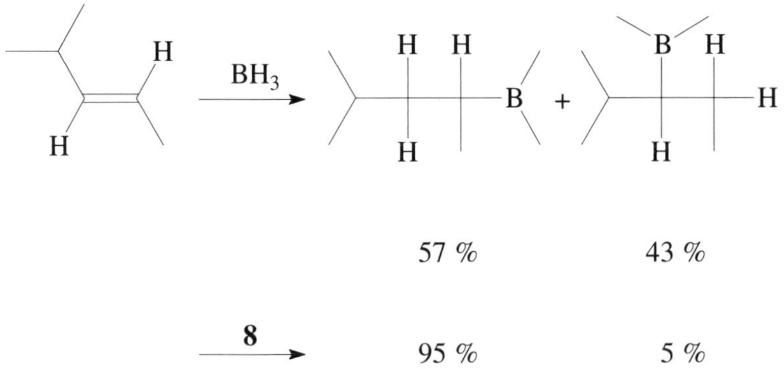

57 %   43 %

$\xrightarrow{\textbf{8}}$ 95 %   5 %

Since borane BH_3 reacts with only one or two equivalents of a sterically hindered alkene, it is possible to prepare less reactive and more selective borane reagents R_2BH and RBH_2 respectively. In addition to disiamylborane **8** and the xylborane **10**, the 9-borabicyclo[3.3.1]nonane (9-BBN) **14** is an important reagent for hydroboration, since it is stable to air; it is prepared by addition of borane **2** to cycloocta-1,5-diene **13**:

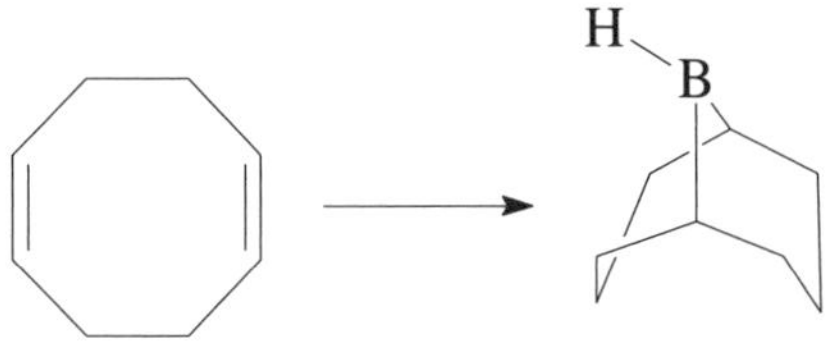

13   **14**

By reaction of borane with two equivalents of α-pinene **15**, the chiral hydroboration reagent diisopinocampheylborane **16** (Ipc_2BH) is formed:

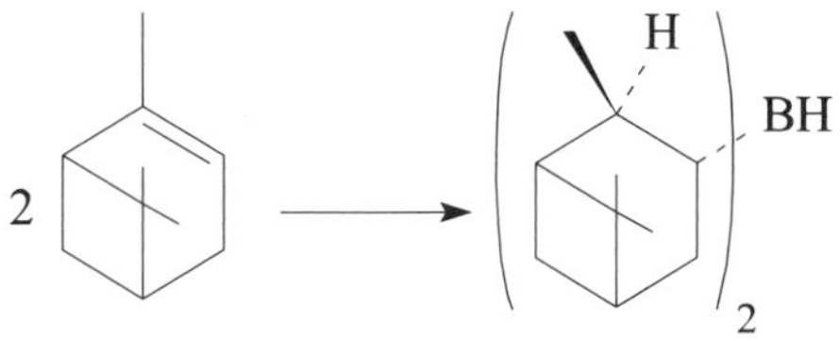

15   **16**

This reagent can be used for the enantioselective hydroboration of Z-alkenes with enantiomeric excess of up to 98%. Other chiral hydroboration reagents have been developed.[6]

The alkylboranes obtained by the hydroboration reaction are versatile intermediates for further transformations. The most important transformation is the
oxidation to yield alcohols **17**; it is usually carried out by treatment with hydroperoxide in alkaline solution. The group R migrates from boron to oxygen with
retention of configuration:

$$R-\overset{\underset{|}{R}}{\underset{|}{B}}+ {}^-OOH \longrightarrow R-\overset{\underset{|}{R}}{\underset{|}{B}}-O-OH \xrightarrow{-OH^-} R-\overset{\underset{|}{R}}{\underset{|}{B}}-OR$$

$$\longrightarrow B(OR)_3 \xrightarrow[\text{NaOH}]{\text{H}_2\text{O}} RO \overset{OH}{\underset{B}{\diagup}} OR + OR^-$$

$$\longrightarrow ROH + RO \overset{O^-}{\underset{B}{\diagup}} OR \longrightarrow 3\ ROH + Na_3BO_3$$

17

The overall result of the sequence hydroboration + oxidation is a regioselective
anti-Markownikoff-addition of water to an alkene. This reaction is an important method in organic synthesis, since it can be made stereoselective and even
enantioselective.

Other important applications for organoboranes[4] include the Michael-like addition reaction to α,β-unsaturated carbonyl compounds, and the alkylation of α-
halogenated carbonyl compounds.

The reaction conditions are mild—e.g. reaction temperatures between 0 °C and
room temperature. Side-reactions like for example rearrangements of the carbon
skeleton are usually not observed.

1. H. C. Brown, B. C. Subba Rao, *J. Am. Chem. Soc.* **1956**, *78*, 5694–5695.
2. G. Zweifel, H. C. Brown, *Org. React.* **1963**, *13*, 1–54.
3. H. Hopf, *Chem. Unserer Zeit* **1970**, *4*, 95–98.
4. A. Suzuki, R. S. Dhillon, *Top. Curr. Chem.* **1986**, *130*, 23–88.
5. A. Pelter, K. Smith, H. C. Brown, *Borane Reagents*, Academic Press, New York,
    1988.
6. H. C. Brown, B. Singaram, *Acc. Chem. Res.* **1988**, *21*, 287–293.

J

Japp-Klingemann Reaction

Arylhydrazones from reaction of β-dicarbonyl compounds with arenediazonium salts

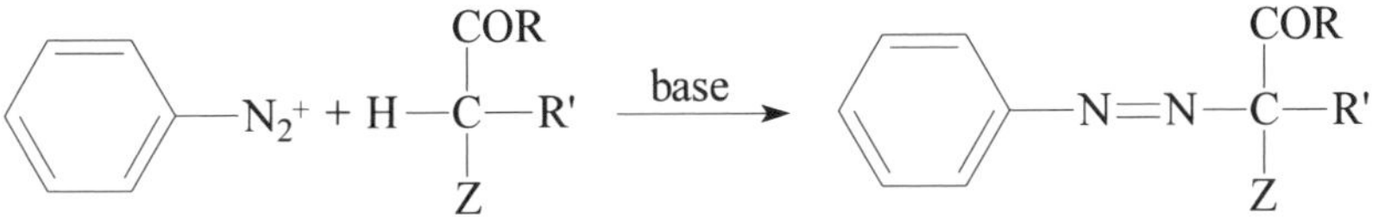

The coupling of arenediazonium compounds **1** to 1,3-dicarbonyl substrates **2** (Z = COR) is known as the *Japp–Klingemann reaction*.[1,2] As suitable substrates, β-ketoacids (Z = COOH) and β-ketoesters (Z = COOR) can be employed. As reaction product an arylhydrazone **4** is obtained.

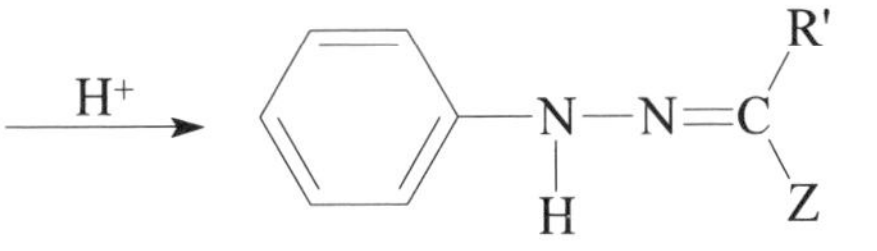

In general the reaction is conducted in aqueous solution under basic conditions—e.g. in the presence of KOH. The 1,3-dicarbonyl substrate is deprotonated to give the corresponding anion **5**, which then couples to the arenediazonium species **1**, to give the diazo compound **3**:

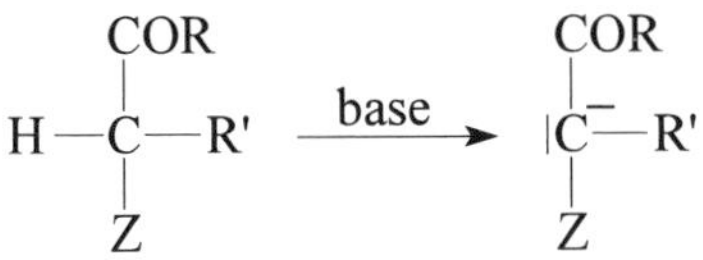

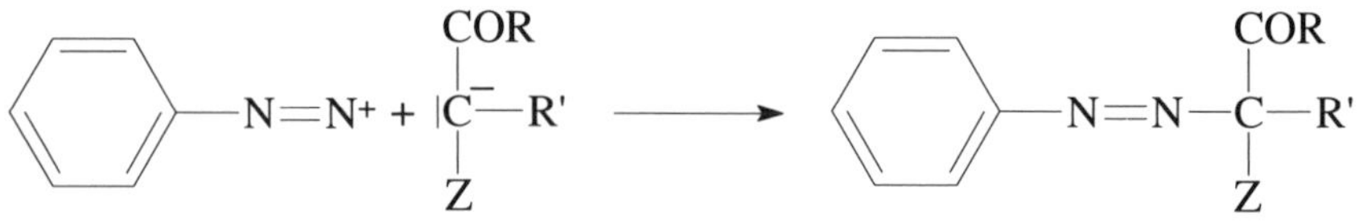

Such diazo compounds **3** however, that contain two electron-withdrawing substituents, are unstable under these reaction conditions. They further react by hydrolytic cleavage of one carbonyl substituent to give an anionic species **6**, that is stabilized by resonance, and which yields the hydrazone **4** upon acidic workup:

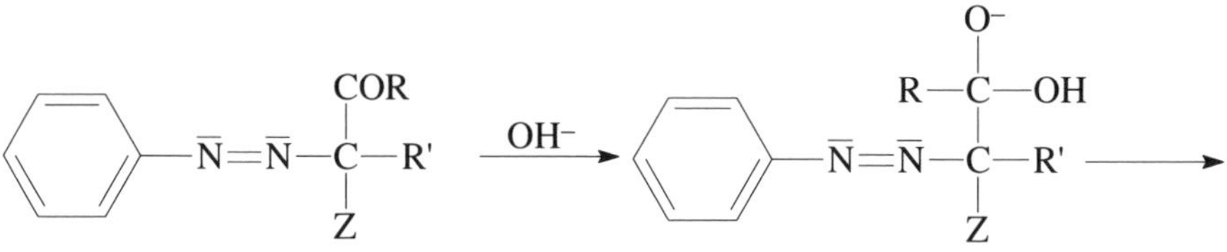

The mechanism outlined above is supported by the fact that various diazo intermediates **3** could be isolated.[3,4]

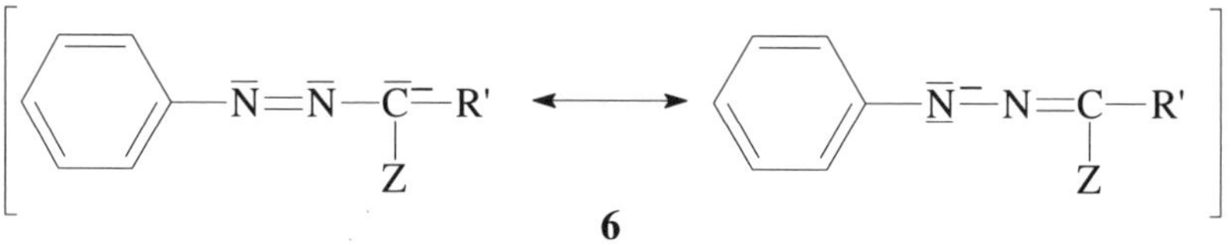

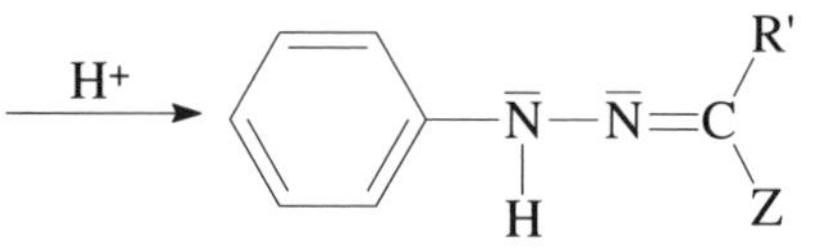

The Japp–Klingemann reaction is a special case of the aliphatic diazo coupling. For a successful reaction, the dicarbonyl substrate **2** should bear a sufficiently CH-acidic hydrogen.

In addition to β-diketones, β-ketoacids and β-ketoesters, cyanoacetic ester and related compounds are suitable starting materials. The arylhydrazones **4** thus obtained are of great importance as starting materials for the *Fischer indole synthesis*, as well as for the preparation of other *N*-heterocycles.[5]

1. F. R. Japp, F. Klingemann, *Justus Liebigs Ann. Chem.* **1888**, *247*, 190–225.
2. R. R. Phillips, *Org. React.* **1959**, *10*, 143–178.
3. O. Dimroth, *Ber. Dtsch. Chem. Ges.* **1908**, *41*, 4012–4028.
4. L. Kalb, F. Schweizer, H. Zellner, E. Berthold, *Ber. Dtsch. Chem. Ges.* **1926**, *59*, 1860–1870.
5. M. Kocevar, D. Kolman, H. Krajnc, S. Polanc, B. Porovne, B. Stanovnik, M. Tisler, *Tetrahedron* **1976**, *32*, 725–729.

Knoevenagel Reaction

Condensation of an aldehyde or ketone with an active methylene compound

$$
\underset{\textbf{1}}{\overset{\displaystyle \diagdown}{\underset{\diagup}{C}}=O} + \underset{\textbf{2}}{H_2C\overset{\displaystyle \diagup COOR}{\diagdown COOR}} \xrightarrow{\text{base}} \underset{\textbf{3}}{\overset{\displaystyle \diagdown}{\underset{\diagup}{C}}=C\overset{\displaystyle \diagup COOR}{\diagdown COOR}} \xrightarrow[-CO_2]{\text{hydrolysis}} \underset{\textbf{4}}{\overset{\displaystyle \diagdown}{\underset{\diagup}{C}}=C\overset{\displaystyle \diagup COOH}{\diagdown H}}
$$

The prototype of a *Knoevenagel reaction*[1,2] shown in the scheme above is the condensation of an aldehyde or ketone **1** with a malonic ester **2**, to yield an α,β-unsaturated carboxylic ester **4**.

The term Knoevenagel reaction however is used also for analogous reactions of aldehydes and ketones with various types of CH-acidic methylene compounds. The reaction belongs to a class of carbonyl reactions, that are related to the *aldol reaction*. The mechanism[3] is formulated by analogy to the latter. The initial step is the deprotonation of the CH-acidic methylene compound **2**. Organic bases like amines can be used for this purpose; a catalytic amount of amine usually suffices. A common procedure, that uses pyridine as base as well as solvent, together with a catalytic amount of piperidine, is called the *Doebner modification*[4] of the Knoevenagel reaction.

The corresponding anion **5**, formed from **2** by deprotonation, subsequently adds to the carbonyl substrate to give the aldol-type intermediate **6**. Loss of water from intermediate **6** leads to a primary α, β-unsaturated condensation product **3**:

$$
\underset{\textbf{2}}{H_2C\overset{\displaystyle \diagup COOR}{\diagdown COOR}} \xrightarrow{\text{base}} \underset{\textbf{5}}{{}^-\!\overline{|}CH\overset{\displaystyle \diagup COOR}{\diagdown COOR}}
$$

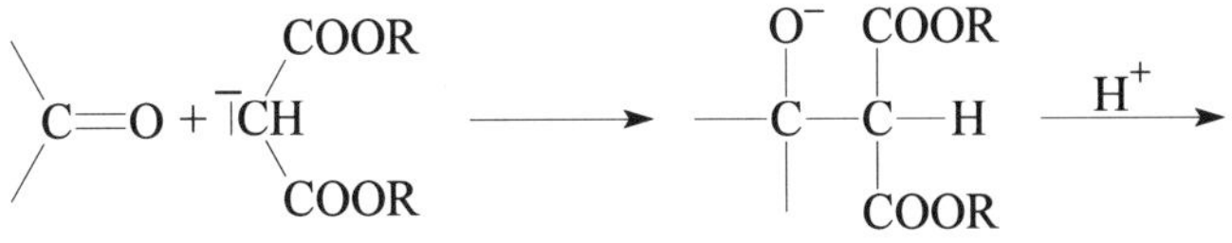

$$\begin{array}{ccc} 1 & & 5 \end{array}$$

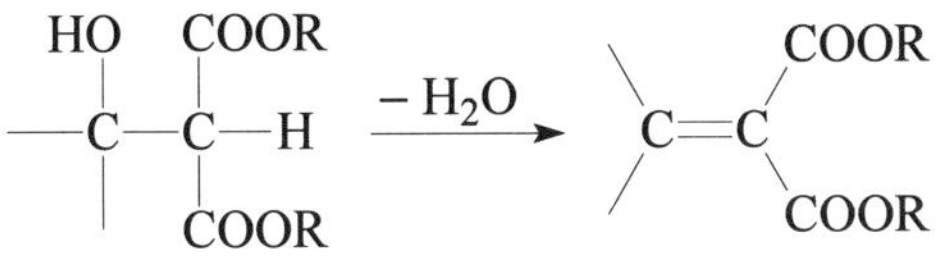

$$\begin{array}{ccc} 6 & & 3 \end{array}$$

Another mechanism has been formulated, which is based on results obtained by Knoevenagel,[5] and which is supported by more recent investigations.[6,7] It involves the formation of an intermediate iminium species **7**:

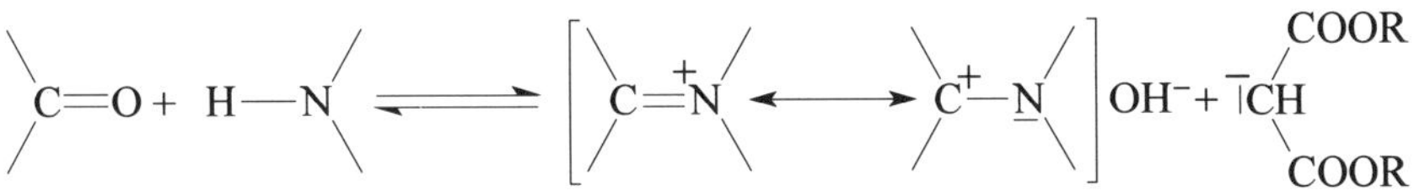

$$\begin{array}{ccc} 1 & 7 & 5 \end{array}$$

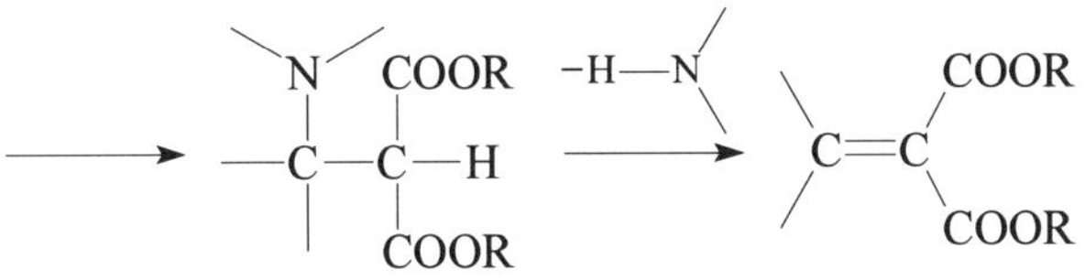

$$3$$

This reaction pathway explains well the applicability of amines as catalysts. There is evidence for each one of the mechanisms outlined above. Because of the wide scope of the reaction, there may be no uniform mechanism that would apply to all cases.

If a Knoevenagel condensation with malonic acid is conducted in refluxing pyridine, a subsequent decarboxylation often occurs. It has been shown that the decarboxylation of α,β-unsaturated diesters **3** under these conditions is slow;[2] the decarboxylation of the corresponding free dicarboxylic acid is formulated as follows:

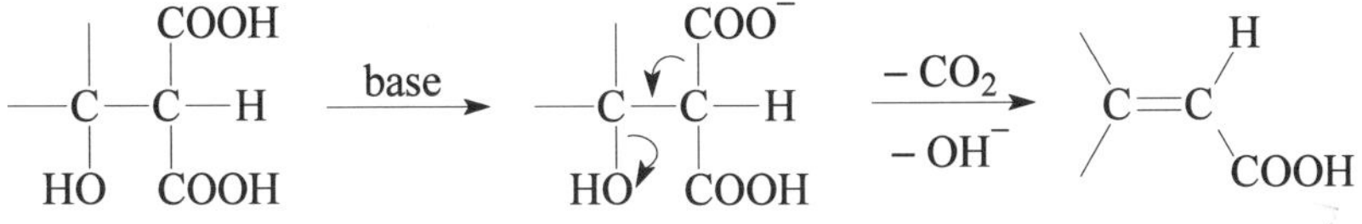

The formation of *bis*-adducts—e.g. **8** by a consecutive *Michael addition reaction*, is observed in some cases. This reaction is formulated as a 1,4-addition of a second molecule of the CH-acidic starting material **2** to the initially formed α,β-unsaturated carbonyl compound **3**:

Virtually any aldehyde or ketone and any CH-acidic methylene compound can be employed in the Knoevenagel reaction; however the reactivity may be limited due to steric effects. Some reactions may lead to unexpected products from side-reactions or from consecutive reactions of the initially formed Knoevenagel product.

Suitable substituents X and Y that can activate the methylene group to become CH-acidic, are electron-withdrawing groups—e.g. carboxy, nitro, cyano and carbonyl groups. Malonic acid as well as cyano acetic acid and derivatives (ester, nitrile, amide) are often used. In general two activating groups X and Y are required to achieve sufficient reactivity; malononitrile $CH_2(CN)_2$ is considered to be the most reactive methylene compound with respect to the Knoevenagel reaction. As would be expected, ketones are less reactive than aldehydes. In addition yield and rate of the condensation reaction are influenced by steric factors.

Because of the mild reaction conditions, and its broad applicability, the Knoevenagel reaction is an important method for the synthesis of α,β-unsaturated carboxylic acids.[2] Comparable methods[8] are the *Reformatsky reaction*, the *Perkin reaction*, as well as the *Claisen ester condensation*. The Knoevenagel reaction is of greater versatility; however the Reformatsky reaction permits the preparation of α,β-unsaturated carboxylic acids that are branched in α-position.

A more recent application of the Knoevenagel reaction is its use in *domino reactions*. The term domino reaction is used for two or more subsequent transformations, where the next reaction step is based on the functionality generated in the preceding step.[9] Such reactions are also called *tandem reactions* or *cascade reactions*.

The following example is a sequence consisting of a Knoevenagel condensation and a subsequent *hetero-Diels–Alder reaction*.[10] An aromatic aldehyde is condensed *in situ* with a β-dicarbonyl compound **9**—i.e. meldrum's acid—to give a 1-oxabuta-1,3-diene derivative **10**, which further undergoes an intramolecular [4 + 2]-cycloaddition reaction with the trisubstituted double bond acting as dienophile:

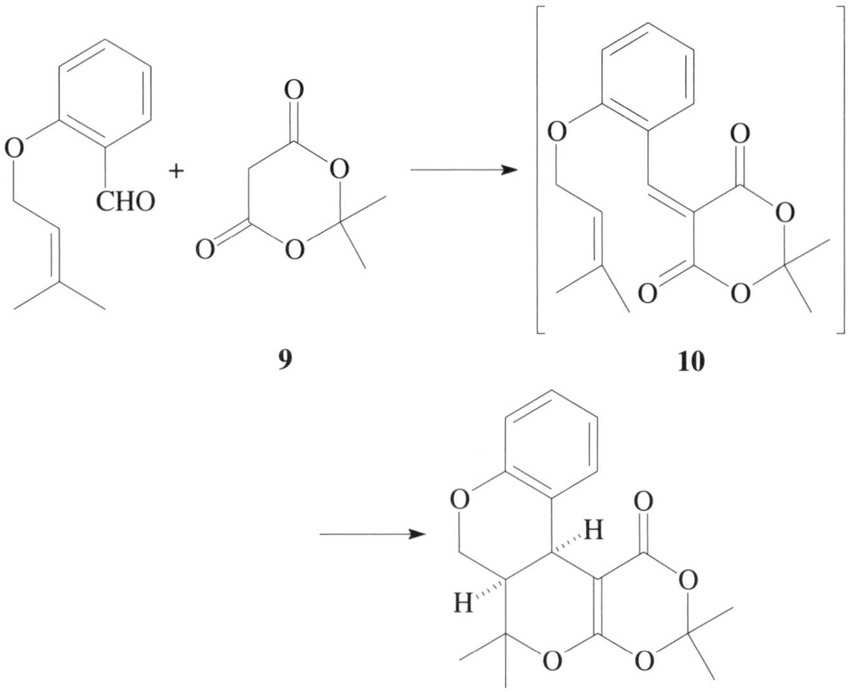

9   **10**

A large number of aldehydes and structurally different CH-acidic methylene compounds can be employed in such a *domino-Knoevenagel + hetero-Diels–Alder reaction.*

1. E. Knoevenagel, *Ber. Dtsch. Chem. Ges.* **1894**, *27*, 2345–2346.
2. G. Jones, *Org. React.* **1967**, *15*, 204–599.
3. A. C. O. Hann, A. Lapworth, *J. Chem. Soc.* **1904**, *85*, 46–56.
4. O. Doebner, *Ber. Dtsch. Chem. Ges.* **1900**, *33*, 2140–2142.
5. E. Knoevenagel, *Ber. Dtsch. Chem. Ges.* **1898**, *31*, 2596–2619.
6. G. Charles, *Bull. Soc. Chim. Fr.* **1963**, 1576–1583.
7. T. I. Crowell, D. W. Peck, *J. Am. Chem. Soc.* **1953**, *75*, 1075–1077.
8. R. L. Shriner, *Org. React.* **1942**, *1*, 1–37.
9. H. Waldmann, *Nachr. Chem. Tech. Lab.* **1992**, *40*, 1133–1140.
10. L. F. Tietze, U. Beifuss, *Angew. Chem.* **1993**, *105*, 137–170;
     Angew. Chem. Int. Ed. Engl. **1993**, *32*, 131.

Knorr Pyrrole Synthesis

Formation of pyrroles by condensation of ketones with α-aminoketones

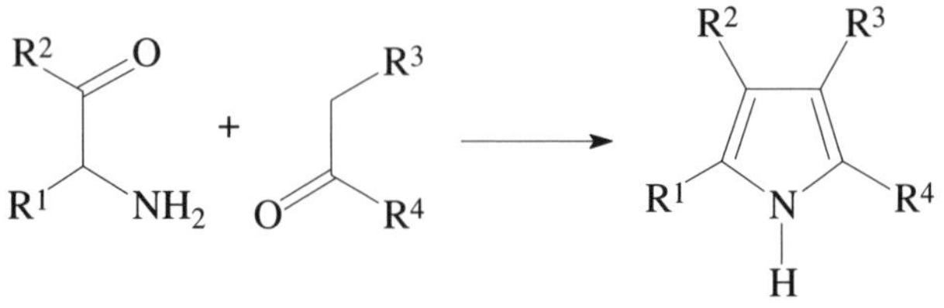

By a condensation reaction of an α-aminoketone **1** with a ketone **2**, a pyrrole **3** can be obtained. This reaction is known as the Knorr pyrrole synthesis.[1,2]

A mechanism has been formulated, starting with a condensation to give the imine **4**, that can tautomerize to the corresponding enamine **5**. The latter can be isolated in some cases, thus supporting the formulated mechanism. A cyclization and subsequent dehydration leads to the imine **6**, which tautomerizes to yield the aromatic pyrrole **3**:

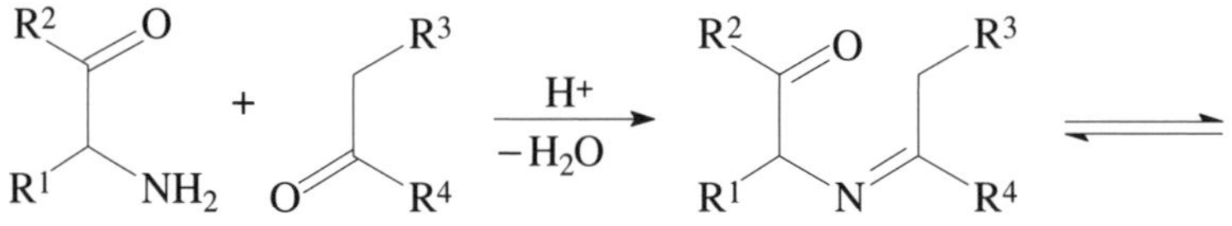

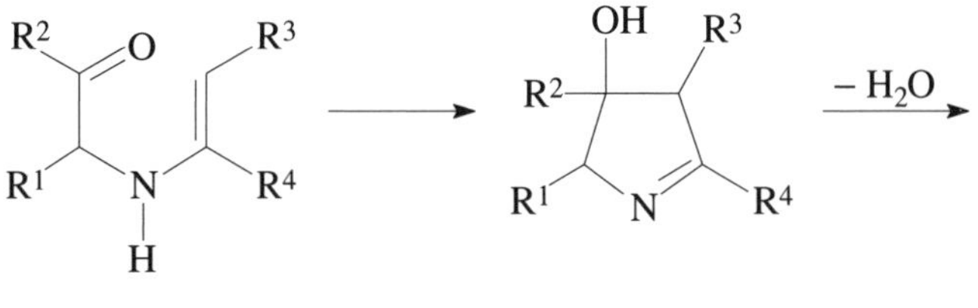

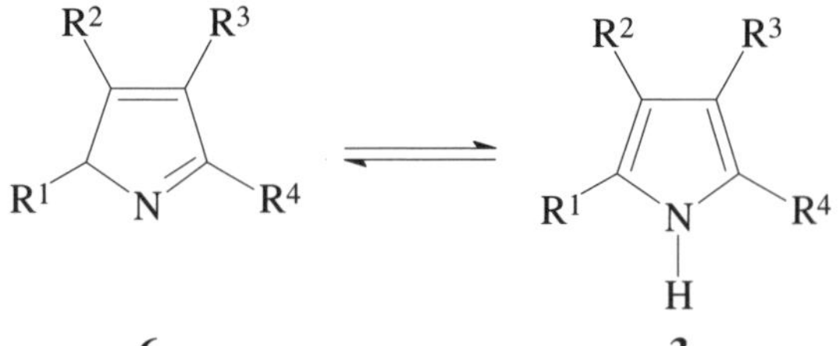

The aminoketone **1**, required as starting material, can be obtained by a *Neber rearrangement* from a *N*-tosylhydrazone. Another route to α-aminoketones starts with the nitrosation of an α-methylene carbonyl compound—often *in situ*—to give the more stable tautomeric oxime **7**, which is then reduced in a subsequent step to yield **1**:

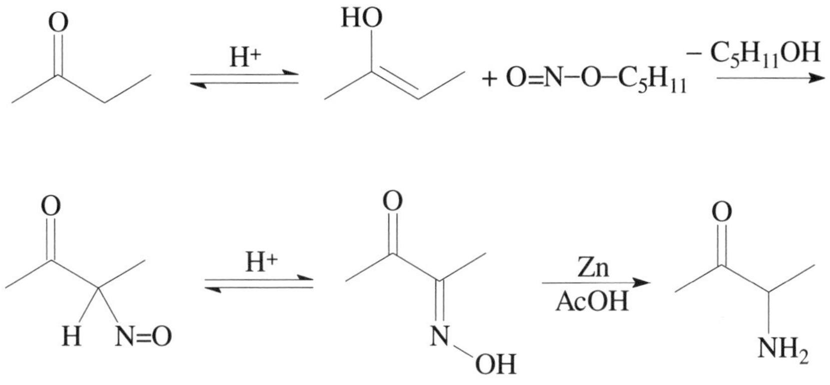

With excess ketone, the preparation of the aminoketone and subsequent condensation to a pyrrole can be conducted in one pot. In a side-reaction α-aminoketones can undergo a self-condensation to give pyrazines **8**:

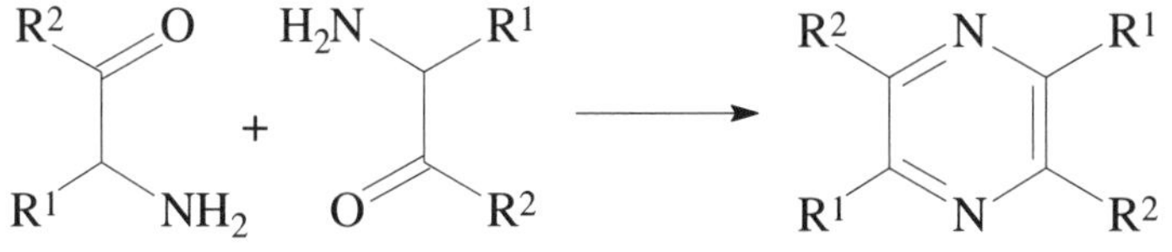

The self-condensation is largely suppressed in reactions with those ketones **2**, that are activated by an electron-withdrawing substituent R^3 or R^4. The carbonyl activity is then increased, and the enamine-intermediate **5** is favored over the imine **4**, by conjugation with the electron-withdrawing group.[3]

Mainly C-substituted pyrroles have been synthesized by application of the Knorr pyrrole synthesis; however N-substituted pyrroles can also be prepared, when starting with secondary aminoketones, e.g. bearing an N-methyl or N-phenyl substituent.

Another important route to pyrroles is offered by the *Paal–Knorr reaction*;[4] where the pyrrole system is formed by condensation of a 1,4-diketone **9** with ammonia:[5]

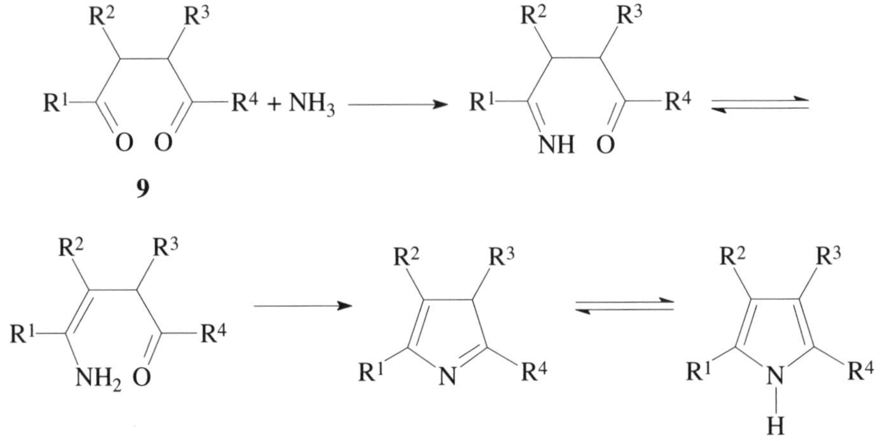

This reaction is of wide scope; it is limited only by the availability of the appropriate 1,4-diketone. 1,4-Diketones are easily accessible, e.g. by the *Nef reaction*.

Methods for the synthesis of pyrroles are of importance, since the pyrrole unit is found in natural products widespread in nature. For example a pyrrole unit is the building block of the porphyrin skeleton, which in turn is the essential structural subunit of chlorophyll and hemoglobin.

1. L. Knorr, *Ber. Dtsch. Chem. Ges.* **1884**, *17*, 1635–1642.
2. R. P. Bean in *The Chemistry of Pyrroles* (Ed.: R. A. Jones), Wiley, New York, **1990**, *Vol. 48/1*, p. 108–113.
3. A. H. Corwin, *Heterocycl. Compd.*, Wiley, New York, **1950**, Vol. 1, p. 287–290.
4. C. Paal, *Ber. Dtsch. Chem. Ges.*, **1885**, *18*, 367–371.
5. R. P. Bean in *The Chemistry of Pyrroles* (Ed.: R. A. Jones), Wiley, New York, **1990**, *Vol. 48/1*, p. 206–220.

Kolbe Electrolytic Synthesis

Electrolysis of carboxylate salts

$$2 \text{ RCOO}^- \xrightarrow[- 2 \text{ CO}_2]{- 2 \text{ e}^-} 2 \text{ R}^\bullet \longrightarrow \text{R} - \text{R}$$

$$\textbf{1} \qquad\qquad\qquad \textbf{2} \qquad\qquad \textbf{3}$$

The anodic oxidation of the carboxylate anion **1** of a carboxylate salt to yield an alkane **3** is known as the *Kolbe electrolytic synthesis*.[1-4] By decarboxylation alkyl radicals **2** are formed, which subsequently can dimerize to an alkane. The initial step is the transfer of an electron from the carboxylate anion **1** to the anode. The carboxyl radical species **4** thus formed decomposes by loss of carbon dioxide. The resulting alkyl radical **2** dimerizes to give the alkane **3**:[4]

The radical mechanism is supported by a number of findings: for instance, when the electrolysis is carried out in the presence of an olefin, the radicals add to the olefinic double bond; styrene does polymerize under those conditions. Side products can be formed by further oxidation of the alkyl radical **2** to an intermediate carbenium ion **5**, which then can react with water to yield an alcohol **6**, or with an alcohol to yield an ether **7**:

$$RCH_2^+ + H_2O \xrightarrow{\ -H^+\ } RCH_2OH$$

$$\ \ \mathbf{5}\phantom{ + H_2O \xrightarrow{\ -H^+\ }}\ \ \ \mathbf{6}$$

$$RCH_2^+ + R'OH \xrightarrow{\ -H^+\ } RCH_2OR'$$

$$\ \ \mathbf{5}\phantom{ + R'OH \xrightarrow{\ -H^+\ }}\ \ \ \mathbf{7}$$

From the mechanism outlined above it follows that the Kolbe electrolytic synthesis is first of all applicable for the preparation of symmetrical target molecules with an even number of carbons. By electrolysis of a mixture of two carboxylates the formation of unsymmetrical products is possible. Generally a statistical mixture of symmetrical and unsymmetrical products is obtained; however if one carboxylic acid, e.g. the less expensive one, is used in large excess, the formation of the symmetrical product from the minor component can be largely suppressed. Such a mixed Kolbe electrolytic synthesis often gives unsatisfactory yields; however this variant leads to the more valuable products. An instructive example is the synthesis of 3,11-dimethyl-2-nonacosanone **10**, which serves as a sex pheromone for the German cockroach. As starting material the 6-methyltetracosanoic acid **8** together with three equivalents of 5-methyl-6-oxoheptanoic acid **9** is used; the product **10** is then obtained in 42% yield:[5]

$$C_{18}H_{37} \overset{H}{\underset{CH_3}{C}} (CH_2)_4 COOH + HOOC (CH_2)_3 \overset{H}{\underset{CH_3}{C}} \overset{O}{C} \underset{CH_3}{}$$

8   **9**

$$\xrightarrow[KOH/MeOH]{electrolysis} \quad C_{18}H_{37} \overset{H}{\underset{CH_3}{C}} (CH_2)_7 \overset{H}{\underset{CH_3}{C}} \overset{O}{C} \underset{CH_3}{}$$

10

Suitable starting materials for the Kolbe electrolytic synthesis are aliphatic carbo-xylic acids that are not branched in α-position. With aryl carboxylic acids the reaction is not successful. Many functional groups are tolerated. The generation of the desired radical species is favored by a high concentration of the carboxylate salt as well as a high current density. Product distribution is further dependend on the anodic material, platinum is often used, as well as the solvent, the temperature and the pH of the solution.[4]

1. H. Kolbe, *Justus Liebigs Ann. Chem.* **1849**, *69*, 257–294.
2. A. K. Vijh, B. E. Conway, *Chem. Rev.* **1967**, *67*, 623–664.
3. H. J. Schäfer, *Angew. Chem.* **1981**, *93*, 978–1000; *Angew. Chem. Int. Ed. Engl.* **1981**, *20*, 911.
4. H. J. Schäfer, *Top. Curr. Chem.* **1990**, *152*, 91–151.
5. W. Seidel, H. J. Schäfer, *Chem. Ber.* **1981**, *113*, 451–456.

Kolbe Synthesis of Nitriles

Nitriles from alkyl halides

$$R-X + CN^- \longrightarrow R-CN + X^-$$

1   **2**

A common method for the preparation of alkyl cyanide **2** is the treatment of corresponding alkyl halides **1** with cyanide.[1-3] The corresponding reaction with aromatic substrates is called the *Rosenmund–von-Braun reaction*.

The *Kolbe synthesis of nitriles* is an important method for the elongation of an alkyl chain by one carbon center (see also the *Arndt–Eistert synthesis*). The nitrile **2** can for example easily be converted to the corresponding carboxylic acid by hydrolysis.

Since the cyanide anion is an ambident nucleophile, isonitriles $R-NC$ may be obtained as by-products. The reaction pathway to either nitrile or isonitrile can be controlled by proper choice of the counter cation for the cyanide anion. With alkali cyanides, a reaction *via* a S_N2-mechanism takes place; the alkyl halide is attacked by cyanide with the more nucleophilic carbon center rather than the nitrogen center, and the alkylnitrile is formed. In contrast, with silver cyanide the reaction proceeds by a S_N1-mechanism, and an isonitrile is formed, since the carbenium intermediate reacts preferentially with the more elctronegative center of the cyanide—i.e. the nitrogen (*Kornblum's rule, HSAB concept*).[4,5]

The reaction works well with primary alkyl halides, especially with allylic and benzylic halides, as well as other alkyl derivatives with good leaving groups. Secondary alkyl halides give poor yields. Tertiary alkyl halides react under the usual reaction conditions by elimination of HX only. Nitriles from tertiary alkyl halides can however be obtained by reaction with trimethylsilyl cyanide **4**:[6]

This variant gives good to very good yields, and it is chemoselective; the primary alkyl halide function in **3** is left unaffected under these reaction conditions.

1. F. Wöhler, J. von Liebig, *Justus Liebigs Ann. Chem.* **1832**, *3*, 267–268.
2. K. Friedrich, K. Wallenfels in *The Chemistry of the Cyano Group* (Ed.: S. Patai), Wiley, New York, **1970**, p. 77–86.
3. K. Friedrich in *The Chemistry of Functional Groups, Supp. C* (Eds. S. Patai, Z. Rappoport), Wiley, New York, **1970**, Vol. 2, p. 1345–1390.
4. N. Kornblum, R. A. Smiley, R. K. Blackwood, D. C. Iffland, *J. Am. Chem. Soc* **1955**, 77, 6269–6280.
5. B. Saville, *Angew. Chem.* **1967**, 79, 966–977; *Angew. Chem. Int. Ed. Engl.* **1967**, 6, 928.
6. M. T. Reetz, I. Chatziiosifidis, *Angew. Chem.* **1981**, 93, 1075–1076; *Angew. Chem. Int. Ed. Engl.* **1981**, 20, 1017.

Kolbe–Schmitt Reaction

Carboxylation of phenolates/synthesis of salicylic acid

Carbon dioxide reacts with phenolates **1** to yield salicylate **2**; with less reactive mono-phenolates, the application of high pressure may be necessary in order to obtain high yields. This reaction, which is of importance for the large scale synthesis of salicylic acid, is called the *Kolbe–Schmitt reaction*.[1–3]

In order to rationalize the *ortho*-selectivity observed in the reaction of sodium phenoxide **1** with carbon dioxide, the formation of a complex **3** is assumed. By that complexation the carbon dioxide becomes polarized, and its electrophilic character is increased. Complex **3** is of suitable geometry for reaction with the activated *ortho*-carbon center:[4]

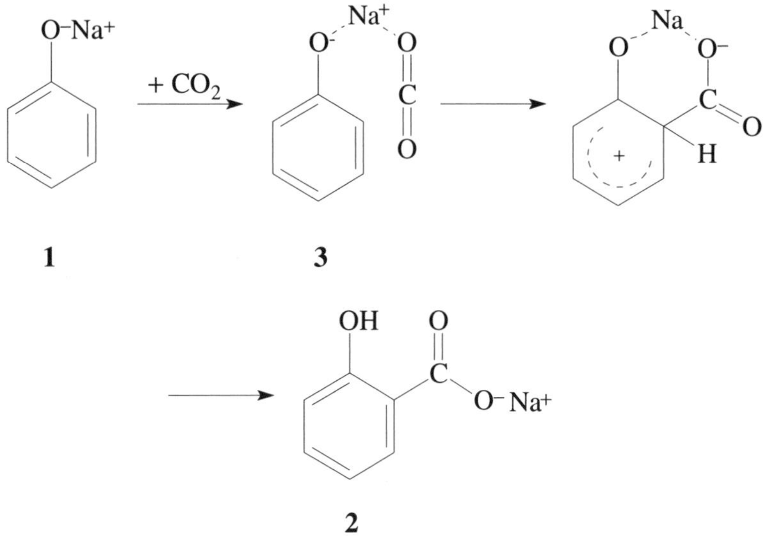

The *para*-substituted product, which is not accessible from complex **3**, can however be obtained from reaction of potassium phenoxide with carbon dioxide.

The Kolbe–Schmitt reaction is limited to phenol, substituted phenols and certain heteroaromatics.[5] The classical procedure is carried out by application of high pressure using carbon dioxide without solvent; yields are often only moderate.[2] In contrast to the minor importance on laboratory scale, the large scale process for the synthesis of salicylic acid is of great importance in the pharmaceutical industry.

1. H. Kolbe, *Justus Liebigs Ann. Chem.* **1860**, *113*, 125–127.
2. A. S. Linsey, H. Jeskey, *Chem. Rev.* **1957**, *57*, 583–620.
3. H. J. Shine, *Aromatic Rearrangements*, Elsevier, New York, **1967**, p. 344–348.
4. I. Hirao, T. Kito, *Bull. Chem. Soc. Jpn.* **1973**, *46*, 3470–3474.
5. H. Henecka, *Methoden Org. Chem. (Houben-Weyl)*, **1952**, Vol. 8, p. 372–377.

L

Leuckart–Wallach Reaction

Reductive alkylation of amines

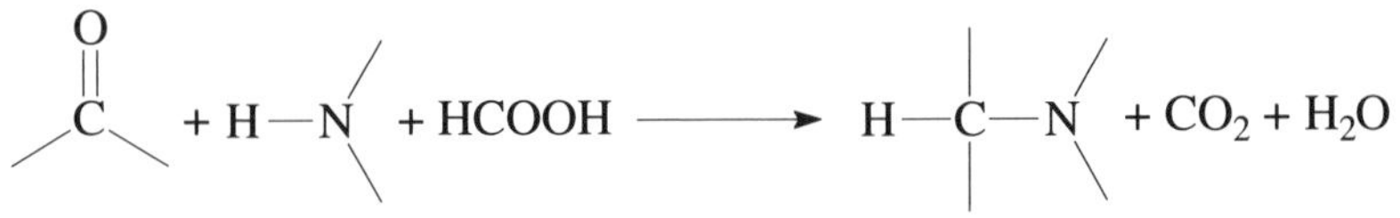

$$\underset{\textbf{1}}{\overset{O}{\underset{\|}{C}}} + \underset{\textbf{2}}{H-N} + \underset{\textbf{3}}{HCOOH} \longrightarrow \underset{\textbf{4}}{H-\overset{|}{\underset{|}{C}}-N} + CO_2 + H_2O$$

By application of the *Leuckart–Wallach reaction*,[1-3] amines **2** can be alkylated with a carbonyl compound **1**; formic acid is used as reductive agent, and is in turn oxidized to give carbon dioxide.

The carbonyl compound **1** is assumed to react first with amine **2** to give the unstable α-aminoalcohol **5** as intermediate,[3] from which an iminium species **6** is formed. The latter is reduced by reaction with formic acid *via* cyclic transition state **7** to yield the alkylated amine:

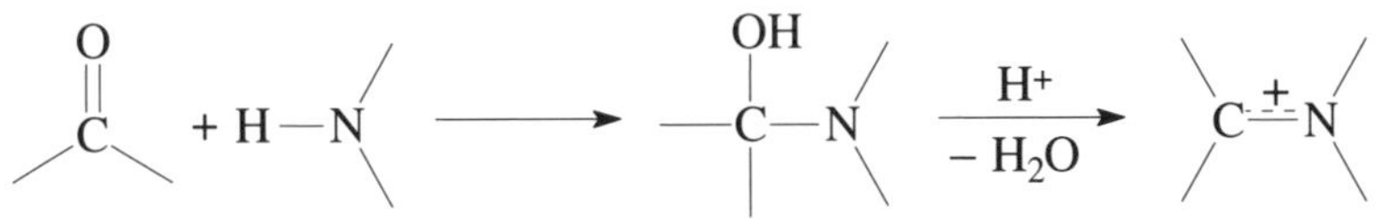

$$\underset{\textbf{1}}{\overset{O}{\underset{\|}{C}}} + \underset{\textbf{2}}{H-N} \longrightarrow \underset{\textbf{5}}{\overset{OH}{\underset{|}{C}}-N} \xrightarrow[-H_2O]{H+} \underset{\textbf{6}}{C\overset{+}{=}N}$$

$$\xrightarrow{HCOOH} \left[\begin{array}{c} C\overset{+}{=}N \\ H \quad\quad H \\ C\overset{=}{-}O \\ O \end{array} \right]^{\ddagger} \longrightarrow H-\overset{|}{\underset{|}{C}}-\overset{+}{\underset{|}{N}}-H + CO_2$$

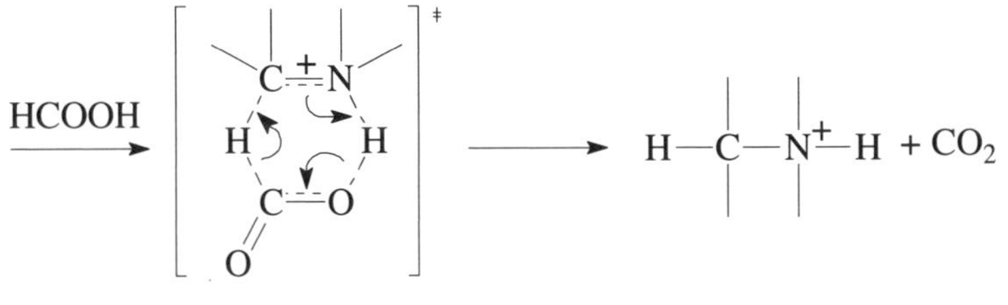

7

A primary or secondary amine can be used, as well as ammonia. With respect to the carbonyl component used, the best results have been obtained with aromatic aldehydes and with high boiling ketones.

The usual procedure is to simply heat a mixture of the starting materials. A common side-reaction is the polyalkylation; it can be suppressed by employing an excess of amine. In addition carbonyl substrates with α-hydrogens may undergo competitive *aldol reactions*; the corresponding reaction products may then undergo a subsequent Leuckart–Wallach reaction.

1. R. Leuckart, *Ber. Dtsch. Chem. Ges.* **1885**, *18*, 2341–2344.
2. M. L. Moore, *Org. React.* **1949**, *5*, 301–330.
3. A. Lukasiewicz, *Tetrahedron* **1963**, *19*, 1789–1799.

Lossen Reaction

Isocyanates from hydroxamic acids

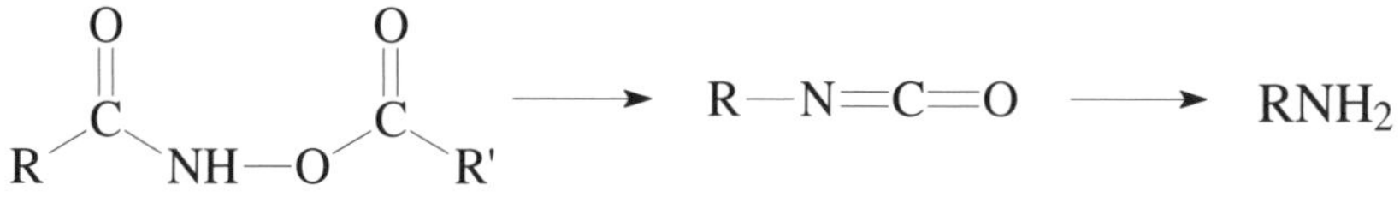

In the *Lossen reaction*[1,2] a hydroxamic acid derivative (usually an O-acyl derivative) is deprotonated by base, and rearranges *via* migration of the group R to give an isocyanate **2**. Under the usual reaction conditions—i.e. aqueous alkaline solution—the isocyanate reacts further to yield the amine **3**. The Lossen reaction is closely related to the *Hofmann rearrangement* and the *Curtius reaction*.

By reaction of the hydroxamic acid derivative **1** with a base, the anionic species **4** is formed. Cleavage of the leaving group from the nitrogen and migration of the group R from the carbon center to the developing electron-deficient nitrogen center are concerted—just like in the Hofmann rearrangement. The isocyanate formally is the final product of the Lossen reaction:

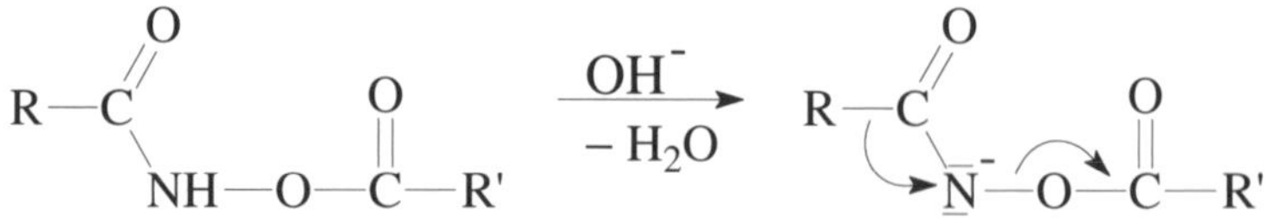

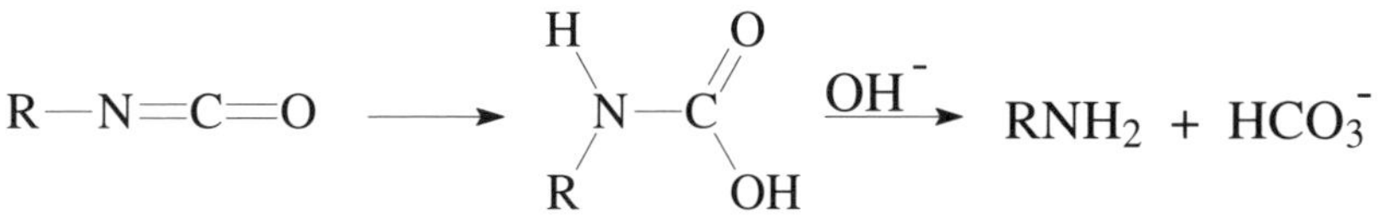

$$R-N{=}C{=}O \longrightarrow \underset{\underset{R}{|}}{N}-\overset{\overset{O}{\parallel}}{\underset{OH}{C}} \xrightarrow{\ OH^-\ } RNH_2 + HCO_3^-$$

2   5   3

In aqueous alkaline solution, the isocyanate is unstable; it reacts by addition of water to give the intermediate carbaminic acid **5**, which subsequently decarboxylates to yield the amine **3**.

Unsubstituted hydroxamic acids do not undergo the Lossen reaction.[3] An activation by an electron-acceptor substituent is necessary—e.g. by an acyl group. Furthermore a carboxylate anion is a much better leaving group as is the hydroxide anion. A substituent R with electro-donating properties can also facilitate the reaction. With substrates containing a chiral group R, the configuration of R is usually retained.

The Lossen reaction is of limited importance in synthetic organic chemistry; one reason for that is the poor availability of the required hydroxamic acid derivatives. Some hydroxamic acids are even unreactive.[4]

1. W. Lossen, *Justus Liebigs Ann. Chem.* **1872**, *161*, 347–362.
2. H. L. Yale, *Chem. Rev.* **1943**, *33*, 209–256.
3. L. Bauer, O. Exner, *Angew. Chem.* **1974**, *86*, 419–428;
    Angew. Chem. Int. Ed. Engl. **1974**, *13*, 376.
4. G. B. Bachmann, J. E. Goldmacher, *J. Org. Chem.* **1964**, *29*, 2576–2579.

M

Malonic Ester Synthesis

Alkylation of malonic esters

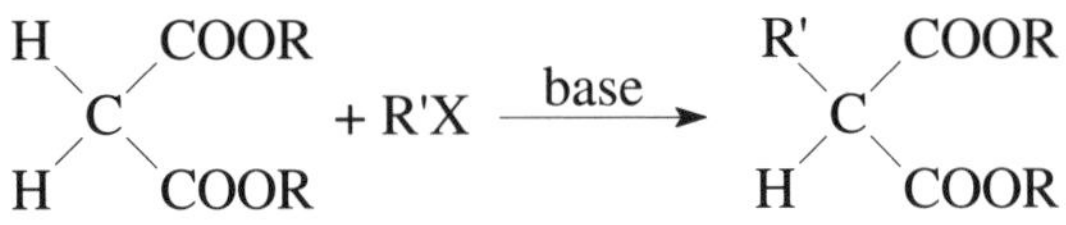

$$\underset{\textbf{1}}{\overset{H}{\underset{H}{>}}C\overset{COOR}{\underset{COOR}{<}}} + \underset{\textbf{2}}{R'X} \xrightarrow{\text{base}} \underset{\textbf{3}}{\overset{R'}{\underset{H}{>}}C\overset{COOR}{\underset{COOR}{<}}}$$

Compounds bearing two strongly electron-withdrawing groups at a methylene or methine group can be deprotonated and subsequently alkylated at that position;[1,2] as alkylating agent an alkyl halide **2** is often used. The most important reaction of this type is known as the *malonic ester synthesis*, where both electron-withdrawing substituents are ester groups; see scheme above **1** → **3**.

The reactive species is the corresponding enolate-anion **4** of malonic ester **1**. The anion can be obtained by deprotonation with a base; it is stabilized by resonance. The alkylation step with an alkyl halide **2** proceeds by a S_N2 reaction:

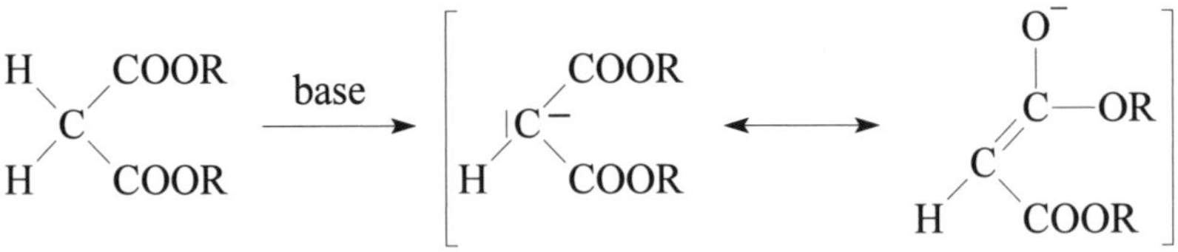

$$\underset{\textbf{1}}{\overset{H}{\underset{H}{>}}C\overset{COOR}{\underset{COOR}{<}}} \xrightarrow{\text{base}} \left[\underset{H}{\overset{|C^-}{>}}\overset{COOR}{\underset{COOR}{<}} \longleftrightarrow \underset{\textbf{4}}{\overset{O^-}{\underset{H}{\overset{C-OR}{\underset{COOR}{<}}}}} \right]$$

$$\underset{\textbf{4}}{\overset{|C^-}{\underset{H}{>}}\overset{COOR}{\underset{COOR}{<}}} + \underset{\textbf{2}}{R'X} \longrightarrow \underset{\textbf{3}}{\overset{R'}{\underset{H}{>}}C\overset{COOR}{\underset{COOR}{<}}} + X^-$$

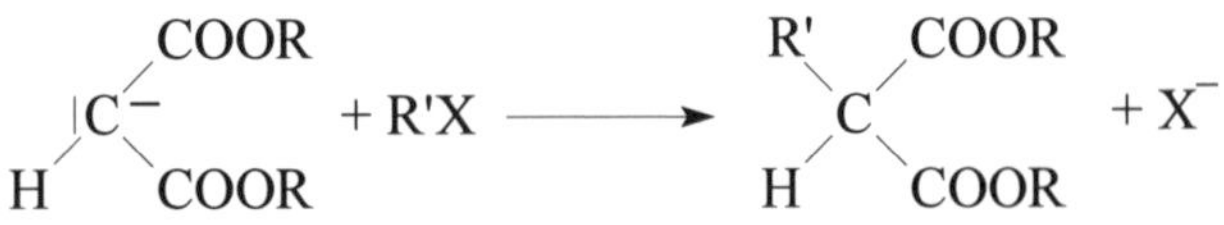

In order to generate a sufficient amount of malonic ester-enolate **4** in the reaction mixture, the corresponding acid BH^+ of the base and the solvent used, have to be less acidic than malonic ester **1**. A metal alkoxide is often used as base—e.g. sodium ethoxide or potassium *t*-butoxide. Usually that alkoxide is employed that corresponds to the ester alkyl, in order to avoid the formation of a product mixture through exchange of the alkoxy groups at the carbonyl center. For the same reason, the corresponding alcohol is used as solvent. Inert solvents are often less suitable, since the starting materials may not be sufficiently soluble, and the strength of the base will usually differ from that in the system RO^-/ROH.

The rate of the alkylation reaction depends on the enolate concentration, since it proceeds by a S_N2-mechanism. If the concentration of the enolate is low, various competitive side-reactions may take place. As expected, among those are E2-eliminations by reaction of the alkyl halide **2** with base. A second alkylation may take place with *mono*-alkylated product already formed, to yield a *bis*-alkyated malonic ester; however such a reaction is generally slower than the alkylation of unsubstituted starting material by a factor of about 10^2. The monoalkylation is in most cases easy to control. Dialkylated malonic esters with different alkyl substituents—e.g. ethyl and isopropyl—can be prepared by a step by step reaction sequence:

With highly reactive alkyl halides, like allylic, benzylic or phenacyl halides, the *bis*-alkylation can be a serious side-reaction. Because of a S_N1-like mechanism in those cases, the effect of enolate concentration on the reaction rate is low, and the resulting monoalkylester **5** may be more acidic than the unsubstituted starting material:

5

In general the *bis*-alkylation can be suppressed by using excess malonic ester **1**. Another side-reaction is the decarbalkoxylation, whereby dialkyl carbonates **6** are formed:

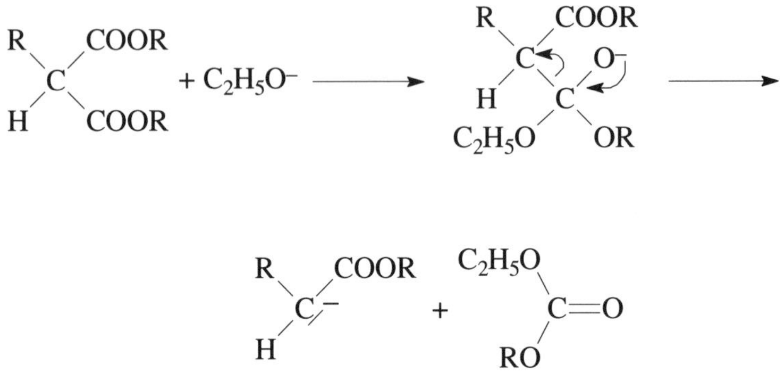

An important application of the potential for bis-alkylation is the use of a dihalide **7** as alkylating agent. This variant allows for the synthesis of cyclic compounds **8**; by this route, mainly five- to seven-membered ring compounds have been prepared:[3]

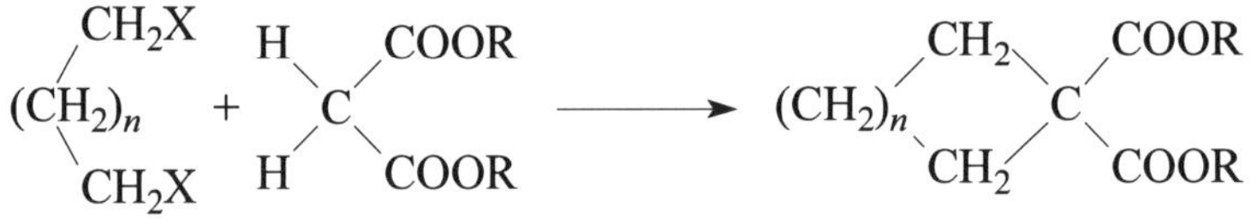

The synthetic importance of the malonic ester synthesis follows from the fact that the substituted malonic ester can easily be hydrolyzed, and subsequently decarboxylates to yield a substituted acetic acid **9**. This route to substituted acetic acids is an important method in organic synthesis:

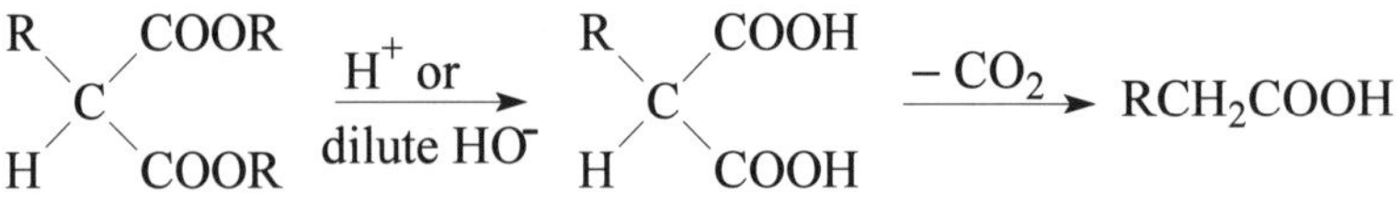

As alkylating agents may for example be used: alkyl halides, dialkyl sulfates, alkyl sulfonates and epoxides. Aryl halides and vinylic halides do not react.

Related and equally important reactions are the *acetoacetic ester synthesis* and the *cyanoacetic ester synthesis.*[2] Here too the initial substituted product can be hydrolyzed and decarboxylated, to yield a ketone **11** (i.e. a substituted acetone) from acetoacetic ester **10**, and a substituted acetonitrile **14** from cyanoacetic ester **13** respectively. Furthermore a substituted acetoacetic ester can be cleaved into a substituted acetic ester **12** and acetate by treatment with strong alkali:

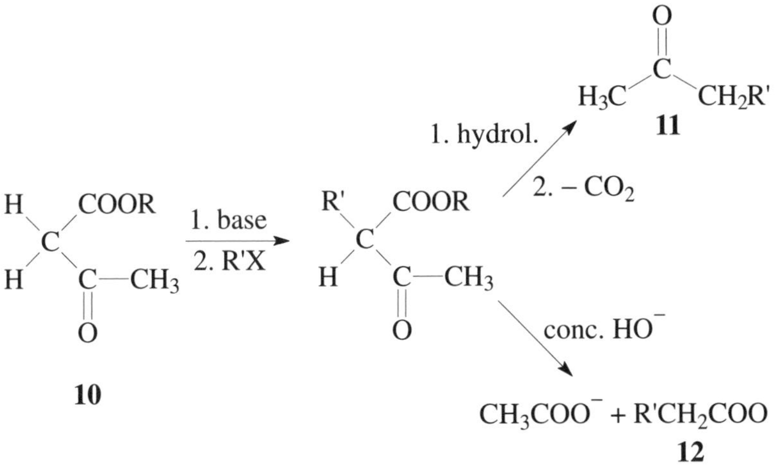

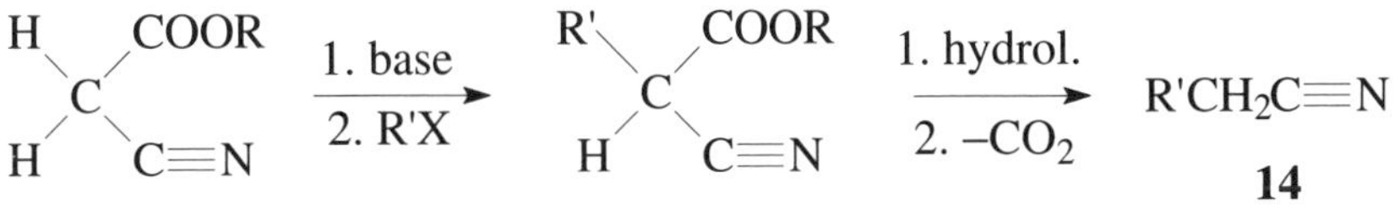

With two acidic CH-centers present in the starting material, the deprotonation and subsequent substitution (e.g. alkylation) takes place at the more acidic position; as shown above, acetoacetic ester will be alkylated at the methylene group rather than the methyl group. A substitution at the less acidic methyl group can be achieved by first converting the acetoacetic ester into its dianion; the reaction with an alkylating agent will then first take place at the more reactive, less acidic (i.e. stronger basic) position.

1. J. Wislicenus, *Justus Liebigs Ann. Chem.* **1877**, *186*, 161–228.
2. A. C. Cope, H. L. Holmes, H. O. House, *Org. React.* **1957**, *9*, 107–331.
3. N. S. Zefirov, T. S. Kuznetsova, S. I. Kozhushkov, L. S. Surmina, Z. A. Rashchupkina, *J. Org. Chem. USSR* **1983**, *19*, 474–480.

Mannich Reaction

Aminomethylation of CH-acidic compounds

$$
\underset{\textbf{1}}{\overset{H}{\underset{H}{\Large{C}}}{=}O} + \underset{\textbf{2}}{NH_3} + \underset{\textbf{3}}{CH{-}\overset{O}{\underset{R}{\Large{C}}}} \longrightarrow \underset{\textbf{4}}{H_2N{-}CH_2{-}\underset{}{\Large{C}}{-}\overset{O}{\underset{R}{\Large{C}}}}
$$

The condensation reaction of a CH-acidic compound—e.g. a ketone **3**—with formaldehyde **1** and ammonia **2** is called the *Mannich reaction*;[1–5] the reaction products **4** are called *Mannich bases*. The latter are versatile building blocks in organic synthesis, and of particular importance in natural products synthesis.

There have been extensive investigations on the reaction mechanism.[6,7] In most cases the reaction proceeds *via* initial nucleophilic addition of ammonia **2** to formaldehyde **1** to give adduct **5**, which is converted into an iminium ion species **6** (note that a resonance structure—an aminocarbenium ion can be formulated) through protonation and subsequent loss of water. The iminium ion species **6** then reacts with the enol **7** of the CH-acidic substrate by overall loss of a proton:

$$
\underset{\textbf{1}}{\overset{H}{\underset{H}{\Large{C}}}{=}O} + \underset{\textbf{2}}{NH_3} \longrightarrow \underset{\textbf{5}}{H{-}\underset{OH}{\overset{NH_2}{\Large{C}}}{-}H} \xrightarrow[-\,H_2O]{H^+}
$$

$$
\left[\underset{}{H{-}\overset{\overset{+}{N}H_2}{\underset{}{\Large{C}}}{-}H} \longleftrightarrow H{-}\overset{NH_2}{\underset{}{\overset{+}{\Large{C}}}}{-}H \right] \underset{\textbf{7}}{+ \ \overset{OH}{\underset{R}{C{=}C}}}
$$

6   **7**

$$
\longrightarrow \underset{\textbf{4}}{H_2N{-}CH_2{-}\underset{}{\Large{C}}{-}\overset{O}{\underset{R}{\Large{C}}}}
$$

Instead of formaldehyde, other aldehydes or ketones may be used—aliphatic as well as aromatic; recently methylene dihalides have been employed with success. The amine component is often employed as hydrochloride; in addition to ammonia, aliphatic amines, hydroxylamine or hydrazine can be used. Aromatic amines usually do not undergo the reaction.

As solvent an alcohol—often ethanol—as well as water or acetic acid can be used. The reaction conditions vary with the substrate; various CH-acidic compounds can be employed as starting materials. The Mannich bases formed in the reaction often crystallize from the reaction mixture, or can be isolated by extraction with aqueous hydrochloric acid.

With an unsymmetrical ketone as CH-acidic substrate, two regioisomeric products can be formed. A regioselective reaction may in such cases be achieved by employing a preformed iminium salt instead of formaldehyde and ammonia. An iminium salt reagent—the *Eschenmoser salt*—has also found application in Mannich reactions.[8,9]

Because of their manifold reactivity, Mannich bases **4** are useful intermediates in organic synthesis. For example the elimination of amine leads to formation of an α, β-unsaturated carbonyl compound **8**:

$$H_2N-CH_2-\overset{\displaystyle |}{\underset{\displaystyle H}{C}}-C\overset{\displaystyle O}{\underset{\displaystyle R}{\diagup}} \quad \xrightarrow[-NH_3]{\Delta} \quad CH_2=\overset{\displaystyle |}{C}-C\overset{\displaystyle O}{\underset{\displaystyle R}{\diagup}}$$

$$\textbf{4} \qquad\qquad\qquad\qquad \textbf{8}$$

Furthermore, a substitution of the amino group is possible. An important application of Mannich bases **4** is their use as alkylating agents:[3]

$$H_2N-CH_2-\overset{\displaystyle |}{\underset{\displaystyle |}{C}}-C\overset{\displaystyle O}{\underset{\displaystyle R}{\diagup}} \quad \xrightarrow{HX} \quad X-CH_2-\overset{\displaystyle |}{\underset{\displaystyle |}{C}}-C\overset{\displaystyle O}{\underset{\displaystyle R}{\diagup}}$$

$$\textbf{4}$$

The reaction with organolithium or organomagnesium reagents **9** leads to formation of β-aminoalcohols **10**:

$$H_2N-CH_2-\overset{\displaystyle |}{\underset{\displaystyle |}{C}}-C\overset{\displaystyle O}{\underset{\displaystyle R}{\diagup}} \quad \xrightarrow[\textbf{9}]{R'M} \quad H_2N-CH_2-\overset{\displaystyle |}{\underset{\displaystyle |}{C}}-\overset{\displaystyle OH}{\underset{\displaystyle R}{C}}-R'$$

$$\textbf{4} \qquad\qquad\qquad\qquad\qquad \textbf{10}$$

The Mannich reactions plays an important role in pharmaceutical chemistry. Many β-aminoalcohols show pharmacological activity. The Mannich reaction can take place under physiological conditions (with respect to pH, temperature, aqueous solution), and therefore can be used in a biomimetic synthesis; e.g. in the synthesis of alkaloids.

11	12	13	14

The synthesis of tropinone **14**, a precursor of atropine and related compounds, is a classical example. In 1917 *Robinson*[10] has prepared tropinone **14** by a Mannich reaction of succindialdehyde **11** and methylamine **12** with acetone **13**; better yields of tropinone were obtained when he used the calcium salt of acetonedicarboxylic acid instead of acetone.

1. C. Mannich, *Arch. Pharm.* **1917**, *255*, 261–276.
2. F. F. Blicke, *Org. React.* **1942**, *1*, 303–341.
3. M. Tramontini, *Synthesis* **1973**, 703–775.
4. G. A. Gevorgyan, A. G. Agababyan, O. L. Mndzhoyan, *Russ. Chem. Rev.* **1984**, *53*, 561–581.
5. M. Tramontini, L. Angiolini, *Tetrahedron* **1990**, *46*, 1791–1837.
    L. Overman, *Acc. Chem Res.* **1992**, *25*, 352–359;
    L. Overman, *Aldrichim. Acta* **1995**, *28*, 107–119.
6. B. Thompson, *J. Pharm. Sci.* **1968**, *57*, 715–733.
7. T. F. Cummings, J. R. Shelton, *J. Org. Chem.* **1960**, *25*, 419–423.
8. J. Schreiber, H. Maag, N. Hashimoto, A. Eschenmoser, *Angew.Chem.* **1971**, *83*, 355–357; *Angew. Chem. Int. Ed. Engl.* **1971**, *10*, 330.
9. N. Holy, R. Fowler, E. Burnett, R. Lorenz *Tetrahedron* **1979**, *35*, 613–618.
10. R. Robinson, *J. Chem. Soc.* **1917**, *111*, 762–768.

McMurry Reaction

Reductive coupling of aldehydes or ketones

1	2

Among the more recent name reactions in organic chemistry, the *McMurry reaction*[1-3] is of particular importance as a synthetic method. It permits the reductive dimerization of aldehydes or ketones **1** to yield alkenes **2** by reaction with low-valent titanium as reducing agent.

The initial step of the coupling reaction is the binding of the carbonyl substrate to the titanium surface, and the transfer of an electron to the carbonyl group.[3] The carbonyl group is reduced to a radical species **3**, and the titanium is oxidized. Two such ketyl radicals can dimerize to form a pinacolate-like intermediate **4**, that is coordinated to titanium. Cleavage of the C—O bonds leads to formation of an alkene **2** and a titanium oxide **5**:

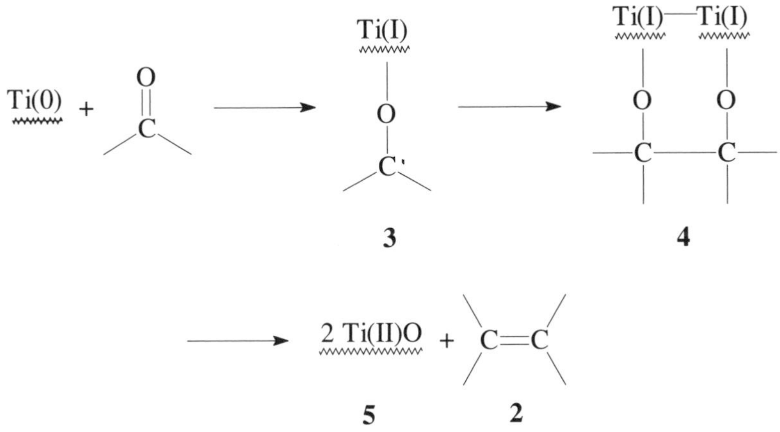

Under appropriate reaction conditions—e.g. at low temperatures—the cleavage of the C—O bonds does not take place, and a vicinal diol can be isolated as product.[4]

The low-valent titanium reagents used are not soluble under the reaction conditions; the McMurry coupling is thus a heterogenous reaction. Low-valent titanium can be prepared from $TiCl_4$ or $TiCl_3$ by reaction with, e.g., lithium aluminum hydride, an alkali metal (Li, Na or K), magnesium or zinc-copper couple. The low-valent titanium reagent has to be prepared freshly prior to reaction. Depending on molar ratio and the reductive agent used, a reagent is obtained that contains titanium of different oxidation states; however Ti(0) appears to be the active species.

The coupling of unsymmetrical ketones leads to formation of stereoisomeric alkenes; the ratio depending on steric demand of substituents R:[2]

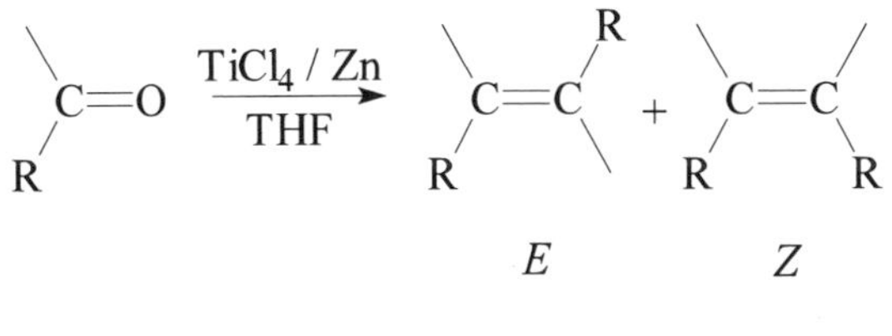

R = *n*-Propyl $E/Z = 3 : 1$
R = *tert*-Butyl $E/Z = 200 : 1$

The intermolecular McMurry reaction is first of all a suitable method for the synthesis of symmetrical alkenes. With a mixture of carbonyl compounds as starting material, the yield is often poor. An exception to this being the coupling of diaryl ketones with other carbonyl compounds, where the mixed coupling product can be obtained in good yield. For example benzophenone and acetone (stoichiometric ratio 1 : 4) are coupled in 94% yield.[5]

The McMurry procedure is a valuable method for the synthesis of highly substituted, strained alkenes; such compounds are difficult to prepare by other methods. Diisopropyl ketone 6 can be coupled to give tetraisopropylethene 7 in 87% yield; attempts to prepare tetra-*t*-butylethene however were not successful.[3,6]

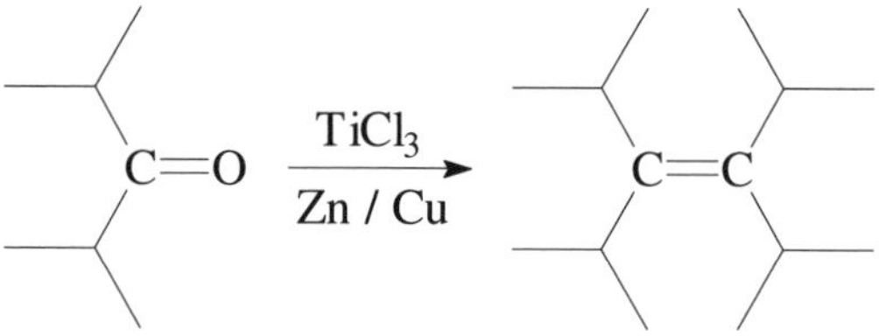

6   **7**

Highly strained cyclic compounds are accessible by an intramolecular variant. An instructive example for the synthetic potential of the McMurry coupling reaction is the synthesis of 3,3-dimethyl-1,2-diphenylcyclopropene 8:[7]

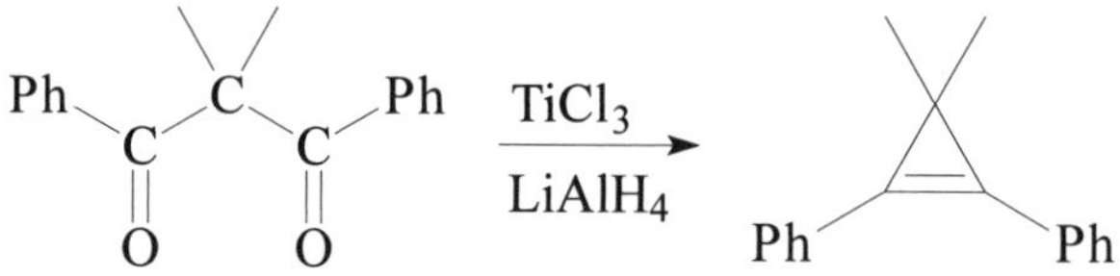

8

Ketoesters 9 can be coupled to give enol ethers 10, which may for example be converted to cycloalkanones by hydrolysis.

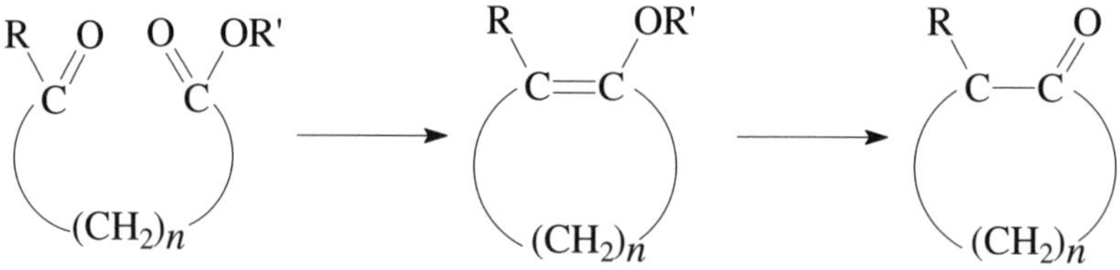

9   **10**   **11**

The McMurry reaction is a valuable tool in organic synthesis. Yields are generally good, even for sterically demanding targets. The optimal ratio of titanium precursor and reducing agent has to be adjusted for a particular reaction. Functional groups that can be reduced by low-valent titanium usually will interfere with the attempted coupling reaction.

1. J. E. McMurry, M. P. Fleming, *J. Am. Chem. Soc.* **1974**, *96*, 4708–4709.
2. D. Lenoir, *Synthesis* **1989**, 883–897.
3. J. E. McMurry, *Chem. Rev.* **1989**, *89*, 1513–1524;
    A. Fürstner, *Angew. Chem.* **1993**, *105*, 171; *Angew. Chem. Int. Ed. Engl.* **1993**, *32*, 164.
    A. Fürstner, B. Bogdanovic, *Angew. Chem.* **1996**, *108*, 2582–2609; *Angew. Chem. Int. Ed. Engl.* **1996**, *35*, 2442–2469.
4. E. J. Corey, R. L. Danheiser, S. Chandrasekaran, *J. Org. Chem.* **1976**, *41*, 260–265.
5. J. E. McMurry, L. R. Krepski, *J. Org. Chem.* **1976**, *41*, 3929–3930.
6. J. E. McMurry, T. Lectka, J. G. Rico, *J. Org. Chem.* **1989**, *54*, 3748–3749.
7. A. L. Baumstark, C. J. McCloskey, K. E. Witt, *J. Org. Chem.* **1978**, *43*, 3609–3611.

Meerwein–Ponndorf–Verley Reduction

Reduction of aldehydes and ketones with aluminum isopropoxide

The reduction of ketones to secondary alcohols and of aldehydes to primary alcohols using aluminum alkoxides is called the *Meerwein–Ponndorf–Verley reduction.*[1-4] The reverse reaction also is of synthetic value, and is called the *Oppenauer oxidation.*[5,6]

The aldehyde or ketone, when treated with aluminum triisopropoxide in isopropanol as solvent, reacts *via* a six-membered cyclic transition state **4**. The aluminum center of the Lewis-acidic reagent coordinates to the carbonyl oxygen, enhancing the polar character of the carbonyl group, and thus facilitating the hydride transfer from the isopropyl group to the carbonyl carbon center. The intermediate mixed aluminum alkoxide **5** presumably reacts with the solvent isopropanol to yield the product alcohol **3** and regenerated aluminum triisopropoxide **2**; the latter thus acts as a catalyst in the overall process:

$$\text{(scheme)} \quad 3 \qquad 2$$

Thus one of the transferred hydrogens comes from the aluminum reagent, and the other one from the solvent. In addition to the mechanism *via* a six-membered cyclic transition state, a radical mechanism is discussed for certain substrates.[7]

In order to shift the equilibrium of the reaction, the low boiling reaction product acetone is continuously removed from the reaction mixture by distillation. By keeping the reaction mixture at a temperature slightly above the boiling point of acetone, the reaction can then be driven to completion.

Other Lewis-acidic alkoxides might also be employed; however aluminum isopropoxide has the advantage to be sufficiently soluble in organic solvents, and acetone as oxidation product can be easily removed for its low boiling point. Recently lanthan isopropoxide[8] has been used with success, and showed good catalytic activity.

The Meerwein–Ponndorf–Verley procedure has largely been replaced by reduction procedures that use lithium aluminum hydride, sodium borohydride or derivatives thereof. The Meerwein–Ponndorf–Verley reduction however has the advantage to be a mild and selective method, that does not affect carbon–carbon double or triple bonds present in the substrate molecule.

The reverse reaction, the so-called *Oppenauer oxidation*, is carried out by treating a substrate alcohol with aluminum tri-*t*-butoxide in the presence of acetone. By using an excess of acetone, the equilibrium can be shifted to the right, yielding the ketone 1 and isopropanol:

$$\text{(scheme)} \quad 1$$

As a synthetic method however the Oppenauer oxidation is of limited importance.

1. H. Meerwein, R. Schmidt, *Justus Liebigs Ann. Chem.* **1925**, *444*, 221–238.
2. W. Ponndorf, *Angew. Chem.* **1926**, *39*, 138–143.
3. A. Verley, *Bull. Soc. Chim,* **1925**, *37*, 537–542.
4. A. L. Wilds, *Org. React,* **1944**, *2*, 178–223.
5. R. V. Oppenauer, *Recl. Trav. Chim. Pays-Bas* **1937**, *56*, 137–144.
6. C. Djerassi, *Org. React.* **1951**, *6*, 207–272.
7. C. G. Screttas, C. T. Cazianis, *Tetrahedron* **1978**, *34*, 933–940.
8. T. Okano, M. Matsuoka, H. Konishi, J. Kiji, *Chem. Lett.* **1987**, 181–184.

Michael Reaction

1,4-Addition to α, β-unsaturated carbonyl compounds

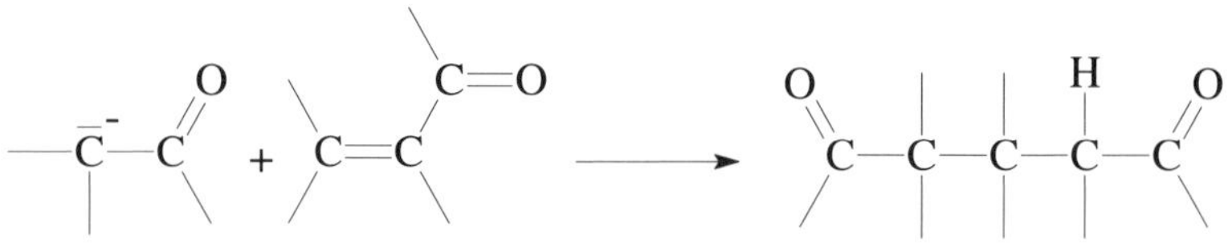

The 1,4-addition of an enolate anion **1** to an α, β-unsaturated carbonyl compound **2**, to yield a 1,5-dicarbonyl compound **3**, is a powerful method for the formation of carbon–carbon bonds, and is called the *Michael reaction* or *Michael addition*.[1,2] The 1,4-addition to an α, β-unsaturated carbonyl substrate is also called a *conjugate addition*. Various other 1,4-additions are known, and sometimes referred to as *Michael-like additions*.

The enolate anion **1** may in principle be generated from any enolizable carbonyl compound **4** by treatment with base; the reaction works especially well with β-dicarbonyl compounds. The enolate **1** adds to the α, β-unsaturated compound **2** to give an intermediate new enolate **5**, which yields the 1,5-dicarbonyl compound **3** upon hydrolytic workup:

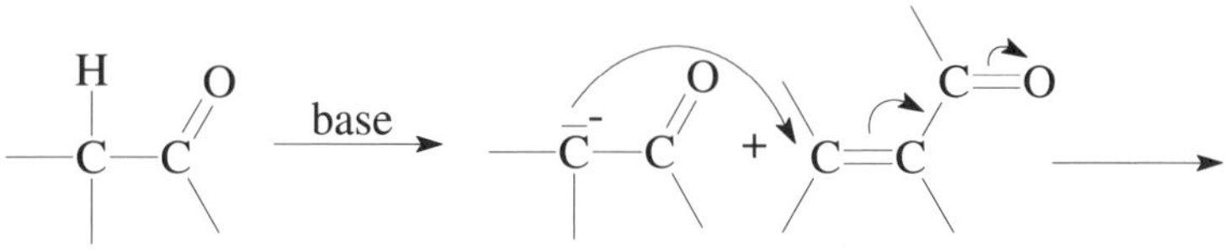

As enolate precursors can be used CH-acidic carbonyl compounds such as malonic esters, cyanoacetic esters, acetoacetic esters and other β-ketoesters, as well as aldehydes and ketones. Even CH-acidic hydrocarbons such as indene and fluorene can be converted into suitable carbon nucleophiles.

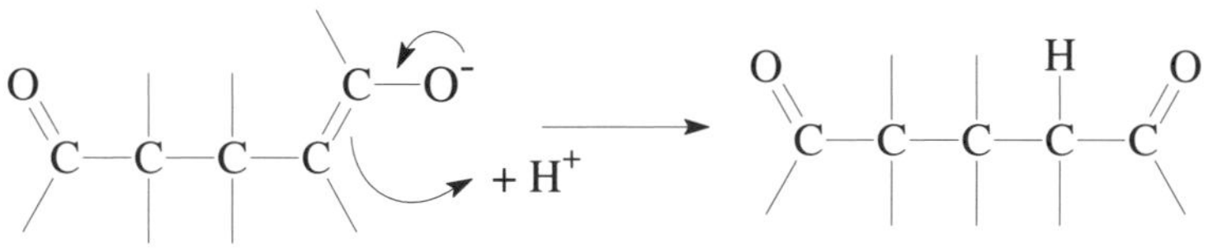

The classical Michael reaction is carried out in a protic organic solvent—e.g. an alcohol—by use of an alkoxide as base—e.g. potassium *t*-butoxide or sodium ethoxide.

The overall process is the addition of a CH-acidic compound to the carbon–carbon double bond of an α, β-unsaturated carbonyl compound. The Michael reaction is of particular importance in organic synthesis for the construction of the carbon skeleton. The above CH-acidic compounds usually do not add to ordinary carbon-carbon double bonds. Another and even more versatile method for carbon–carbon bond formation that employs enolates as reactive species is the *aldol reaction*.

Various competitive reactions can reduce the yield of the desired Michael-addition product. An important side-reaction is the 1,2-addition of the enolate to the C=O double bond (see *aldol reaction, Knoevenagel reaction*); especially with α, β-unsaturated aldehydes, the 1,2-addition product may be formed preferentially, rather than the 1,4-addition product. Generally the 1,2-addition is a kinetically favored and reversible process. At higher temperatures, the thermodynamically favored 1,4-addition products are obtained.

Certain starting materials may give rise to the non-selective formation of regioisomeric enolates, leading to a mixture of isomeric products. Furthermore α, β-unsaturated carbonyl compounds tend to polymerize. The classical Michael procedure (i.e. polar solvent, catalytic amount of base) thus has some disadvantages, some of which can be avoided by use of preformed enolates. The CH-acidic carbonyl compound is converted to the corresponding enolate by treatment with an equimolar amount of a strong base, and in a second step the α, β-unsaturated carbonyl compound is added—often at low temperature. A similar procedure is applied for variants of the *aldol reaction*.

Substituted enolates are usually obtained as a mixture of *E*- and *Z*-isomers; under suitable reaction conditions, one particular isomer may be obtained preferentially:[4]

The stereochemical outcome of the Michael addition reaction with substituted starting materials depends on the geometry of the α, β-unsaturated carbonyl compound as well as the enolate geometry; a stereoselective synthesis is possible.[3,4] Diastereoselectivity can be achieved if both reactants contain a stereogenic center. The relations are similar to the aldol reaction, and for kinetically controlled Michael reactions, the observed selectivities can be rationalized by taking into account the diastereomeric transition states. For diastereoselective reactions, four cases have to be considered, since enolate **1** and acceptor **2** can each exist as E- as well as Z-isomer. If for example a Z-enolate **6** reacts with an E-configurated acceptor **7**, two staggered transition states **8** and **9** can be written, where the reactants are brought together through coordination to the metal center (chelation control):

Transition state **8**, which would lead to the *syn*-addition product **10**, is energetically disfavored because of mutual steric hindrance of groups R^2 and R^4. The predominant formation of *anti*-products **11** is generally observed, and is believed to proceed via the favored transition state **9**. The actual diastereoselectivity strongly depends on the steric demand of R^2 and R^4. Analogous considerations apply to the other combinations, and can be summarized as follows. An E-configured enolate reacts with an E-configured acceptor to give the *syn*-product; Z-enolate and E-acceptor react to yield preferentially the *anti*-product. For reactions with a Z-configured acceptor the opposite diastereoselectivities are observed.

With the use of chiral reagents a differentiation of enantiotopic faces is possible, leading to an enantioselective reaction. The stereoselective version of the Michael addition reaction can be a useful tool in organic synthesis, for instance in the synthesis of natural products.

An interesting feature is the sometimes observed pressure dependence of the reaction.[5] The Michael addition of dimethyl methylmalonate **12** to the bicyclic ketone **13** does not occur under atmospheric pressure, but can be achieved at 15 Kbar in 77% yield:

The Michael reaction is of great importance in organic synthesis.

1. A. Michael, *J. Prakt. Chem.* **1887**, *36*, 113–114.
2. E. D. Bergman, D. Gunsburg, R. Rappo, *Org. React.* **1959**, *10*, 179–560.
3. D. A. Oare, C. H. Heathcock, *Topics Stereochem.* **1989**, *19*, 227–407.
4. C. H. Heathcock in *Modern Synthetic Methods 1992* (Ed.: R. Scheffold), VHCA, Basel **1992**, p. 1–103.
5. W. G. Dauben, J. M. Gerdes, G. C. Look, *Synthesis* **1986**, 532–535.

Mitsunobu Reaction

Esterification of an alcohol with carboxylic acid in the presence of dialkyl azodicarboxylate and triphenylphosphine

The major application of the *Mitsunobu reaction*[1–3] is the conversion of a chiral secondary alcohol **1** into an ester **3** with concomitant inversion of configuration at the secondary carbon center. In a second step the ester can be hydrolyzed to yield the inverted alcohol **4**, which is enantiomeric to **1**. By using appropriate nucleophiles, alcohols can be converted to other classes of compounds—e.g. azides, amines or ethers.

The mechanistic pathway[4–6] can be divided into three steps: 1. formation of the activating agent from triphenylphosphine and diethyl azodicarboxylate (DEAD) or diisopropyl azodicarboxylate (DIAD); 2. activation of the substrate alcohol **1**; 3. a bimolecular nucleophilic substitution (S_N2) at the activated carbon center.

In an initial step triphenylphosphine adds to diethyl azodicarboxylate **5** to give the zwitterionic adduct **6**, which is protonated by the carboxylic acid **2** to give intermediate salt **7**. The alcohol reacts with **7** to the alkoxyphosphonium salt **8** and the hydrazine derivative **9**, and is thus activated for a S_N2-reaction:

$$Ph_3P + EtO_2CN=NCO_2Et \longrightarrow \underset{\underset{Ph_3P^+}{|}}{EtO_2CN-N^-CO_2Et} \overset{R^1COOH}{\longrightarrow}$$

$$\textbf{5} \qquad\qquad\qquad \textbf{6}$$

$$\underset{\underset{R^1COO^-}{\underset{Ph_3P^+\ \ H}{|\quad\ |}}}{EtO_2CN-N-CO_2Et} \quad \overset{ROH}{\longrightarrow} \quad \begin{array}{c} Ph_3P^+ -OR \quad R^1COO^- \\ \textbf{8} \\ \underset{H\ \ H}{\underset{|\ \ \ |}{EtO_2CN-NCO_2Et}} \end{array}$$

$$\textbf{7} \qquad\qquad\qquad\qquad \textbf{9}$$

The final step is the nucleophilic displacement of the oxyphosphonium group by the carboxylate anion *via* a S_N2-mechanism, yielding ester **3** with inverted configuration at the stereogenic center, and triphenylphosphine oxide. A hydrolysis of the ester **3** will leave the new configuration unchanged, and yield the inverted alcohol **4**:

$$\underset{\textbf{8}}{\overset{Ph_3P^+O \quad H \quad R^1COO^-}{\underset{R^2 \qquad R^3}{\diagdown\diagup}}} \longrightarrow \underset{\textbf{3}}{\overset{\overset{O}{\overset{||}{}}H \quad OCR^1}{\underset{R^2 \qquad R^3}{\diagdown\diagup}}} + O=PPh_3 \longrightarrow \underset{\textbf{4}}{\overset{H \quad OH}{\underset{R^2 \qquad R^3}{\diagdown\diagup}}}$$

Recent mechanistic studies have shown that the many combinations of alcohols, carboxylic acids and solvents cannot be correctly described by a uniform mechanism. In certain cases the reaction appears to involve a pentavalent dialkoxyphosphorane **10** as an intermediate, which is in equilibrium with oxyphosphonium salt **8**:[4,6]

$$\underset{\textbf{10}}{(RO)_2PPh_3} \overset{H^+}{\rightleftharpoons} \underset{\textbf{8}}{ROP^+Ph_3} + ROH$$

In summary the Mitsunobu reaction can be described as a condensation of an alcohol **1** and a nucleophile—NuH—**11**, where the reagent triphenylphosphine is oxidized to triphenylphosphine oxide and the azodicarboxylate reagent **12** is reduced to a hydrazine derivative **13**:

$$PPh_3 + \underset{\textbf{12}}{RO_2CN=NCO_2R} + \underset{\textbf{1}}{R'OH} + \underset{\textbf{11}}{HNu} \longrightarrow$$

$$O=PPh_3 + RO_2CN-N-CO_2R + R'Nu$$
$$\quad\quad\quad\quad | \quad\ |$$
$$\quad\quad\quad\quad H \quad H$$

13

Alkyl aryl ethers and enol ethers are also accessible by the Mitsunobu method.[2] Cyclic ethers can be obtained by an intramolecular variant, which is especially suitable for the synthesis of three- to seven-membered rings:

$$HO(CH_2)_nOH \xrightarrow[PPh_3]{DEAD} (CH_2)_n \quad O$$
$$n = 2-6$$

The conversion of an alcohol to an amine can be achieved in a one-pot reaction;[2] the alcohol **1** is treated with hydrazoic azid (HN_3), excess triphenylphosphine and diethyl azodicarboxylate (DEAD). The initial Mitsunobu product, the azide **14**, further reacts with excess triphenylphosphine to give an iminophosphorane **15**. Subsequent hydrolytic cleavage of **15** yields the amine—e.g. as hydrochloride **16**:

$$ROH \xrightarrow[HN_3]{DEAD\,/\,PPh_3} R-N_3 \xrightarrow{PPh_3} R-N=PPh_3 \xrightarrow[H_2O]{HCl} R-N^+H_3Cl^-$$

1   **14**   **15**   **16**

Suitable starting materials for the Mitsunobu reaction are primary and secondary alcohols. Tertiary alcohols are less suitable since these are bad substrates for a S_N2-mechanism.

A variety of nucleophiles can be employed—e.g. carboxylic acids, phenols, imides, thiols, thioamides, and even β-ketoesters as carbon nucleophiles. Of major importance however is the esterification as outlined above, and its use for the clean inversion of configuration of a chiral alcohol.

1. O. Mitsunobu, *Bull. Chem. Soc. Jpn.* **1967**, *40*, 4235–4238.
2. D. L. Hughes, *Org. React.* **1992**, *42*, 335–656.
3. O. Mitsunobu, *Synthesis* **1981**, 1–28.
4. D. L. Hughes, R. A. Reamer, J. J. Bergan, E. J. J. Grabowski, *J. Am. Chem. Soc.* **1988**, *110*, 6487–6491.
5. D. Crich, H. Dyker, R. J. Harris, *J. Org. Chem.* **1989**, *54*, 257–259.
6. D. J. Camp, I. D. Jenkins, *J. Org. Chem.* **1989**, *54*, 3045–3049 and 3049–3054.

Nazarov Cyclization

Cyclization of divinyl ketones to yield cyclopentenones

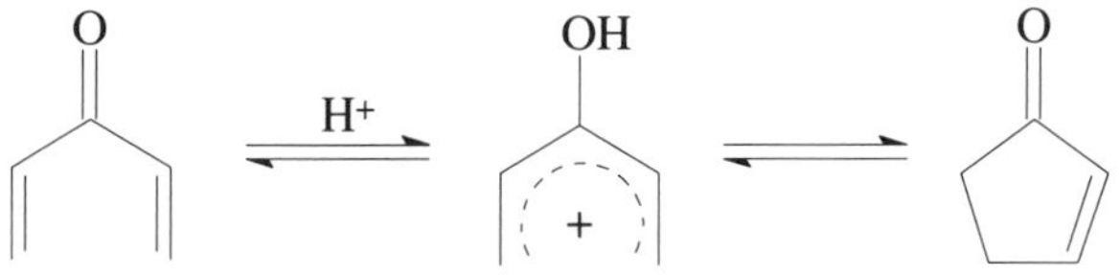

Upon treatment of a divinyl ketone **1** with a protic acid or a Lewis acid, an electrocyclic ring closure can take place to yield a cyclopentenone **3**. This reaction is called the *Nazarov cyclization*.[1,2] Protonation at the carbonyl oxygen of the divinyl ketone **1** leads to formation of a hydroxypentadienyl cation **2**, which can undergo a thermally allowed, conrotatory electrocyclic ring closure reaction to give a cyclopentenyl cation **4**. Through subsequent loss of a proton a mixture of isomeric cyclopentenones **5** and **6** is obtained:

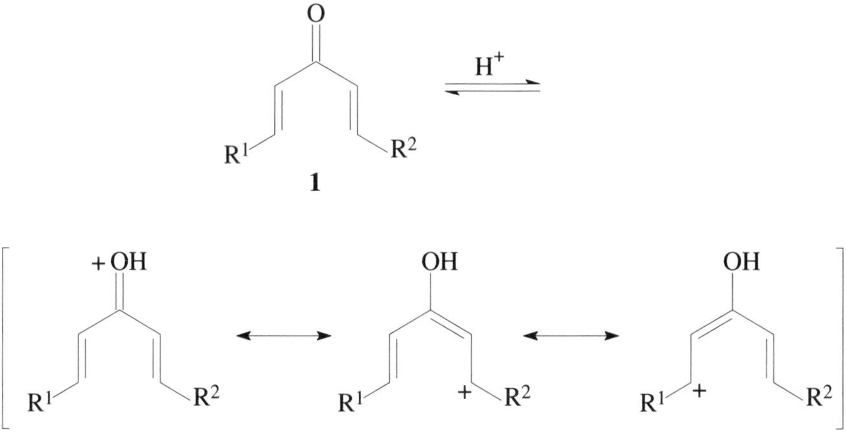

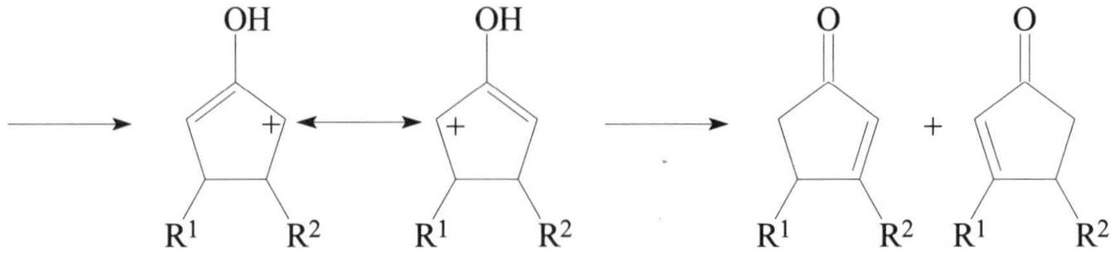

With the use of trimethylsilyl-substituted starting materials after *Denmark et al.*,[4] the disadvantageous formation of a mixture of isomers can be avoided. The vinyl silane derivatives react by loss of the TMS group in the last step:

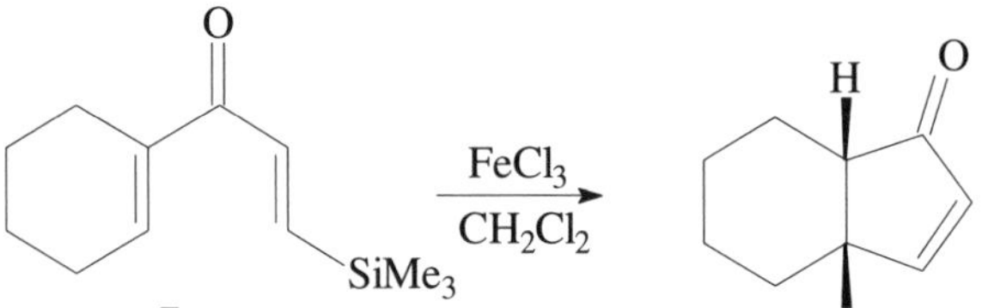

A variant of the Nazarov reaction is the cyclization of allyl vinyl ketones **8**. These will first react by double bond isomerization to give divinyl ketones, and then cyclize to yield a cyclopentenone **9** bearing an additional methyl substituent:[2]

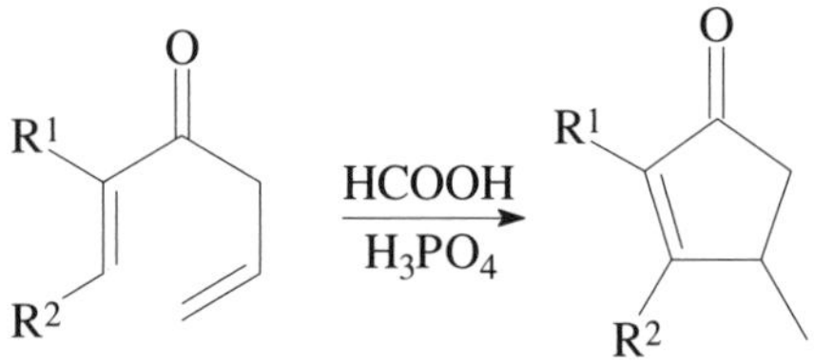

For the preparation of divinyl ketones, as required for the Nazarov reaction, various synthetic routes have been developed.[3,5] A large variety of substituted divinyl ketones, including vinylsilane derivatives, can thus be prepared. The Nazarov cyclization, and especially the vinylsilane variant, has found application for the synthesis of complex cyclopentanoids.

1.   I. N. Nazarov, *Usp. Khim.* **1949**, *18*, 377–401.
2.   C. Santelli-Rouvier, M. Santelli, *Synthesis* **1983**, 429–442.
3.   J. Mulzer, H.-J. Altenbach, M. Braun, K. Krohn, H.-U. Reissig, *Organic Synthesis Highlights*, VCH, Weinheim, **1991**, p. 137–140.
4.   S. E. Denmark, T. K. Jones, *J. Am. Chem. Soc.* **1982**, *104*, 2642–2645.
5.   R. M. Jacobson, G. P. Lahm, J. W. Clader, *J. Org. Chem.* **1980**, *45*, 395–405.

Neber Rearrangement

α-Amino ketones from ketoxime tosylates

$$R-CH_2-C \overset{NOSO_2Ar}{\diagdown} \quad \longrightarrow \quad R-CH-C \overset{NH_2 \quad O}{\underset{R'}{}}$$

$$\underset{R'}{} $$

1   **2**

A ketoxime tosylate **1** can be converted into an α-amino ketone **2** *via* the *Neber rearrangement*[1,2] by treatment with a base—e.g. using an ethoxide or pyridine. Substituent R is usually aryl, but may as well be alkyl or H; substituent R' can be alkyl or aryl, but not H.

The following mechanism is generally accepted, since azirine **3**, that has been identified as intermediate, can be isolated:[3,4]

$$R-CH_2-C \overset{NOSO_2Ar}{\diagdown} \quad \overset{base}{\longrightarrow} \quad R-\bar{C}H-C \overset{N-OSO_2Ar}{\underset{R'}{}}$$

1

$$\longrightarrow \quad R \overset{N}{\underset{H}{\diagup}} \overset{}{\diagdown} R' \quad \overset{H_2O}{\longrightarrow} \quad R-CH-C \overset{NH_2 \quad O}{\underset{R'}{}}$$

3   **2**

The ketoxime derivatives, required as starting materials, can be prepared from the appropriate aromatic, aliphatic or heterocyclic ketone. Aldoximes (where R' is H) do not undergo the rearrangement reaction, but rather an elimination of toluenesulfonic acid to yield a nitrile. With ketoxime tosylates a *Beckmann rearrangement* may be observed as a side-reaction.

Unlike the Beckmann rearrangement, the outcome of the Neber rearrangement does not depend on the configuration of the starting oxime derivative: *E*- as well as *Z*-oxime yield the same product. If the starting oxime derivative contains two different α-methylene groups, the reaction pathway is not determined by the configuration of the oxime, but rather by the relative acidity of the α-methylene protons; the more acidic proton is abstracted preferentially.[2]

An α-amino ketone, obtained by the Neber rearrangement, can be further converted into an oxime tosylate, and then subjected to the Neber conditions; α,α'-diamino ketones can be prepared by this route.

The Neber rearrangement has for example found application in natural product synthesis.

1. P. W. Neber, A. Burgard, *Justus Liebigs Ann. Chem.* **1932**, *493*, 281–285.
2. C. O'Brien, *Chem. Rev.* **1964**, *64*, 81–89.
3. D. J. Cram, M. J. Hatch, *J. Am. Chem. Soc.* **1953**, *75*, 33–38.
4. M. J. Hatch, D. J. Cram, *J. Am. Chem. Soc.* **1953**, *75*, 38–44.

Nef Reaction

Carbonyl compounds from nitro alkanes

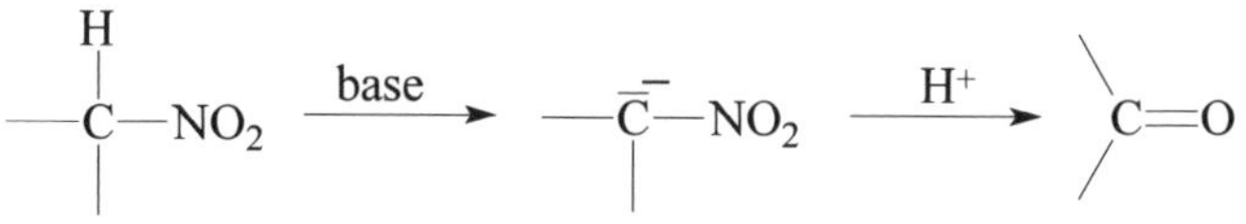

The conversion of a primary or secondary nitro alkane **1** to a carbonyl compound **3** *via* an intermediate nitronate **2** is called the *Nef reaction*.[1,2] Since carbonyl compounds are of great importance in organic synthesis, and nitro alkanes can on the other hand be easily prepared, the Nef reaction is an important tool in organic chemistry.

The mechanism of the Nef reaction has been thoroughly investigated.[2] Initial step is the abstraction of a proton from the α-carbon of the nitro alkane **1**, leading to nitronate anion **2**, which is then stepwise protonated at both negatively charged oxygens. Subsequent hydrolytic cleavage yields the carbonyl compound **3** together with dinitrogen oxide and water:

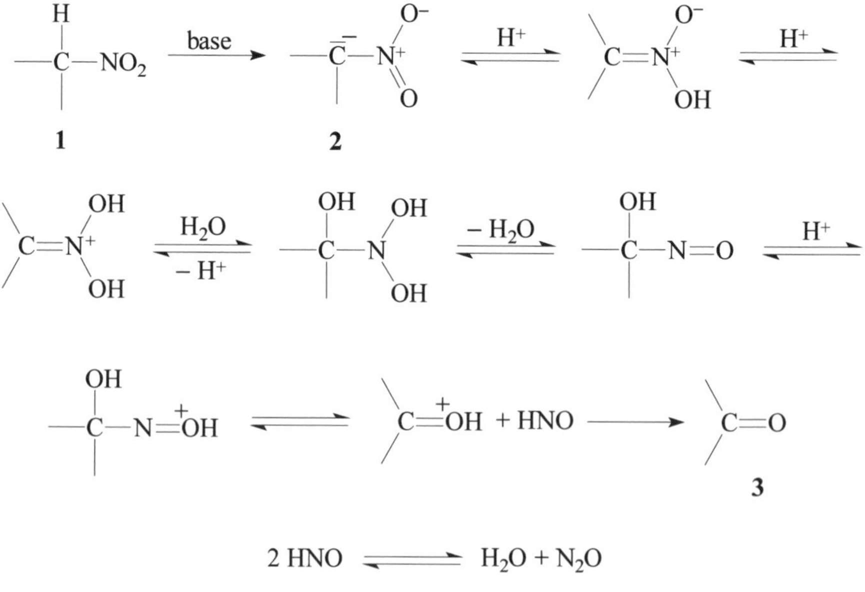

$$2\ HNO \rightleftharpoons H_2O + N_2O$$

Various side-reactions may complicate the course of the Nef reaction. Because of the delocalized negative charge, the nitronate anion **2** can react at various positions with an electrophile; addition of a proton at the α-carbon reconstitutes the starting nitro alkane. **1**. The nitrite anion can act as leaving group, thus leading to elimination products.

The required nitro compounds are easy to prepare, and are useful building blocks for synthesis. Treatment with an appropriate base—e.g. aqueous alkali—leads to formation of nitronates **2**. Various substituted nitro compounds, such as nitro-ketones, -alcohols, -esters and -nitriles are suitable starting materials.

The Nef reaction has for example been applied for the 1,2-transposition of carbonyl groups:[3]

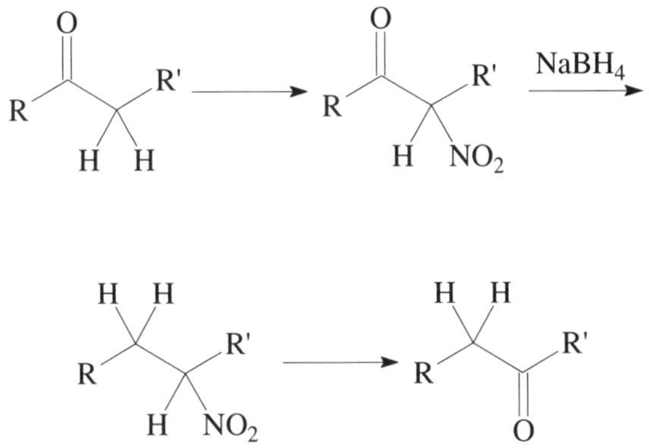

Another important feature of the Nef reaction is the possible use of a $CH-NO_2$ function as an *umpoled* carbonyl function. A proton at a carbon α to a nitro group is acidic, and can be abstracted by base. The resulting anionic species has a nucleophilic carbon, and can react at that position with electrophiles. In contrast the carbon center of a carbonyl group is electrophilic, and thus reactive towards nucleophiles. 1,4-Diketones **4** can for example be prepared from α-acidic nitro compounds by a *Michael addition*/Nef reaction sequence:[4]

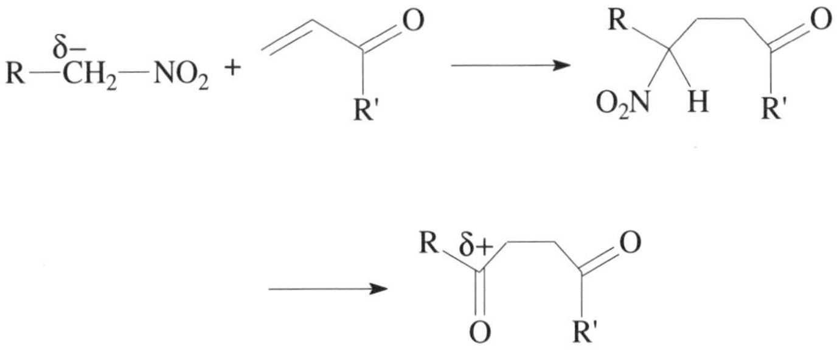

For starting materials containing base- and/or acid-sensitive functional groups, modified procedures have been developed—e.g. using oxidizing agents.[2]

1. J. U. Nef, *Justus Liebigs Ann. Chem.* **1894**, *280*, 263–291.
2. H. W. Pinnick, *Org. React.* **1990**, *38*, 655–792.
3. A. Hassner, J. M. Larkin, J. E. Dowd, *J. Org. Chem.* **1968**, *33*, 1733–1739.
4. O. W. Lever Jr., *Tetrahedron.* **1976**, *32*, 1943–1971.

Norrish Type I Reaction

Photochemical cleavage of aldehydes and ketones

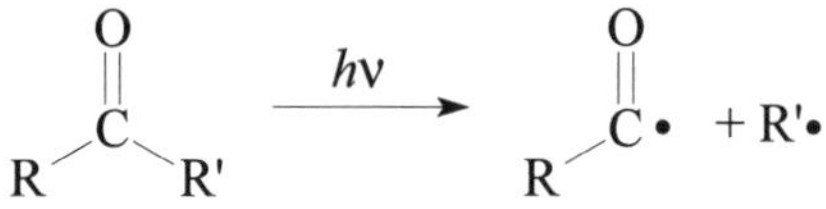

Carbonyl compounds can undergo various photochemical reactions; among the most important are two types of reactions that are named after *Norrish*.[1] The term *Norrish type I fragmentation*[1-4] refers to a photochemical reaction of a carbonyl compound **1** where a bond between carbonyl group and an α-carbon is cleaved homolytically. The resulting radical species **2** and **3** can further react by decarbonylation, disproportionation or recombination, to yield a variety of products.

By absorption of a photon of light, a ketone or aldehyde molecule **1** can be converted into a photoactivated species; it is promoted to the singlet excited (S_1)-state **4**, from which it can reach the triplet excited (T_1)-state **5** by *intersystem crossing*. The homolytic Norrish type I cleavage may occur from either or both states, and leads to formation of an acyl radical **2** and an allyl radical **3**. Aromatic ketones generally undergo the photolytic cleavage from the triplet excited state, since the intersystem crossing is usually fast in those cases.

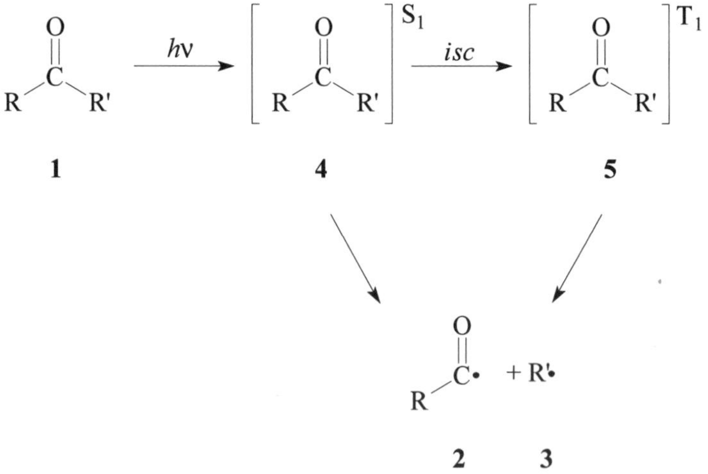

With unsymmetrical ketones two different bonds are available for photolytic cleavage; the actual cleavage pathway depends on the relative stability of the possible radical species R• and R'•.

The radical pair **2/3** can undergo various subsequent reactions: the most obvious is the recombination to the starting carbonyl compound **1**. The acyl radical **2** can undergo a fragmentation by loss of CO to the radical **6**, which can further react with radical **3** to yield the hydrocarbon **7** (i.e. R-R'). Cleavage of CO from **2** and subsequent combination of **6** and **3** usually is a fast process taking place in a solvent cage, which largely prevents formation of symmetrical hydrocarbons (R-R or R'-R'). If the acyl radical **2** bears an α-hydrogen, this hydrogen can be abstracted by radical **3**, resulting in formation of a ketene **8** and hydrocarbon **9**:

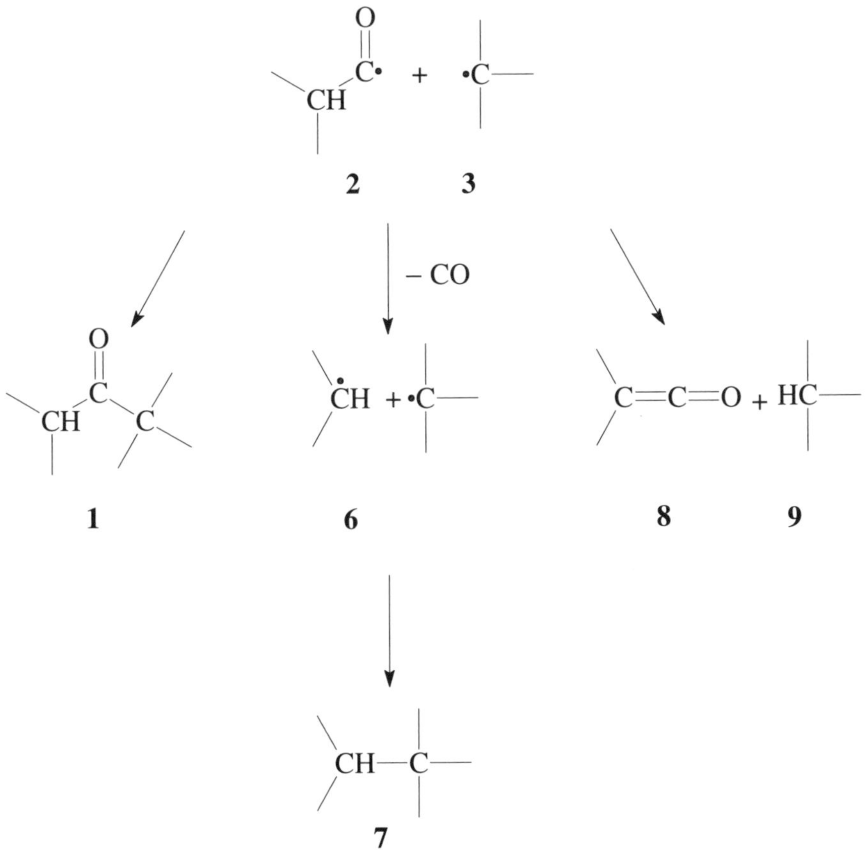

The acyl radical **2** can abstract a β-hydrogen from the radical **3**, to give an aldehyde **10** and an alkene **11**:

3   **10**   **11**

Since the quantum yield of the Norrish type I reaction is generally low, it has been assumed that the initial homolytic cleavage is a reversible process. Evidence came from an investigation by *Barltrop et al.*[5] which has shown that *erythro*-2,3-dimethylcyclohexanone **12** isomerizes to *threo*-2,3-dimethylcyclohexanone **13** upon irradiation:

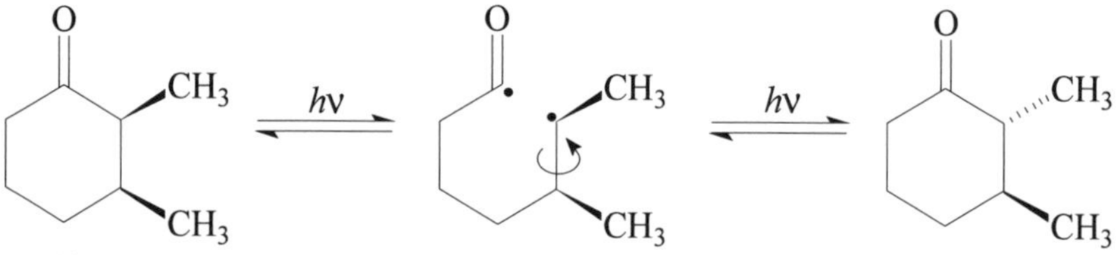

12   **13**

The photolytic cleavage of cyclic ketones **14** leads to formation of a diradical species, that can undergo analogously the various reactions outlined above. The decarbonylation followed by intramolecular recombination yields a ring-contracted cycloalkane **15**:

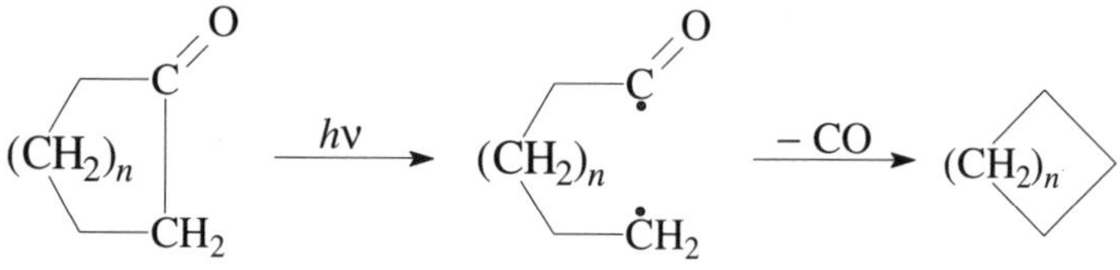

14   **15**

With strained cycloketones the type I-cleavage gives better yields, and can be used as a preparative method. For example photolysis of the bicyclic ketone **16** gives diene **17** in good yield:[6]

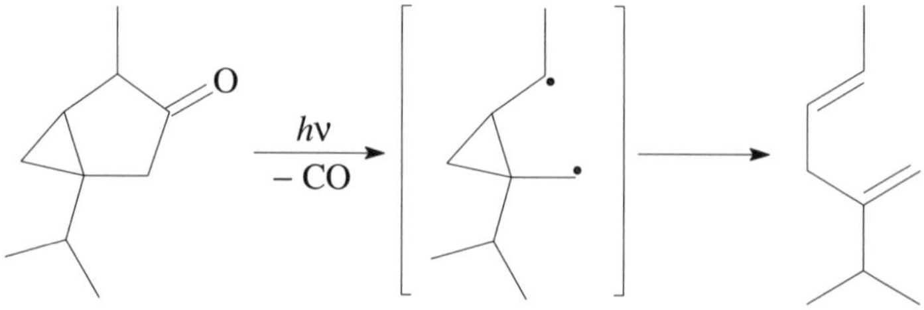

16   **17**

In general however the various possible reaction pathways give rise to formation of a mixture of products. The type I-cleavage reaction is only of limited synthetic importance, but rather an interfering side-reaction—e.g. with an attempted *Paterno–Büchi reaction*, or when an aldehyde or ketone is used as sensitizer in a [2 + 2]-*cycloaddition reaction*.

1. R. G. W. Norrish, *Trans. Faraday Soc.* **1937**, *33*, 1521–1528.
2. J. N. Pitts, Jr., J. K. S. Wan in *The Chemistry of the Carbonyl Group* (Ed.: S. Patai), Wiley, New York, **1966**, p. 823–916.
3. J. S. Swenton, *J. Chem. Educ.* **1969**, *46*, 217–226.
4. J. M. Coxon, B. Halton, *Organic Photochemistry*, Cambridge University Press, London, **1974**, p. 58–78.
5. J. A. Barltrop, J. D. Coyle, *Chem. Commun.* **1969**, 1081–1082.
6. J. Kopecky, *Organic Photochemistry*, VCH, Weinheim, **1991**, p. 119–122.

Norrish Type II Reaction

Photochemical reaction of aldehydes or ketones bearing γ-hydrogens

An aldehyde or ketone **1** bearing a γ-hydrogen atom can upon irradiation undergo an intramolecular hydrogen shift by the so-called *Norrish type II reaction*.[1–4] The resulting diradical species **2** can undergo a subsequent ring closure reaction to yield a cyclobutanol **3**, or suffer fragmentation to yield an enol **4** and an alkene **5**.

Photoactivated aldehyde or ketone molecules with γ-hydrogens can undergo the intramolecular hydrogen abstraction from the singlet excited (S_1)-state as well as the triplet excited (T_1)-state. This reaction proceeds via a cyclic six-membered transition state. The resulting 1,4-diradical species **2** can further react either by ring closure to give a cyclobutanol **3** or by carbon–carbon bond cleavage to give an enol **4** and an alkene **5**; enol **4** will subsequently tautomerize to carbonyl compound **6**:

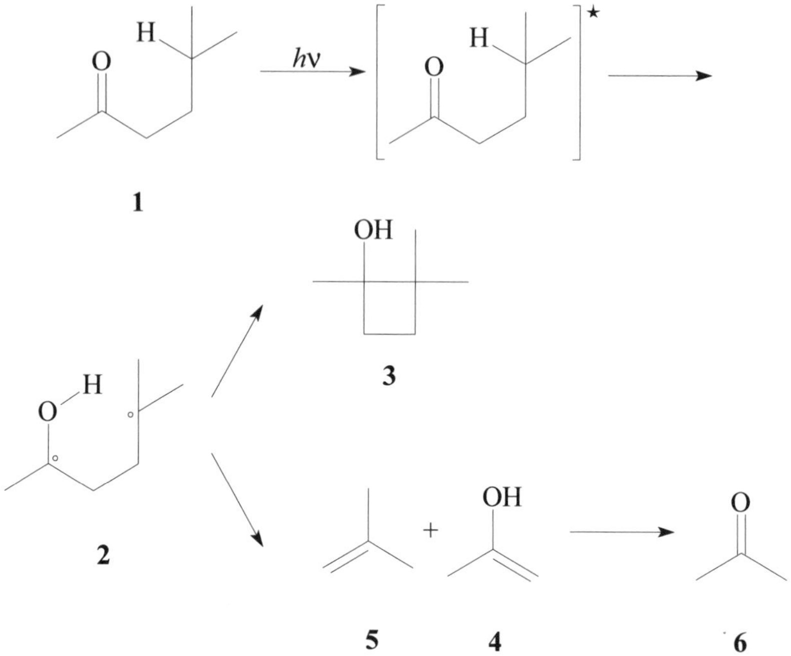

The fragmentation/cyclization ratio is determined by the relative orientation of the respective molecular orbitals, and thus by the conformation of diradical species **2**.[5] The quantum yield with respect to formation of the above products is generally low; the photochemically initiated 1,5-hydrogen shift from the γ-carbon to the carbonyl oxygen is a reversible process, and may as well proceed back to the starting material. This has been shown to be the case with optically active ketones **7**, containing a chiral γ-carbon center; an optically active ketone **7** racemizes upon irradiation to a mixture of **7** and **9**:

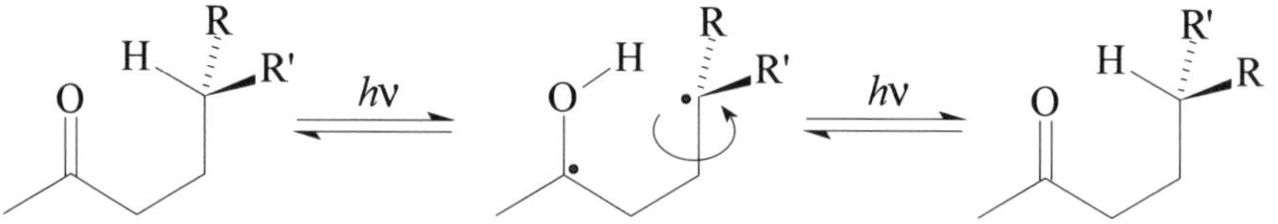

As a side reaction, the *Norrish type I reaction* is often observed. The stability of the radical species formed by α-cleavage determines the Norrish type I/Norrish type II ratio. For example aliphatic methyl ketones **10** react by a Norrish type II-mechanism, while aliphatic *tert*-butyl ketones **11** react preferentially by a Norrish type I-mechanism.

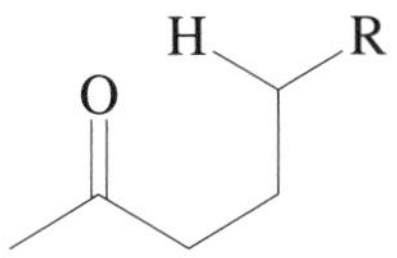

10 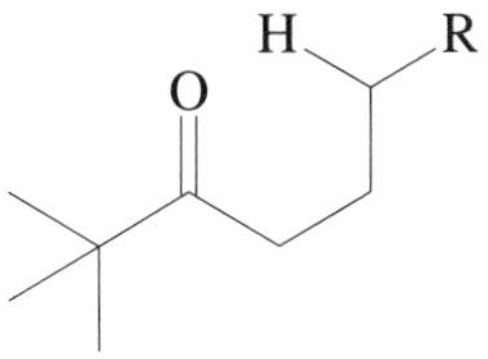   11

There are only a few examples for a preparative use of this reaction;[5] of more importance have so far been mechanistic aspects.

1. R. G. W. Norrish, *Trans. Faraday Soc.* **1937**, *33*, 1521–1528.
2. J. N. Pitts, Jr., J. K. S. Wan in *The Chemistry of the Carbonyl Group* (Ed.: S. Patai), Wiley, New York, **1966**, p. 823–916.
3. P. J. Wagner, *Acc. Chem. Res.* **1971**, *4*, 168–177.
4. J. M. Coxon, B. Halton, *Organic Photochemistry*, Cambridge University Press, London, **1974**, p. 58–78.
5. J. Kopecky, *Organic Photochemistry*, VCH, Weinheim, **1991**, p. 123–125.

O

Ozonolysis

Cleavage of a carbon–carbon double bond by reaction with ozone

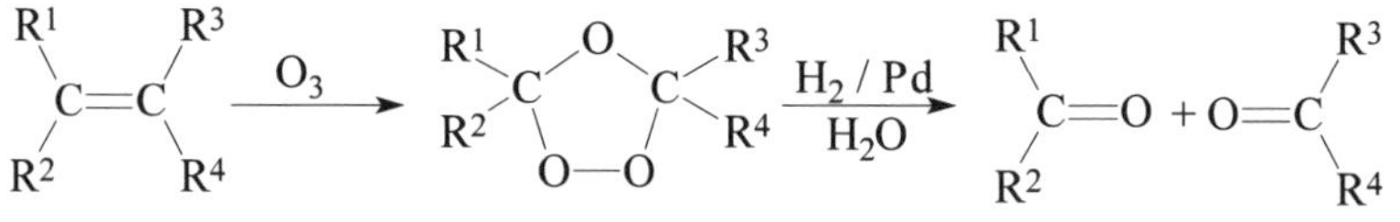

$$R^1R^2C{=}CR^3R^4 \xrightarrow{\ O_3\ } \underset{O{-}O}{R^1R^2C\!\!\begin{smallmatrix}O\end{smallmatrix}\!\!CR^3R^4} \xrightarrow[H_2O]{H_2/Pd} R^1R^2C{=}O + O{=}CR^3R^4$$

1	**2**	**3**	**4**

Harries[1] has introduced at the beginning of this century *ozonolysis*[2,3,4] as a method for the cleavage of carbon–carbon double bonds. The reaction proceeds *via* several intermediates to yield carbonyl compounds **3** and **4**. The following mechanism, which has been proposed by *Criegee*[2] and which is named after him, is generally accepted. Initial step is a *1,3-dipolar cycloaddition* reaction of ozone and alkene **1**, to give the *primary ozonide* **5** (also called *molozonide*). The initial ozonide **5** is unstable under the reaction conditions, and decomposes by a cycloreversion to give a carbonyl oxide **6** together with carbonyl compound **3**. Carbonyl oxide **6** again is a 1,3-dipolar species—isoelectronic to ozone—and rapidly undergoes a cycloaddition to the C=O double bond of **3**, to give the *secondary ozonide* **2**. The actual reaction sequence leading from **1** to **2** thus is: cycloaddition–cycloreversion–cycloaddition:

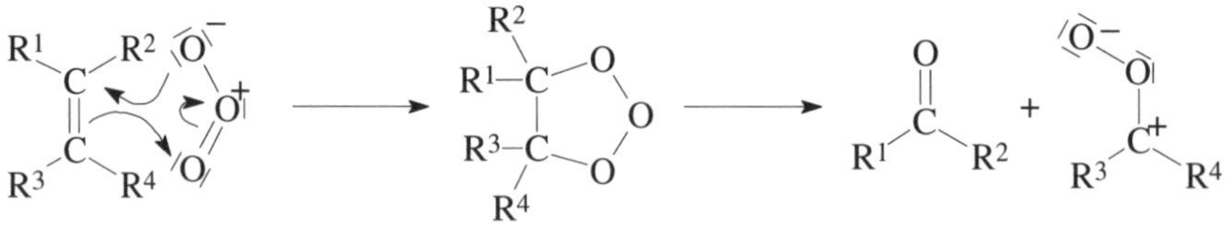

1	**5**	**3**	**6**

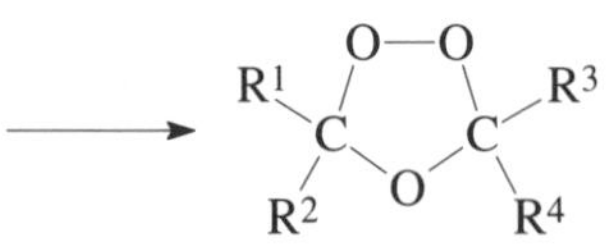

2

The ozonides **2** are reactive compounds that decompose violently upon heating. Nevertheless numerous ozonides have been isolated and studied by spectroscopic methods. With polar solvents, trapping products of carbonyl oxide **6** can be obtained. An external aldehyde, when added to the reaction mixture, will react with **6** to give a new secondary ozonide. Upon subsequent hydrolysis of ozonide **2**, different products may be obtained, depending on the particular reaction conditions. Under oxidative conditions an initially formed aldehyde will be oxidized to yield a carboxylic acid. In most cases however a reducing agent is added during hydrolytic workup, in order to avoid subsequent reactions with hydrogen peroxide. When the ozonide **2** is treated with lithium aluminum hydride, alcohols are obtained as products.

The reaction of ozone with an aromatic compound is considerably slower than the reaction with an alkene. Complete ozonolysis of one mole of benzene with workup under non-oxidative conditions will yield three moles of glyoxal. The selective ozonolysis of particular bonds in appropriate aromatic compounds is used in organic synthesis, for example in the synthesis of a substituted biphenyl **8** from phenanthrene **7**:[5]

In general however, ozonolysis is of limited synthetic importance. For quite some time ozonolysis has been an important tool for structure elucidation in organic chemistry, but has lost its importance when spectroscopic methods were fully developed for that purpose. The identification of the aldehydes and/or ketones obtained by ozonolysis of unsaturated compounds allowed for conclusions about the structure of the starting material, but has practically lost its importance since then.

Ozone has received increased attention for its occurence and function in the Earth's atmosphere.[6,7] For example the decreasing ozone concentration in the stratospheric ozone layer, becoming most obvious with the Antarctic ozone hole, and on the other hand the increasing ozone concentration in the ground level layer of the atmosphere during summer that is related to air pollution.[6] In the latter case ozone has manifold effects on man, animals and plants; for example it contributes to the dying of forests, but also to the degradation of organic environmental chemicals.[7]

1. C. Harries, *Justus Liebigs Ann. Chem.* **1905**, *343*, 311–374.
2. R. Criegee, *Angew. Chem.* **1975**, *87*, 765–771; *Angew. Chem. Int. Ed. Engl.* **1975**, *14*, 745.
3. R. L. Kuczkowski, *Chem. Soc. Rev.* **1992**, *21*, 79–83.
4. W. Sander, *Angew. Chem.* **1990**, *102*, 362–372; *Angew. Chem. Int. Ed. Engl.* **1990**, *29*, 344.
5. T.-J. Ho, C.-S. Shu, M.-K. Yeh, F.-C. Chen, *Synthesis* **1987**, 795–797.
6. F. Zabel, *Chem. Unserer Zeit* **1987**, *21*, 141–150.
   P. Crutzen, *Angew. Chem.* **1996**, *108*, 1878–1898;
   Angew. Chem. Int. Ed. Engl. **1996**, *35*, 1758–1777 (*Nobel lecture*).
   M. J. Molina, *Angew. Chem.* **1996**, *108*, 1900–1907;
   Angew. Chem. Int. Ed. Engl. **1996**, *35*, 1778–1785 (*Nobel lecture*).
   F. S. Rowland, *Angew. Chem.* **1996**, *108*, 1908–1921;
   Angew. Chem. Int. Ed. Engl. **1996**, *35*, 1786–1798 (*Nobel lecture*).
7. H. K. Lichtenthaler, *Naturwiss. Rundsch.* **1984**, *37*, 271–277.

P

Paterno–Büchi Reaction

Cycloaddition of a carbonyl compound to an alkene

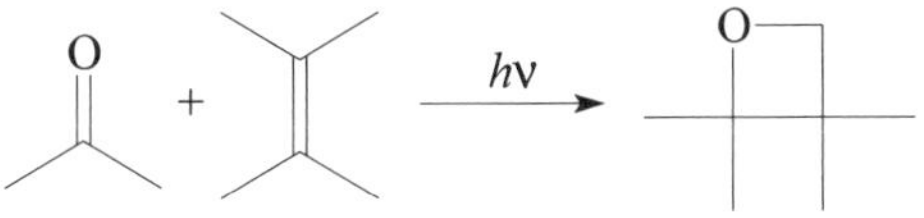

1   **2**   **3**

The photochemical cycloaddition of a carbonyl compound **1** to an alkene **2** to yield an oxetane **3**, is called the *Paterno–Büchi reaction*.[1,2] This reaction belongs to the more general class of photochemical [2 + 2]-cycloadditions, and is just as these, according to the Woodward–Hofmann rules,[3] photochemically a symmetry-allowed process, and thermally a symmetry-forbidden process.

The irradiation is usually carried out with light of the near UV region, in order to activate only the $n \rightarrow \pi^*$ transition of the carbonyl function,[4] thus generating excited carbonyl species. Depending on the substrate, it can be a singlet or triplet excited state. With aromatic carbonyl compounds, the reactive species are usually in a T_1-state, while with aliphatic carbonyl compounds the reactive species are in a S_1-state. An excited carbonyl species reacts with a ground state alkene molecule to form an *exciplex*, from which in turn diradical species can be formed—e.g. **4** and **5** in the following example:

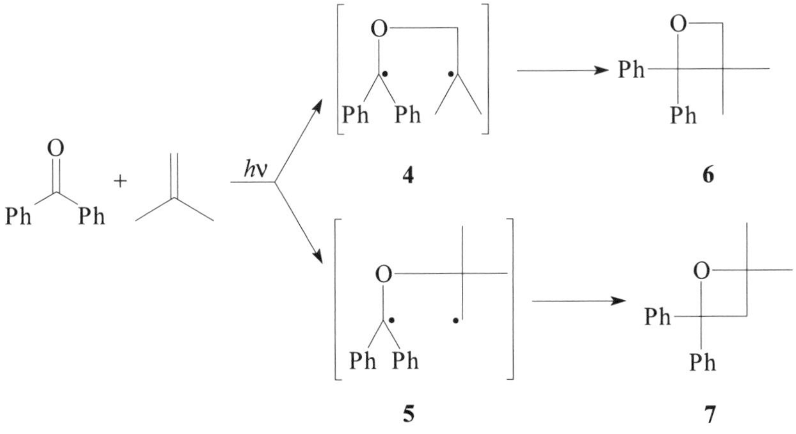

Diradical species **4** is more stable than diradical **5**, and the oxetane **6** is thus formed preferentially; oxetane **7** is obtained as minor product only. Evidence for diradical intermediates came from trapping experiments,[5] as well as spectroscopic investigations.[6]

In addition to the intermolecular Paterno–Büchi reaction, the intramolecular variant has also been studied;[2] the latter allows for the construction of bicyclic structures in one step. For example the diketone **8** reacts quantitatively to the bicyclic ketone **9**:

Although the Paterno–Büchi reaction is of high synthetic potential, its use in organic synthesis is still not far developed.[2] In recent years some promising applications in the synthesis of natural products have been reported.[8] The scarce application in synthesis may be due to the non-selective formation of isomeric products that can be difficult to separate—e.g. **6** and **7**—as well as to the formation of products by competitive side-reactions such as *Norrish type-I-* and *type-II fragmentations*.

1. G. Büchi, C. G. Inman, E. S. Lipinsky, *J. Am. Chem. Soc.* **1954**, *76*, 4327–4331.
2. I. Ninomiya, T. Naito, *Photochemical Synthesis*, Academic Press, New York, **1989**, p. 138–151.
3. R. B. Woodward, R. Hoffmann, *The Conservation of Orbital Symmetry*, Academic Press, New York, **1970**.
4. M. Demuth, G. Mikhail, *Synthesis* **1989**, 145–162.
5. W. Adam, U. Kliem, V. Lucchini, *Tetrahedron Lett.* **1986**, *27*, 2953–2956.
6. S. C. Freilich, K. S. Peters, *J. Am. Chem. Soc.* **1985**, *107*, 3819–3822.
7. R. Bishop, N. K. Hamer, *Chem. Commun.* **1969**, 804.
8. J. Mulzer, H.-J. Altenbach, M. Braun, K. Krohn, H.-U. Reissig, *Organic Synthesis Highlights*, VCH, Weinheim, **1991**, p. 105–110.

Pauson–Khand Reaction

Synthesis of cyclopentenones by a formal [2 + 2 + 1]-cycloaddition

The reaction of an alkyne **1** and an alkene **2** in the presence of dicobaltoctacarbonyl to yield a cyclopentenone **3** is referred to as the *Pauson–Khand reaction*.[1–4]
Formally it is a [2 + 2 + 1]-cycloaddition reaction. The dicobaltoctacarbonyl acts
as coordinating agent as well as a source of carbon monoxide.

Initial step is the formation of a dicobalthexacarbonyl-alkyne complex **5** by
reaction of alkyne **1** with dicobaltoctacarbonyl **4** with concomitant loss of two
molecules of CO. Complex **5** has been shown to be an intermediate by indepen-
dent synthesis. It is likely that complex **5** coordinates to the alkene **2**. Insertion of
carbon monoxide then leads to formation of a cyclopentenone complex **6**, which
decomposes into dicobalthexacarbonyl and cyclopentenone **3**:[2]

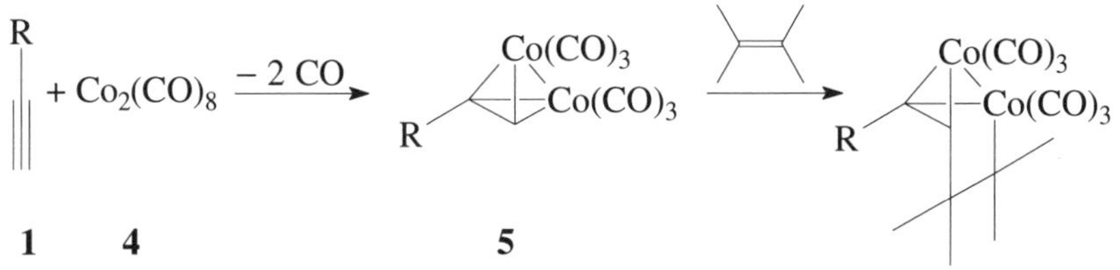

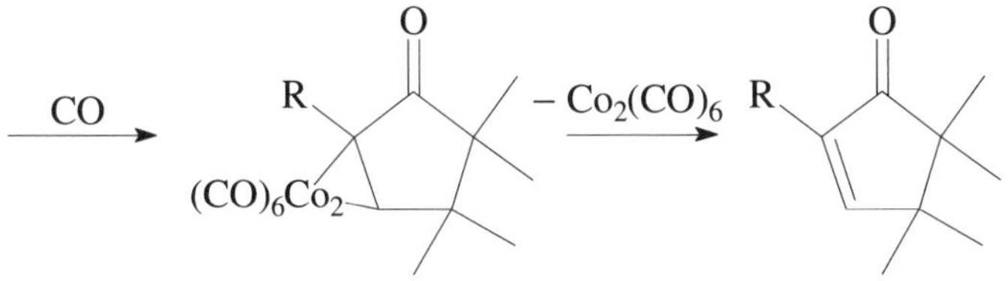

From reaction of simple alkynes and alkenes four regioisomeric products **7a–d**
may be formed:

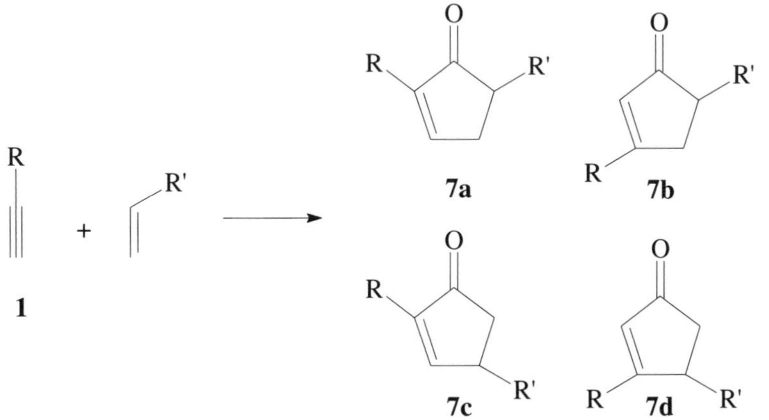

Products **7a** and **7c**, with the substituent R α to the carbonyl group, are by far predominantly formed.[5,6] This regioselectivity is a result of the preferential approach of the alkene **2** to the dicobalthexacarbonyl-alkyne complex **5** from the side opposite to the substituent R of the original alkyne. The actual incorporation of the alkene however is less selective with respect to the orientation of the olefinic substituent R′, thus leading to a mixture of isomers **7a** and **7c**.

The reaction with bicyclic alkenes—e.g. norbornadiene **8**—preferentially yields the *exo*-product **9**:

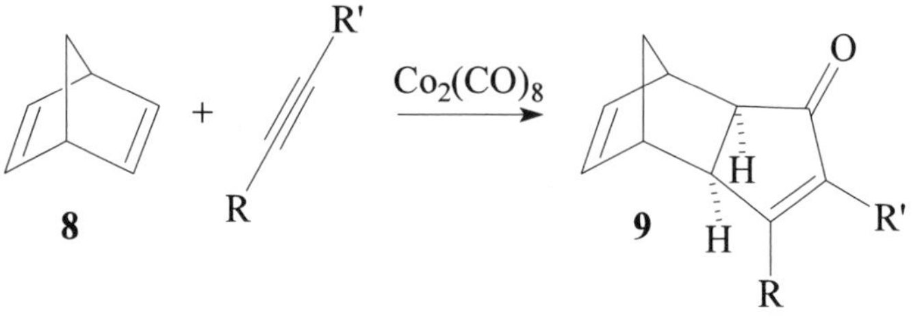

An example for the synthetic potential is the formation of a fenestrane skeleton **11** from the open-chain compound **10** by a cascade of two consecutive *intramolecular Pauson–Khand reactions*; the yield in this case is however only 9%:[7]

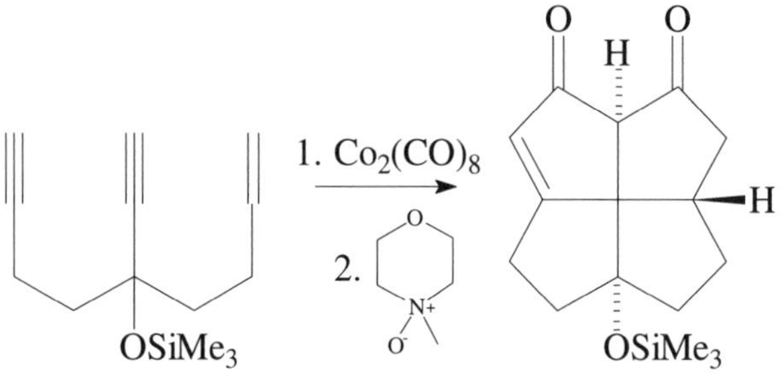

The Pauson–Khand reaction was originally developed using strained cyclic alkenes, and gives good yields with such substrates. Alkenes with sterically demanding substituents and acyclic as well as unstrained cyclic alkenes often are less suitable substrates. An exception to this is ethylene, which reacts well. Acetylene as well as simple terminal alkynes and aryl acetylenes can be used as triple-bond component.

1. I. U. Khand, G. R. Knox, P. L. Pauson, W. E. Watts, M. I. Forman, *J.Chem. Soc., Perkin Trans. 1*, **1973**, 977–981.
2. N. E. Schore, *Org. React.* **1991**, *40*, 1–90.
3. P. L. Pauson, *Tetrahedron* **1985**, *41*, 5855–5860.
4. J. Mulzer, H.-J. Altenbach, M. Braun, K. Krohn, H.-U. Reissig, *Organic Synthesis Highlights*, VCH, Weinheim, **1991**, p. 140–144.

5. S. E. MacWhorter, V. Sampath, M. M. Olmstead, N. E. Schore, *J. Org. Chem.* **1988**, *53*, 203–205.
6. M. E. Krafft, *J. Am. Chem. Soc.* **1988**, *110*, 968–970.
7. L. F. Tietze, U. Beifuss, *Angew. Chem.* **1993**, *105*, 137–170; *Angew. Chem. Int. Ed. Engl.* **1993**, *32*, 131.

Perkin Reaction

Condensation of aromatic aldehydes with carboxylic anhydrides

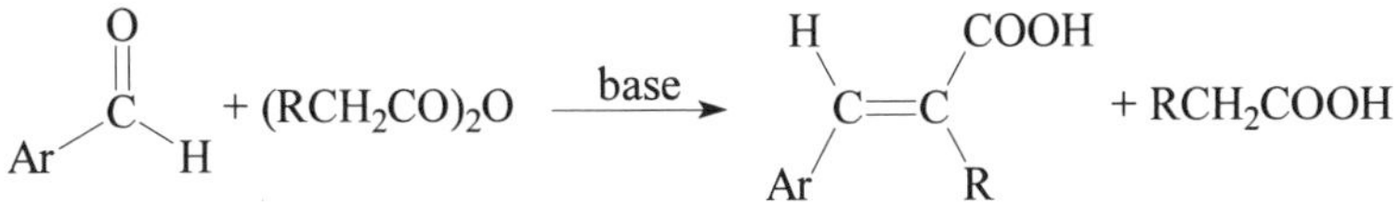

The aldol-like reaction of an aromatic aldehyde **1** with a carboxylic anhydride **2** is referred to as the *Perkin reaction.*[1,2] As with the related *Knoevenagel reaction*, an α,β-unsaturated carboxylic acid is obtained as product; the β-aryl derivatives **3** are also known as cinnamic acids.

The reaction mechanism involves deprotonation of the carboxylic anhydride **2** to give anion **4**, which then adds to aldehyde **1**. If the anhydride used bears two α-hydrogens, a dehydration takes place already during workup; a β-hydroxy carboxylic acid will then not be isolated as product:

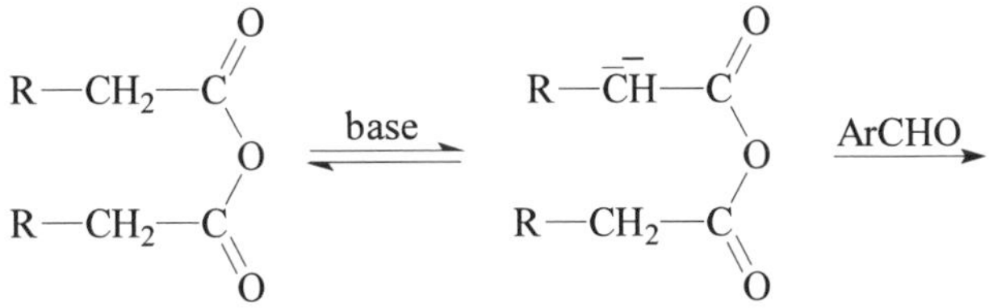

If the starting anhydride bears only one α-hydrogen, the dehydration step cannot take place, and a β-hydroxy carboxylic acid is obtained as the reaction product.

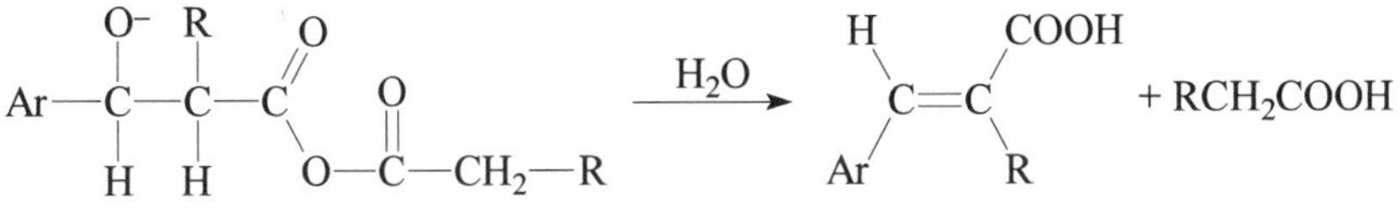

In principle the formation of a mixture of *E*- and *Z*-isomers is possible; however the preferential formation of the *E*-isomer is generally observed.

The general procedure is to heat a mixture of aldehyde **1** and carboxylic anhydride **2** together with a base to a temperature of 170–200 °C for several hours. As base the sodium salt of the carboxylic acid corresponding to the anhydride is most often used.

A variant of the Perkin reaction is the *Erlenmeyer–Plöchl–azlactone synthesis*.[3–5] By condensation of an aromatic aldehyde **1** with an N-acyl glycine **5** in the presence of sodium acetate and acetic anhydride, an azlactone **6** is obtained *via* the following mechanism:

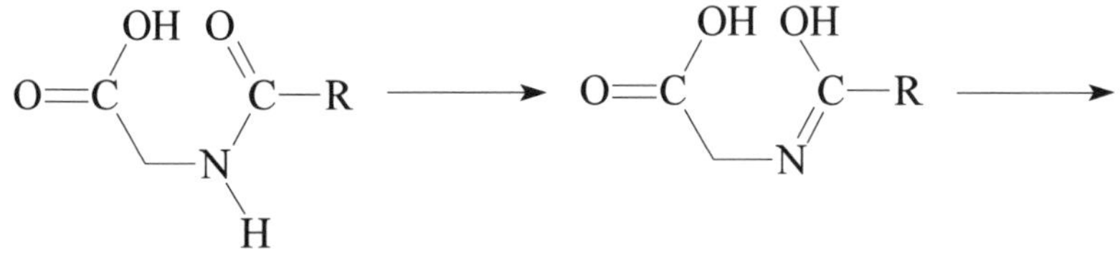

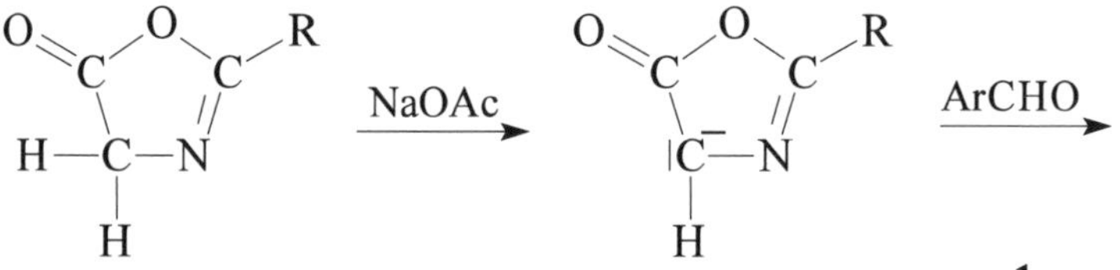

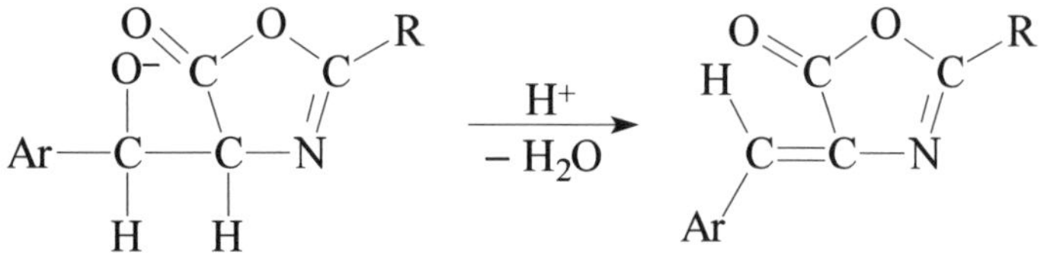

Azlactones like **6** are mainly used as intermediates in the synthesis of α-amino acids and α-keto acids. The Erlenmeyer–Plöchl reaction takes place under milder conditions than the Perkin reaction.

1. W. H. Perkin, *J. Chem. Soc.* **1877**, *31*, 388–427.
2. J. R. Johnson, *Org. React.* **1942**, *1*, 210–266.
3. E. Erlenmeyer, *Justus Liebigs Ann. Chem.* **1893**, *275*, 1–3.

4.   J. Plöchl, *Ber. Dtsch. Chem. Ges.* **1883**, *16*, 2815–2825.
5.   H. E. Carter, *Org. React.* **1946**, *3*, 198–239.

Peterson Olefination

Synthesis of alkenes from ketones or aldehydes

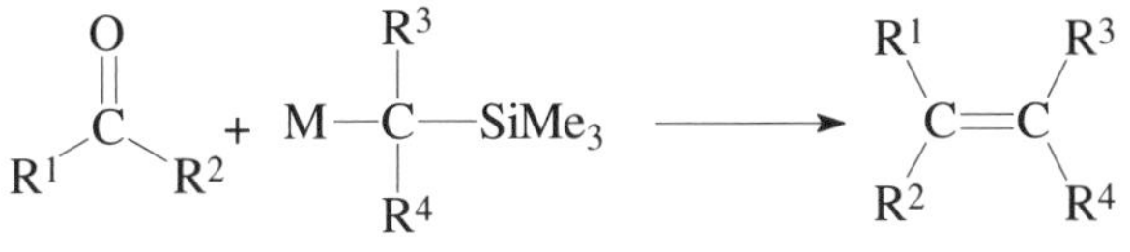

The *Peterson olefination*[1-3] can be viewed as a silicon variant of the *Wittig reaction*, the well-known method for the formation of carbon–carbon double bonds. A ketone or aldehyde **1** can react with an α-silyl organometallic compound **2**—e.g. with M = Li or Mg—to yield an alkene **3**.

The Peterson olefination is a quite modern method in organic synthesis; its mechanism is still not completely understood.[2,4,5] The α-silyl organometallic reagent **2** reacts with the carbonyl substrate **1** by formation of a carbon–carbon single bond to give the diastereomeric alkoxides **4a** and **4b**; upon hydrolysis the latter are converted into β-hydroxysilanes **5a** and **5b**:

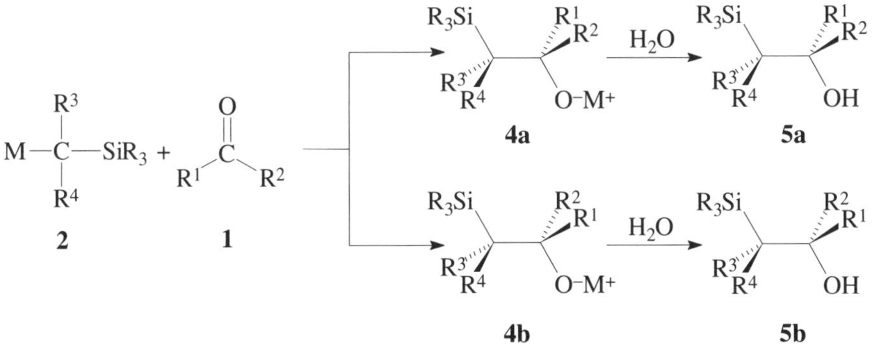

By application of the most common procedure—i.e by using an α-silylated organolithium or magnesium reagent—the β-hydroxysilane **5a/5b** can be isolated. However in the case of M = Na or K, the alkoxide oxygen in **4a/4b** is of strong ionic character, and a spontaneous elimination step follows to yield directly the alkene **3**.

The next step of the Peterson olefination allows for the control of the *E/Z*-ratio of the alkene to be formed by proper choice of the reaction conditions. Treatment of β-hydroxysilanes **5** with a base such as sodium hydride or potassium hydride leads to preferential *syn*-elimination to give alkene **3a** as major product. In contrast treatment with acid leads to preferential *anti*-elimination by an E_2-mechanism, then yielding alkene **3b** as major product:

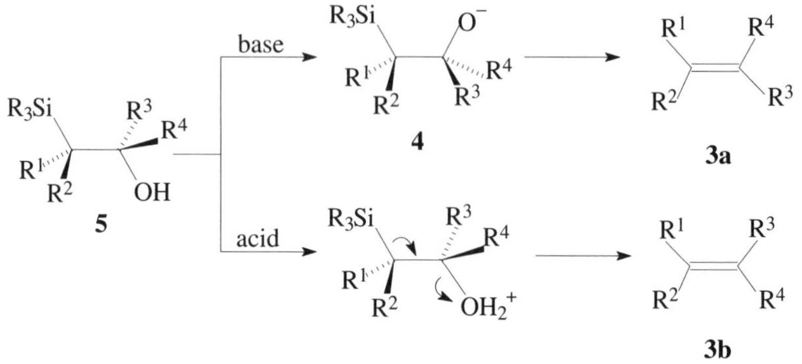

Whether the formation of alkene **3** proceeds directly from alkoxide **4** or *via* a penta-coordinated silicon-species **6**, is not rigorously known. In certain cases—e.g. for β-hydroxydisilanes (R^3 = SiMe₃) that were investigated by *Hrudlik et al.*[4]—the experimental findings suggest that formation of the carbon–carbon bond is synchronous to formation of the silicon–oxygen bond:

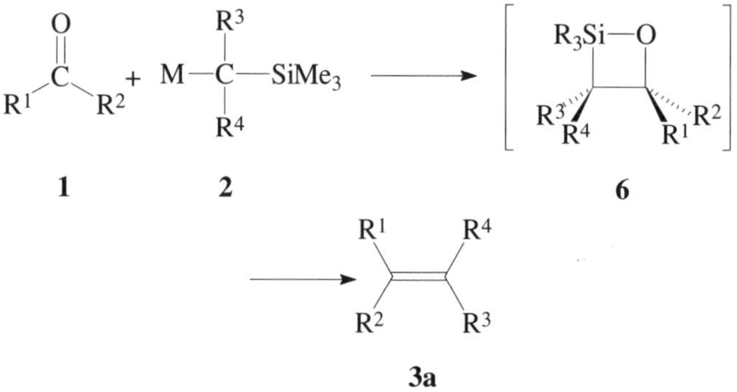

For the purpose of stereoselective synthesis the selective elimination at the stage of the β-hydroxysilane **5** is not a problem; the diastereoselective preparation of the desired β-hydroxysilane however is generally not possible. This drawback can be circumvented by application of alternative reactions to prepare the β-hydroxysilane;[2] however these methods do not fall into the category of the Peterson reaction.

The Peterson olefination presents a valuable alternative to the Wittig reaction. It has the advantage to allow for a simple control of the alkene geometry. Its applicability in synthesis depends on the availability of the required silanes.[2]

1. D. J. Peterson, *J. Org. Chem.* **1968**, *33*, 780–784.
2. D. J. Ager, *Org. React.* **1990**, *38*, 1–223.
3. D. J. Ager, *Synthesis* **1984**, 384–398.

4. P. F. Hudrlik, E. L. O. Agwaramgbo, A. M. Hudrlik, *J. Org. Chem.* **1989**, *54*, 5613–5618.
5. A. R. Bassindale, R. J. Ellis, J. C.-Y. Lau, P. G. Taylor, *J. Chem. Soc., Perkin Trans. 2*, **1986**, 593–597.

Pinacol Rearrangement

Rearrangement of vicinal diols

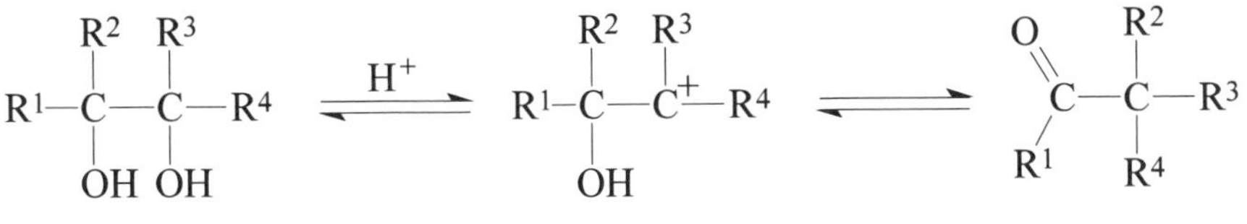

A vicinal diol **1**, when treated with a catalytic amount of acid, can rearrange to give an aldehyde or ketone **3** by migration of an alkyl or aryl group. The prototype of this reaction is the rearrangement of pinacol ($R^1 = R^2 = R^3 = R^4 = CH_3$) to yield pinacolone. The *pinacol rearrangement reaction*[1,2] can be viewed as a special case of the *Wagner–Meerwein rearrangement*.

In the initial step one hydroxy group is protonated, and thus converted into a good leaving group—i.e. water.[3] Subsequent loss of water from the molecule proceeds in such a way that the more stable carbenium ion species **2** is formed. The next step is a 1,2-shift of a group R to the tertiary carbenium center to give a hydroxycarbenium ion species **4**:

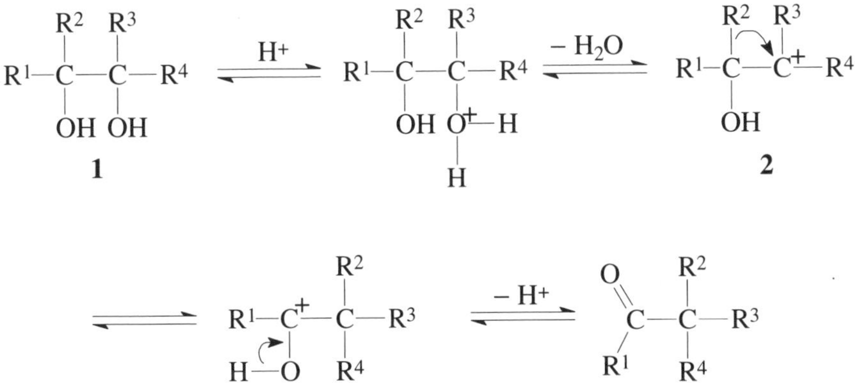

The reaction is strictly intramolecular; the migrating group R is never completely released from the substrate. The driving force is the formation of the more stable rearranged carbenium ion **4**, that is stabilized by the hydroxy substituent. The electron-deficiency of the carbenium ion center in **4** is to some extent compensated by an electron pair from the adjacent oxygen center—a resonance structure—a protonated carbonyl group $C=OH^+$—can be written. Loss of a proton yields the stable carbonyl compound **3**. The elimination of water to give an alkene may be observed as a side-reaction. Substituents R^1, R^2, R^3, R^4 can be alkyl or aryl; single substituents can even be hydrogen. Reaction with an unsymmetrical diol as starting material may give rise to formation of a mixture of products. The order of migration is $R_3C > R_2CH > RCH_2 > CH_3 > H$. The product ratio may also depend on the acid used.

The pinacol rearrangement reaction is of limited synthetic importance; although it can be a useful alternative to the standard methods for synthesis of aldehydes and ketones.[4] Especially in the synthesis of ketones with special substitution pattern—e.g. a spiro ketone like **5**—the pinacol rearrangement demonstrates its synthetic potential:[5]

5

The required vicinal diols are in general accessible by standard methods.[6] Pinacol itself can be obtained by dimerization of acetone. For the rearrangement reaction concentrated or dilute sulfuric acid is often used as catalyst.

1. R. Fittig, *Justus Liebigs. Ann. Chem.* **1859**, *110*, 17–23.
2. C. J. Collins, J. F. Eastham in *The Chemistry of the Carbonyl Group* (Ed.: S. Patai), Wiley, New York, **1966**, p. 762–767.
3. C. J. Collins, *J. Am. Chem. Soc.* **1955**, *77*, 5517–5523.
4. D. Dietrich, *Methoden Org. Chem. (Houben-Weyl)*, **1973**, Vol. 7/2a, p. 1016–1034.
5. E. Vogel, *Chem. Ber.* **1952**, *85*, 25–29.
6. Th. Wirth, *Angew. Chem.* **1996**, *108*, 65–67;
   Angew. Chem. Int. Ed. Engl. **1996**, *35*, 61–63.

Prilezhaev Reaction

Epoxidation of alkenes

1   **2**   **3**   **4**

The *Prilezhaev reaction*[1-4] is a rarely used name for the epoxidation of an alkene **1** by reaction with a peracid **2** to yield an oxirane **3**. The epoxidation of alkenes has been further developed into an enantioselective method, that is named after *Sharpless*.

The hydroxy oxygen of a peracid has a higher electrophilicity as compared to a carboxylic acid. A peracid **2** can react with an alkene **1** by transfer of that particular oxygen atom to yield an oxirane (an epoxide) **3** and a carboxylic acid **4**. The reaction is likely to proceed *via* a transition state as shown in **5** (butterfly mechanism),[5] where the electrophilic oxygen adds to the carbon-carbon π-bond and the proton simultaneously migrates to the carbonyl oxygen of the acid:[3,6]

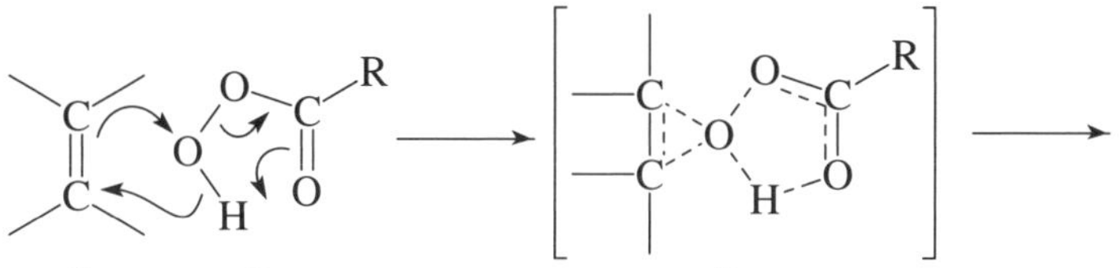

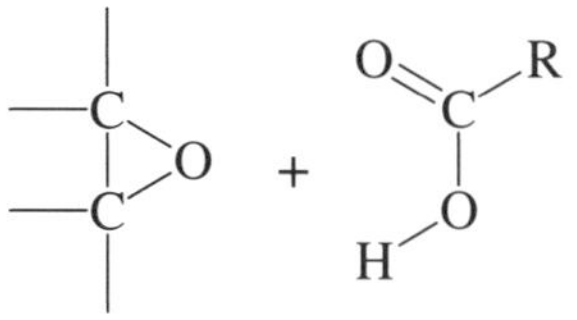

The mechanism formulated above is in agreement with the experimental findings, that stereospecifically a *syn*-addition takes place; the stereochemical relation between substituents in the alkene **1** is retained in the oxirane **3**.

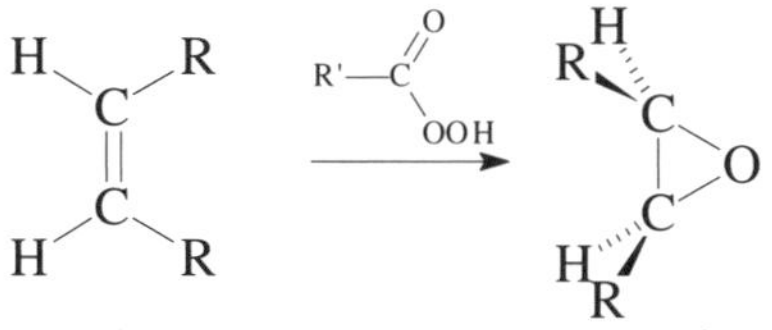

Allenes **6** also react with peracids; allene oxides **7** are formed, or even a spiro dioxide **8** can be obtained by reaction with a second equivalent of peracid:

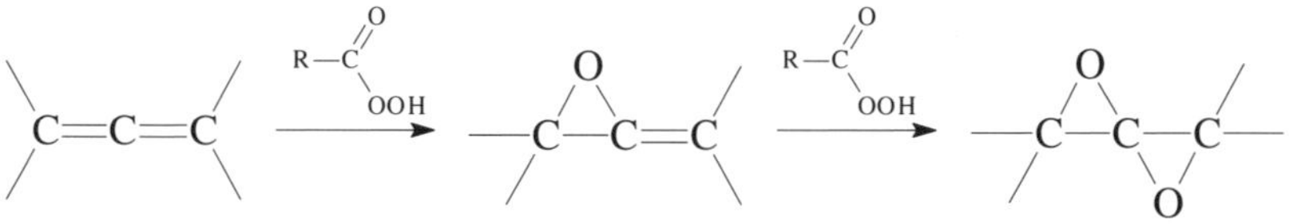

Oxiranes **3** are versatile intermediates in organic synthesis. The ring-opening reaction with a nucleophile is of wide scope. By this route alcohols, vicinal diols, ethers and many other classes of compounds can be prepared. With Grignard reagents (see *Grignard reaction*) the ring-opening proceeds with concomitant formation of a carbon–carbon bond.

m-Chloroperbenzoic acid is often used as epoxidation reagent; it is commercially available, quite stable and easy to handle. Various other peracids are unstable, and have to be prepared immediately prior to use.

The separation of the reaction products—i.e. the oxidation product and the carboxylic acid—can usually be achieved by extraction with mild aqueous base.

A modern reagent, that has found increased application, is *dimethyldioxirane*; it is prepared *in situ* by oxidation of acetone with potassium peroxomonosulfate $KHSO_5$.[7]

The epoxidation reaction usually takes place under mild conditions and with good to very good yield. Functional groups that are sensitive to oxidation should not be present in the starting material; with carbonyl groups a *Baeyer–Villiger reaction* may take place.

1. N. Prilezhaev, *Ber. Dtsch. Chem. Ges.* **1909**, *42*, 4811–4815.
2. B. Plesnicar in *The Chemistry of Peroxides* (Ed.: S. Patai), Wiley, New York, **1983**, p. 521–584.
3. C. Berti, *Top. Stereochem.* **1973**, *7*, 93–251.
4. B. Plesnicar in *Oxidation in Organic Chemistry, Vol. C* (Ed.: W. S. Trahanovsky), Academic Press, New York, **1978**, p. 211–252.
5. K. W. Woods, P. Beak, *J. Am. Chem. Soc.* **1991**, *113*, 6281–6283.
6. V. G. Dryuk, *Russ. Chem. Rev.* **1985**, *54*, 986–1005.
7. W. Adam, R. Curci, J. O. Edwards, *Acc. Chem. Res.* **1989**, *22*, 205–211.

Prins Reaction

Addition of formaldehyde to alkenes

| 1 | 2 | 3 | 4 | 5 |

The acid-catalyzed addition of an aldehyde—often formaldehyde **1**—to a carbon–carbon double bond can lead to formation of a variety of products. Depending on substrate structure and reaction conditions, a 1,3-diol **3**, allylic alcohol **4** or a 1,3-dioxane **5** may be formed. This so-called *Prins reaction*[1–4] often leads to a mixture of products.

The initial step is the protonation of the aldehyde—e.g. formaldehyde—at the carbonyl oxygen. The hydroxycarbenium ion **6** is thus formed as reactive species, which reacts as electrophile with the carbon–carbon double bond of the olefinic substrate by formation of a carbenium ion species **7**. A subsequent loss of a proton from **7** leads to formation of an allylic alcohol **4**, while reaction with water, followed by loss of a proton, leads to formation of a 1,3-diol **3**:[3,4]

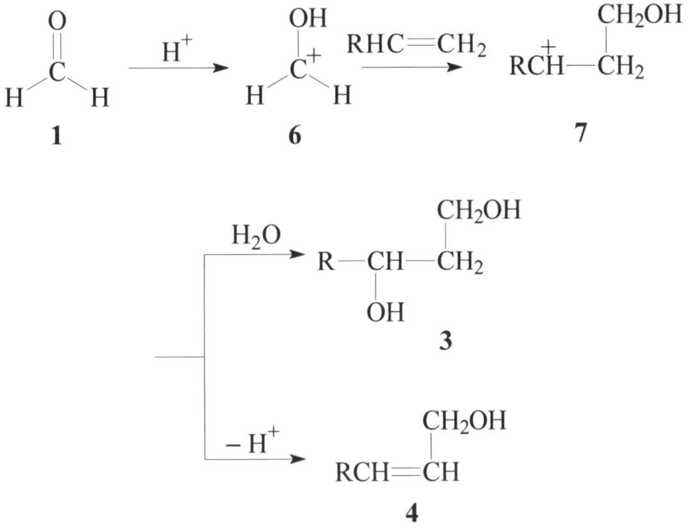

The Prins reaction often yields stereospecifically the *anti*-addition product; this observation is not rationalized by the above mechanism. Investigations of the sulfuric acid-catalyzed reaction of cyclohexene **8** with formaldehyde in acetic acid as solvent suggest that the carbenium ion species **7** is stabilized by a neighboring-group effect as shown in **9**. The further reaction then proceeds from the face opposite to the coordinating OH-group:[3,4]

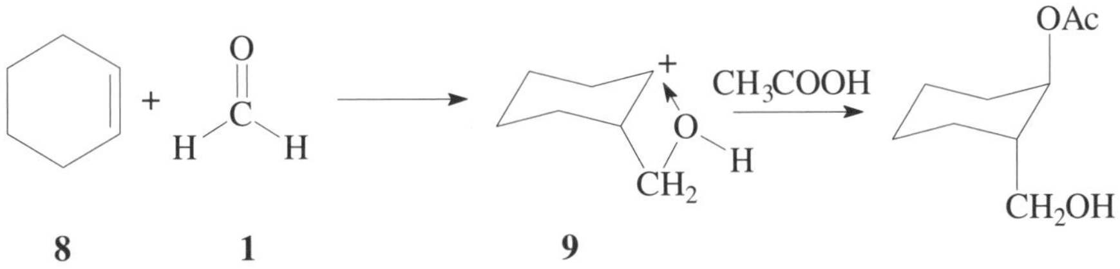

The *anti*-selective addition is not always observed; in some cases the *syn*-addition product predominates, and sometimes there is no selectivity observed at all. Obviously not all substrates are likely to react *via* a four-membered ring intermediate like **9**.

In the presence of excess formaldehyde, the carbenium ion species **7** can further react to give a 1,3-dioxane **5**. If only one equivalent of formaldehyde is used however, 1,3-diol **3** is formed as the major product:

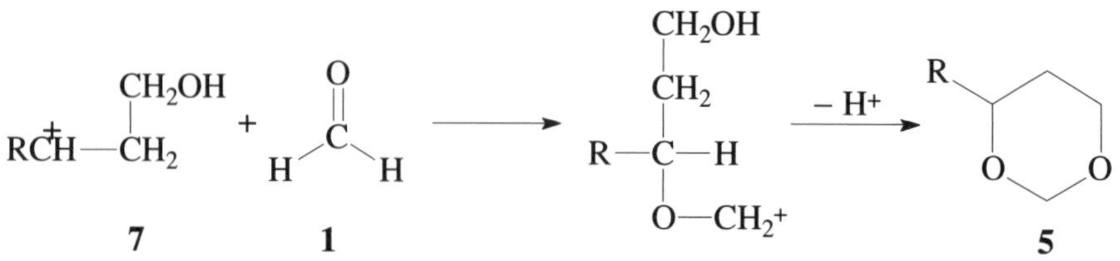

The formation of complex mixtures of products by a Prins reaction can be a problem. An example is the reaction of aqueous formaldehyde with cyclohexene **8** under acid catalysis:

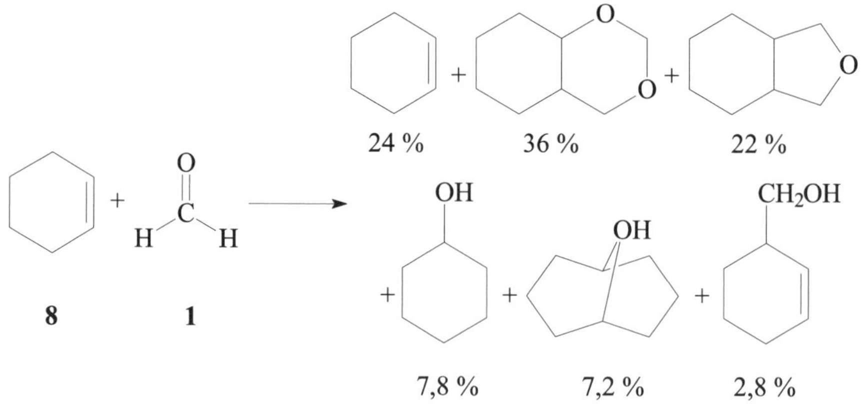

Under appropriate conditions 1,3-dioxanes can be obtained in moderate to good yields. Below 70 °C the acid-catalyzed condensation of alkenes with aldehydes yields 1,3-dioxanes as major products, while at higher temperatures the hydrolysis of dioxanes to diols is observed.

As a catalyst sulfuric acid is most often used; phosphoric acid, boron trifluoride or an acidic ion exchange resin have also found application. 1,1-disubstituted alkenes are especially suitable substrates, since these are converted to relatively stable tertiary carbenium ion species upon protonation. α,β-unsaturated carbonyl compounds do not react as olefinic component.

1. H. J. Prins, *Chem. Weekbl.* **1919**, *16*, 1072–1073.
2. D. R. Adams, S. P. Bhatnagar, *Synthesis* **1977**, 661–672.
3. H. Griegel, W. Sieber, *Monatsh. Chem.* **1973**, *104*, 1008–1026, 1027–1033.
4. V. I. Isagulyants, T. G. Khaimova, V. R. Melikyan, S. V. Pokrovskaya, *Russ. Chem. Rev.* **1968**, *37*, 17–25.

R

Ramberg–Bäcklund Reaction

Conversion of α-halosulfones to alkenes

$$R^1\!-\!CH_2\!-\!\overset{\overset{\displaystyle O}{\|}}{\underset{\underset{\displaystyle O}{\|}}{S}}\!-\!\underset{\underset{\displaystyle X}{|}}{CH}\!-\!R^2 \quad \xrightarrow{\text{base}} \quad \overset{R^1}{\underset{H}{\diagdown}}C\!=\!C\overset{R^2}{\underset{H}{\diagup}}$$

1   2

Treatment of an α-halosulfone **1** with base leads to extrusion of sulfur dioxide and formation of an alkene **2**. This reaction is referred to as the *Ramberg–Bäcklund reaction*;[1,2] it usually yields a mixture of *E*- and *Z*-isomers of the alkene.

An α-halosulfone **1** reacts with a base by deprotonation at the α′-position to give a carbanionic species **3**. An intramolecular nucleophilic substitution reaction, with the halogen substituent taking the part of the leaving group, then leads to formation of an intermediate episulfone **4** and the halide anion. This mechanism is supported by the fact that the episulfone **4** could be isolated.[3] Subsequent extrusion of sulfur dioxide from **4** yields the alkene **2**:

$$R^1\!-\!CH_2\underset{\underset{O}{\diagdown}\underset{}{S}\underset{O}{\diagup}}{\diagdown}\overset{\overset{X}{|}}{CH}\!-\!R^2 \quad \xrightarrow{\text{base}} \quad R^1\!-\!\bar{C}H\underset{\underset{O}{\diagdown}\underset{}{S}\underset{O}{\diagup}}{\diagdown}\overset{X}{CH}\!-\!R^2 \quad \longrightarrow$$

1   3

$$R^1\!-\!\underset{\underset{O}{\diagdown}\underset{}{S}\underset{O}{\diagup}}{\overset{\overset{H}{|}}{C}\!-\!\overset{\overset{H}{|}}{C}}\!-\!R^2 \quad \xrightarrow{-\,SO_2} \quad R^1\!-\!CH\!=\!CH\!-\!R^2$$

4   2

The Ramberg–Bäcklund reaction has been used for the preparation of strained unsaturated ring compounds that are difficult to obtain by other methods. A recent example is the synthesis of ene-diyne **5**[4] that has been used as starting material for a *Bergman cyclization*:

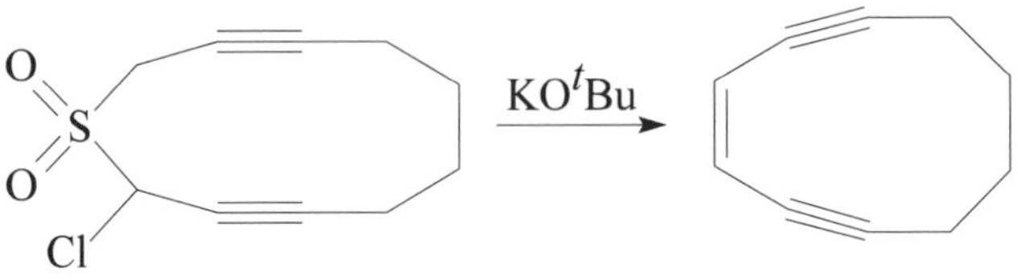

5

The α-halosulfone, required for the Ramberg–Bäcklund reaction, can for example be prepared from a sulfide by reaction with thionyl chloride (or with *N*-chlorosuccinimide) to give an α-chlorosulfide, followed by oxidation to the sulfone—e.g. using *m*-chloroperbenzoic acid. As base for the Ramberg–Bäcklund reaction have been used: alkoxides—e.g. potassium *t*-butoxide in an etheral solvent, as well as aqueous alkali hydroxide. In the latter case the use of a phase-transfer catalyst may be of advantage.[5]

1. L. A. Paquette, *Org. React.* **1977**. *25*, 1–71.
2. F. G. Bordwell, E. Doomes, *J. Org. Chem.* **1974**, *39*, 2526–2531.
3. A. G. Sutherland, R. J. K. Taylor, *Tetrahedron Lett.* **1989**, *30*, 3267–3270.
4. K. C. Nicolaou, W.-M. Dai, *Angew. Chem.* **1991**, *103*, 1453–1481;
   Angew. Chem. Int. Ed. Engl. **1991**, *30*, 1387.
5. G. D. Hartman, R. D. Hartman, *Synthesis* **1982**, 504–506.

Reformatsky Reaction

Synthesis of β-hydroxy esters

$$\text{X-CH}_2\text{CO}_2\text{Et} \xrightarrow{\ \text{Zn}\ } \text{XZn-CH}_2\text{CO}_2\text{Et}$$

1   **2**

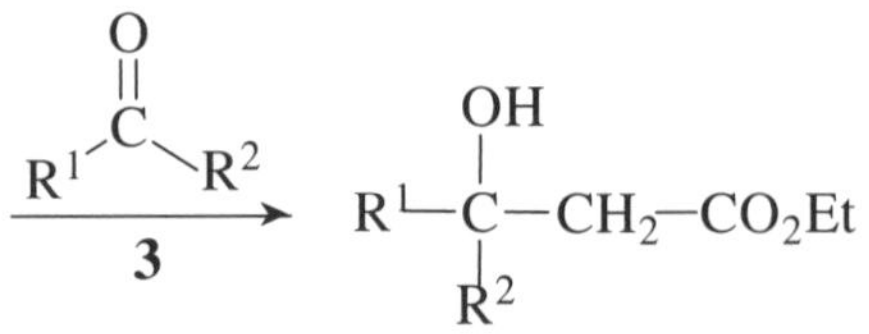

4

The classical *Reformatsky reaction*[1–4] consists of the treatment of an α-halo ester **1** with zinc metal and subsequent reaction with an aldehyde or ketone **3**. Nowadays the name is used generally for reactions that involve insertion of a metal into a carbon–halogen bond and subsequent reaction with an electrophile. Formally the Reformatsky reaction is similar to the *Grignard reaction*.

By reaction of an α-halo ester **1** with zinc metal in an inert solvent such as diethyl ether, tetrahydrofuran or dioxane, an organozinc compound **2** is formed (a Grignard reagent-like species). Some of these organozinc compounds are quite stable; even a structure elucidation by x-ray analysis is possible in certain cases:

$$X\text{-}CH_2CO_2Et + Zn \longrightarrow XZn\text{-}CH_2CO_2Et \quad \xrightarrow[\textbf{3}]{\underset{R^1}{\overset{\displaystyle \underset{\|}{\overset{O}{C}}}{}}\!R^2}$$

$$\textbf{1} \qquad\qquad\qquad \textbf{2}$$

$$R \overset{OZnX}{\underset{R^2}{\overset{|}{\underset{|}{C}}}}-CH_2-CO_2Et \xrightarrow{H^+} R \overset{OH}{\underset{R^2}{\overset{|}{\underset{|}{C}}}}-CH_2-CO_2Et$$

$$\textbf{4}$$

The reaction with a carbonyl substrate **3** is similar to a Grignard reaction. Hydrolytic workup then yields the β-hydroxy ester **4**. Sometimes product **4** easily eliminates water to yield directly an α,β-unsaturated ester.

The organozinc compound **2** is less reactive than an organomagnesium compound; the addition to an ester carbonyl group is much slower than the addition to an aldehyde or ketone. Nevertheless the addition of **2** to the carbonyl group of unreacted α-halo ester **1** is the most frequently observed side-reaction:

$$XCH_2CO_2Et + Zn \longrightarrow XZnCH_2CO_2Et + XCH_2CO_2Et \longrightarrow$$

$$\textbf{1} \qquad\qquad\qquad \textbf{2} \qquad\qquad\qquad \textbf{1}$$

$$XCH_2\overset{OZnX}{\underset{OEt}{\overset{|}{\underset{|}{C}}}}CH_2CO_2Et \longrightarrow XCH_2\overset{O}{\overset{\|}{C}}CH_2CO_2Et + EtOZnX$$

The carbonyl substrate **3** to be reacted with the organozinc compound **2** can be an aldehyde or ketone that may contain additional functional groups. With a vinylogous halo ester—i.e. a γ-halocrotyl ester—the corresponding γ-crotylzinc derivative is formed.

With special techniques for the activation of the metal—e.g. for removal of the oxide layer, and the preparation of finely dispersed metal—the scope of the Reformatsky reaction has been broadened, and yields have been markedly improved.[4,5] The attempted activation of zinc by treatment with iodine or dibromomethane, or washing with dilute hydrochloric acid prior to use, often is only moderately successful. Much more effective is the use of special alloys—e.g. zinc-copper couple, or the reduction of zinc halides using potassium (the so-called *Rieke procedure*[6]) or potassium graphite.[5] The application of ultrasound has also been reported.[7]

1. S. Reformatsky, *Ber. Dtsch. Chem. Ges.* **1887**, *20*, 1210–1211.
2. R. L. Shriner, *Org. React.* **1946**, *1*, 423–460.
3. M. W. Rathke, *Org. React.* **1975**, *22*, 423–460.
4. A. Fürstner, *Synthesis* **1989**, 571–590.
5. A. Fürstner, *Angew. Chem.* **1993**, *105*, 171–197; *Angew. Chem. Int. Ed. Engl.* **1993**, *32*, 164.
6. R. D. Rieke, S. J. Uhm, *Synthesis* **1975**, 452–453.
7. B. H. Han, P. Boudjouk, *J. Org. Chem.* **1982**, *47*, 5030–5032.

Reimer–Tiemann Reaction

Formylation of aromatic substrates with chloroform

The formylation of a phenol **1** with chloroform in alkaline solution is called the *Reimer–Tiemann reaction.*[1–3] It leads preferentially to formation of an *ortho*-formylated phenol—e.g. salicylic aldehyde **2** —while with other formylation reactions, e.g. the *Gattermann reaction*, the corresponding *para*-formyl derivative is obtained as a major product. The Reimer–Tiemann reaction is mainly used for the synthesis of *o*-hydroxy aromatic aldehydes.

The actual formylation process is preceded by the formation of dichlorocarbene **3** as the reactive species. In strongly alkaline solution, the chloroform is deprotonated; the resulting trichloromethide anion decomposes into dichlorocarbene and a chloride anion:

$$CHCl_3 + OH^- \underset{-H_2O}{\overset{-H_2O}{\rightleftharpoons}} CCl_3^- \xrightarrow{-Cl^-} :CCl_2$$
$$3$$

In alkaline solution, the phenol **1** is deprotonated to the phenolate **4**, which reacts at the *ortho*-position with dichlorocarbene **3**. The initial addition reaction product **5** isomerizes to the aromatic *o*-dichloromethyl phenolate **6**, which under the reaction conditions is hydrolyzed to the *o*-formyl phenolate.[4]

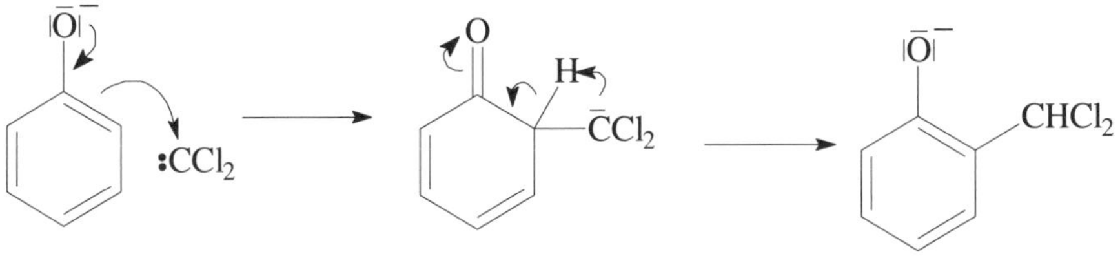

4   **3**   **5**   **6**

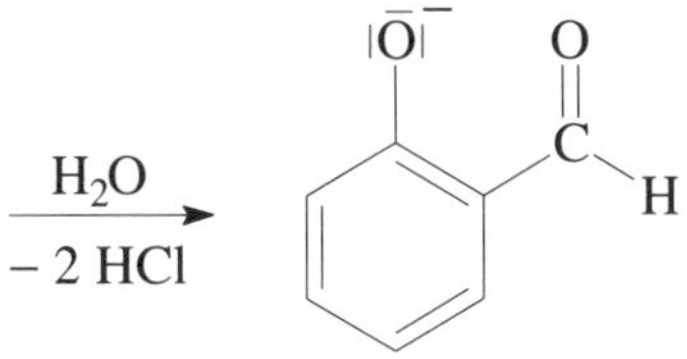

$$\xrightarrow[-2\,HCl]{H_2O}$$

The applicability of the Reimer–Tiemann reaction is limited to the formylation of phenols and certain reactive heterocycles like pyrroles and indoles. Yields are usually below 50%. In contrast to other formylation procedures, the Reimer–Tiemann reaction is *ortho*-selective; it is therefore related to the *Kolbe–Schmitt reaction.*

By a modified procedure using polyethyleneglycol as complexing agent a *para*-selective reaction can be achieved.[5]

As with other two-phase reactions, the application of ultrasound may lead to shorter reaction times and improved yields.[6]

1. K. Reimer, *Ber. Dtsch. Chem. Ges.* **1876**, *9*, 423–424.
2. H. Wynberg, E. W. Meijer, *Org. React.* **1982**, *28*, 1–36.
3. G. Simchen, *Methoden Org. Chem (Houben-Weyl)*, **1983**, Vol. E3, p. 16–19.
4. E. A. Robinson, *J. Chem. Soc.* **1961**, 1663–1671.
5. R. Neumann, Y. Sasson, *Synthesis* **1986**, 569–570.
6. J. C. Cochran, M. G. Melville, *Synth. Commun.* **1990**, *20*, 609–616.

Robinson Annulation

Annulation of a cyclohexenone ring

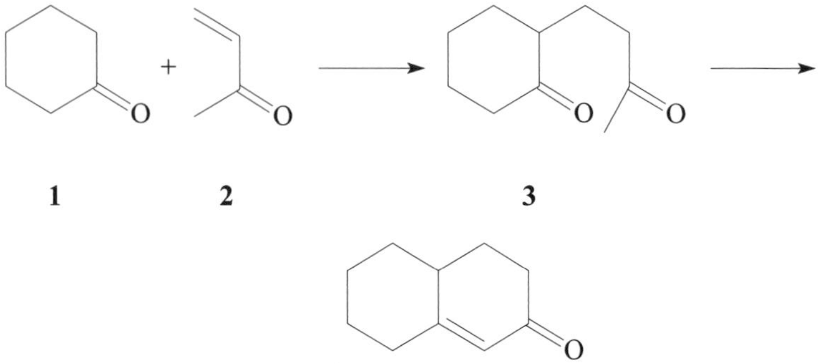

The reaction of a cyclic ketone—e.g. cyclohexanone **1**—with methyl vinyl ketone **2** resulting in a ring closure to yield a bicyclic α,β-unsaturated ketone **4**, is called the *Robinson annulation*.[1-3] This reaction has found wide application in the synthesis of terpenes, and especially of steroids. Mechanistically the Robinson annulation consists of two consecutive reactions, a *Michael addition* followed by an *Aldol reaction*. Initially, upon treatment with a base, the cyclic ketone **1** is deprotonated to give an enolate, which undergoes a conjugate addition to the methyl vinyl ketone, i.e. a Michael addition, to give a 1,5-diketone **3**:

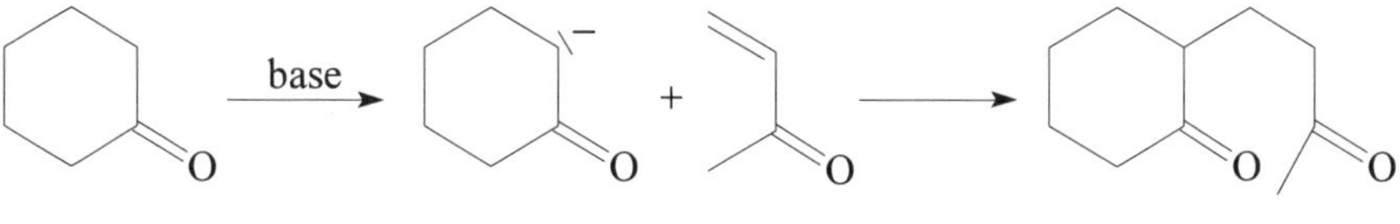

The next step is an intramolecular aldol reaction leading to closure of a six-membered ring. Subsequent dehydration yields the bicyclic enone **4**:

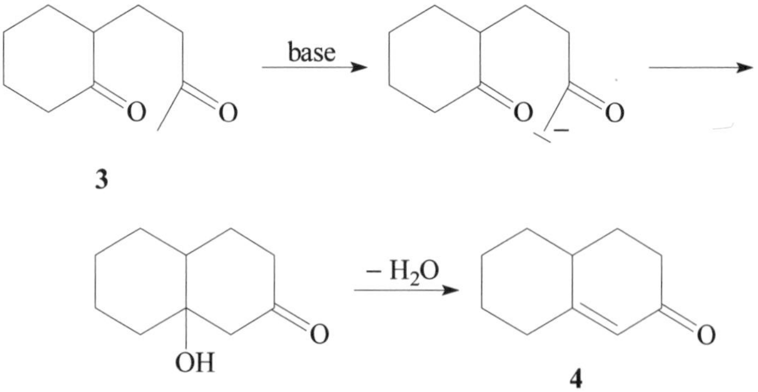

Methyl vinyl ketone **2** tends to polymerize, especially in the presence of a strong base; the yield of annulation product is therefore often low. A methyl vinyl ketone precursor, e.g. **6**, is often employed, from which the Michael acceptor **2** is generated *in situ*, upon treatment with a base. The quaternary ammonium salt **6** can be obtained by reaction of the tertiary amine **5**, which in turn is prepared from acetone, formaldehyde and diethylamine in a *Mannich reaction*.

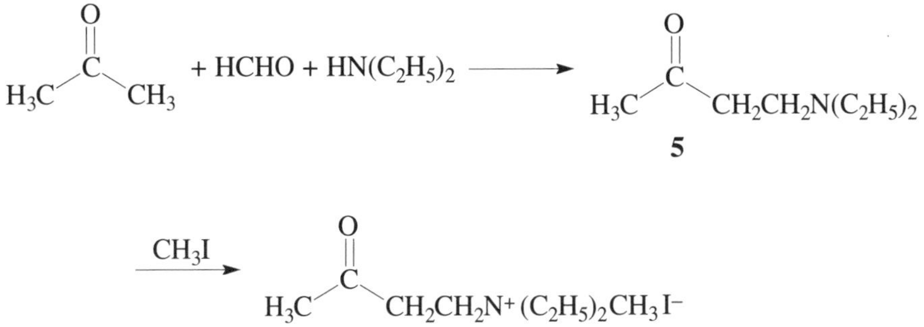

Besides a polymerization of the Michael acceptor, a double alkylation of the starting ketone, by reaction with a second Michael acceptor molecule, may take place as a side reaction, and thus further reduce the yield. The polymerization of the enone **2** as well as the double alkylation of the starting ketone can be avoided by application of a modern procedure for the Robinson annulation that uses an organotin triflate as catalyst.[4]

When 3-butyne-2-one **7** is used as a Michael acceptor component, a 2,5-cyclohexadienone, e.g. **8**, is obtained as the annulation product:[5]

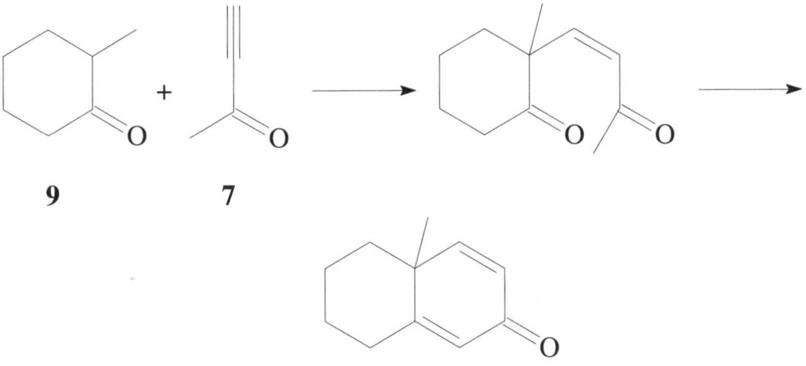

From a stereochemical point of view the Robinson annulation can be a highly complex reaction, since the configuration at five stereogenic sp^3-carbon centers is influenced during formation of the initial annulation product; the subsequent dehydration however, that usually takes place, reduces the number of stereogenic centers to three.

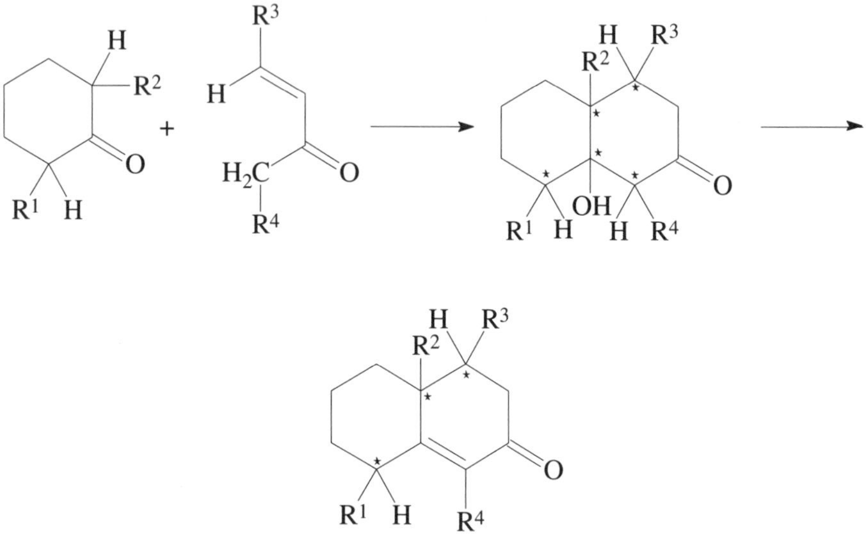

Since most often the selective formation of just one stereoisomer is desired, it is of great importance to develop highly selective methods. For example the second step, the aldol reaction, can be carried out in the presence of a chiral auxiliary—e.g. a chiral base—to yield a product with high enantiomeric excess. This has been demonstrated for example for the reaction of 2-methylcyclopenta-1,3-dione with methyl vinyl ketone in the presence of a chiral amine or α-amino acid. By using either enantiomer of the amino acid proline—i.e. (S)-$(-)$-proline or (R)-$(+)$-proline—as chiral auxiliary, either enantiomer of the annulation product 7a-methyl-5,6,7,7a-tetrahydroindan-1,5-dione could be obtained with high enantiomeric excess.[6] α-Substituted ketones, e.g. 2-methylcyclohexanone **9**, usually add with the higher substituted α-carbon to the Michael acceptor:

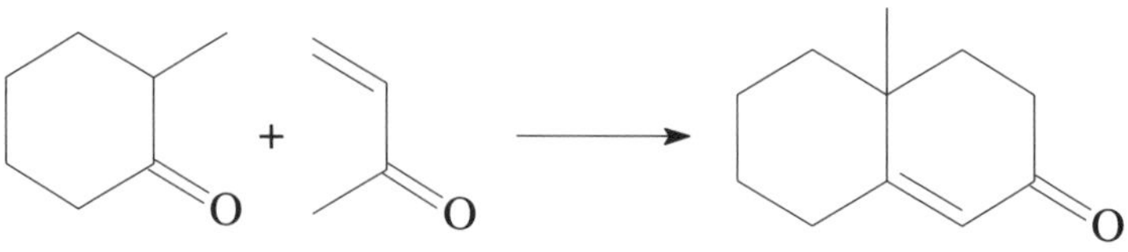

Exceptions to this rule may be a result of steric hindrance. However when the *Stork enamine method* is applied, for example with enamine **10**, the less substituted α-carbon becomes connected to the Michael acceptor:

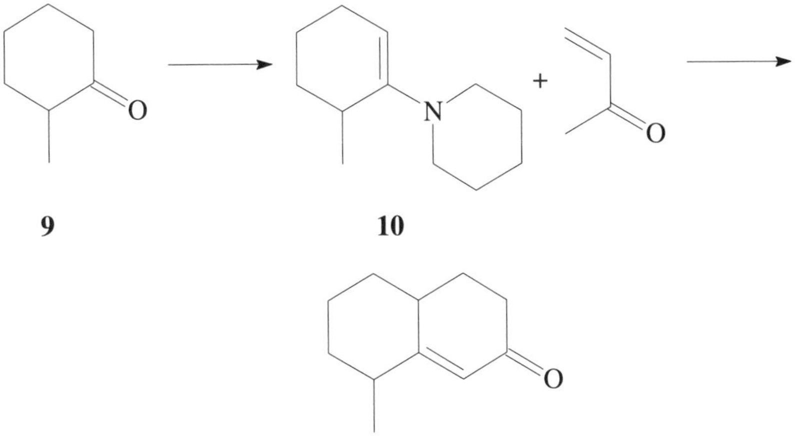

9   **10**

The best method to achieve a high regioselectivity is the use of preformed enolates. A double annulation reaction is possible if, for example, a diketone such as **11** is used as starting material. The product of the Michael addition **12** can undergo two subsequent aldol condensation reactions to yield the tricyclic dienone **13**:[2]

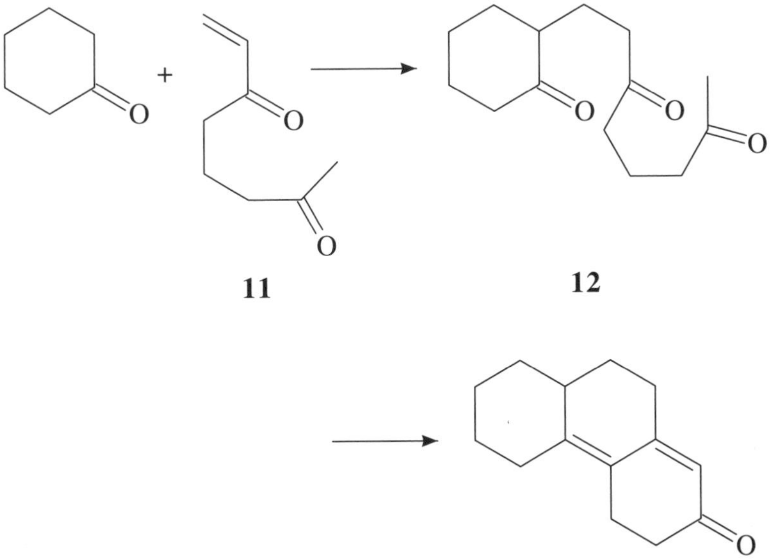

11   **12**

13

Since an annulated six-membered carbocycle is a common structural element of natural products, the Robinson annulation is an important reaction in organic synthesis.

1. W. S. Rapson, R. Robinson, *J. Chem. Soc.* **1935**, 1285–1291.
2. R. E. Gawley, *Synthesis* **1976**, 777–794.
3. M. E. Jung, *Tetrahedron* **1976**, *32*, 3–31.
4. T. Sato, Y. Wakahara, J. Otera, H. Nozaki, *Tetrahedron Lett.* **1990**, *31*, 1581–1584.
5. R. B. Woodward, G. Singh, *J. Am. Chem. Soc.* **1950**, *72*, 494–500.
6. U. Eder, G. Sauer, R. Wiechert, *Angew. Chem.* **1971**, *83*, 492–493;
    Angew. Chem. Int. Ed. Engl. **1971**, *10*, 496–497.
    Z. G. Hajos, D. R. Parrish, *J. Org. Chem.* **1974**, *39*, 1615.
    C. Agami, *Bull. Soc. Chim. Fr.* **1988**, 499–507.

Rosenmund Reduction

Aldehydes by reduction of acyl chlorides

The name *Rosenmund reduction*[1-3] is used for the catalytic hydrogenation of an acyl chloride **1** to yield an aldehyde **2**.

The reaction mechanism differs from that of other catalytic hydrogenations that also are carried out in the presence of palladium as catalyst, e.g. that of olefins. Presumably an organopalladium species is formed as an intermediate, which then reacts with the hydrogen:[6]

By continuously passing hydrogen gas through the reaction mixture, the hydrogen chloride that is formed in the reaction can be removed. Better yields (around 90%) may be obtained by adding a base to the reaction mixture in order to remove the hydrogen chloride.[7]

As catalyst for the Rosenmund reaction palladium on a support, e.g. palladium on barium sulfate, is most often used. The palladium has to be made less active in order to avoid further reduction of the aldehyde to the corresponding alcohol. Such a poisoned catalyst is obtained for example by the addition of quinoline and sulfur. Recent reports state that the reactivity of the catalyst is determined by the morphology of the palladium surface.[4,5]

A number of side-reactions may be observed with the Rosenmund reduction, which however can be avoided by proper reaction conditions. A poorly deactivated catalyst will lead to reduction of aldehyde **2** to the alcohol **4**, or even to the corresponding hydrocarbon **5**:

$$
\underset{\substack{\\ \mathbf{2}}}{\overset{O}{\underset{R}{\parallel}}} \quad \xrightarrow{H_2} \quad \underset{\mathbf{4}}{RCH_2OH} \quad \xrightarrow[-H_2O]{H_2} \quad \underset{\mathbf{5}}{RCH_3}
$$

Reaction of the acyl chloride **1** with alcohol **4** thus formed leads to formation of an ester **6**:

$$
\underset{\mathbf{1}}{R-C(=O)-Cl} \;+\; \underset{\mathbf{4}}{RCH_2OH} \;\longrightarrow\; \underset{\mathbf{6}}{R-C(=O)-OCH_2R}
$$

Small amounts of water present will lead to partial hydrolysis of the acyl chloride to give the carboxylic acid **7**, which then may further react with the acyl chloride to give a carboxylic anhydride **8**:

$$
\underset{\mathbf{1}}{R-C(=O)-Cl} \xrightarrow[-HCl]{H_2O} \underset{\mathbf{7}}{R-C(=O)-OH} \;+\; \underset{\mathbf{1}}{R-C(=O)-Cl} \xrightarrow{-HCl} \underset{\mathbf{8}}{R-C(=O)-O-C(=O)-R}
$$

The Rosenmund reduction is usually applied for the conversion of a carboxylic acid into the corresponding aldehyde *via* the acyl chloride. Alternatively a carboxylic acid may be reduced with lithium aluminum hydride to the alcohol, which in turn may then be oxidized to the aldehyde. Both routes require the preparation of an intermediate product; and each route may have its advantages over the other, depending on substrate structure.

1. M. Saytzeff, *J. Prakt. Chem.* **1873**, *6*, 128–135.
2. K. W. Rosenmund, *Ber. Dtsch. Chem. Ges.* **1918**, *51*, 585–593.
3. E. Mosettig, R. Mozingo, *Org. React.* **1948**, *4*, 362–377.
4. W. F. Maier, S. J. Chettle, R. S. Rai, G. Thomas, *J. Am. Chem. Soc.* **1986**, *108*, 2608–2616.
5. P. N. Rylander, H. Greenfield, R. L. Augustine, *Catalysis of Organic Reactions*, Marcel Dekker, New York, **1988**, p. 221–224.
6. O. Bayer, *Methoden Org. Chem. (Houben-Weyl)* **1954** Vol. 7/1, p. 285–291.
7. A. W. Burgstahler, L. O. Weigel, C. G. Shaefer, *Synthesis* **1976**, 767–768.

S

Sakurai Reaction

Conjugate addition of an allylsilane to an α,β-unsaturated ketone

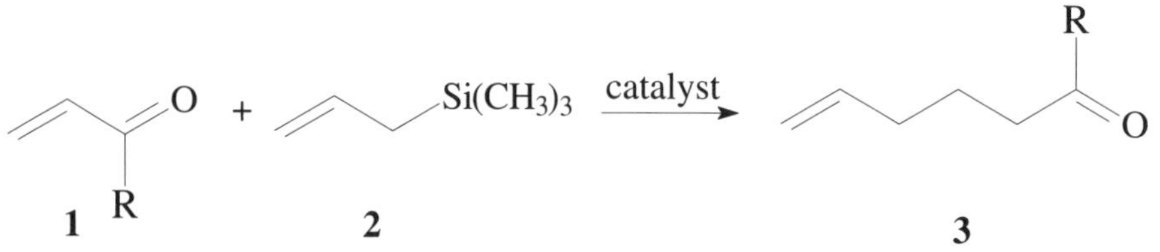

An allylsilane—e.g. allyltrimethylsilane **2**—can upon treatment with fluoride or in the presence of catalytic amounts of a Lewis acid undergo a nucleophilic conjugate addition to an α,β-unsaturated ketone **1**; as reaction product a δ,ε-unsaturated ketone **3** is obtained. This conjugate addition reaction is called the *Sakurai reaction*;[1,2] of particular interest from a synthetic point of view is the intramolecular variant.[3]

When a Lewis acid, e.g. titanium tetrachloride, coordinates to the carbonyl oxygen of an α,β-unsaturated carbonyl compound, the β-carbon center becomes more positively polarized. The allylsilane adds as a nucleophile with its γ-carbon to the β-carbon of the α,β-unsaturated carbonyl substrate.[4,5] This carbon–carbon single-bond forming step is the rate determining step. Cleavage of the trimethylsilyl group from the intermediate carbenium ion **5** leads to formation of a new carbon–carbon double bond. After hydrolytic workup the δ,ε-unsaturated ketone **3** is obtained:

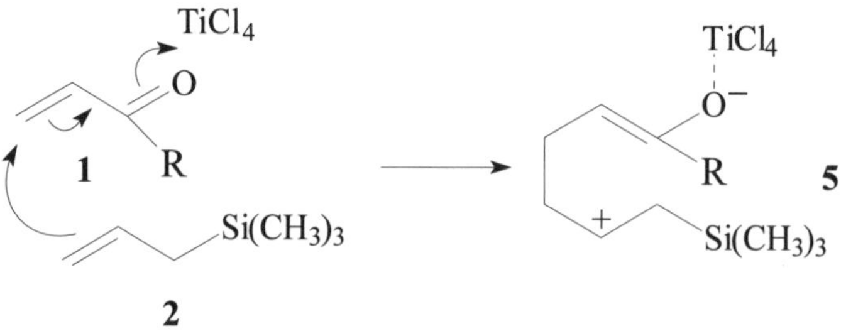

$$\xrightarrow{\text{H}_2\text{O}}$$

R

3

The intramolecular Sakurai reaction allows for the synthesis of functionalized bicyclic systems.[3] By proper choice of the reaction conditions, especially of the Lewis acid or fluoride reagent used, high stereoselectivity can be achieved, which is an important aspect for its applicability in natural products synthesis.

Propargylsilanes can also be employed in the Sakurai reaction. For example the enone **6**, containing a propargylsilane side chain, undergoes an intramolecular Sakurai reaction, catalyzed by an acidic ion-exchange resin—e.g. Amberlyst-15—to give stereoselectively the bicyclic product **7** in good yield:[6]

$$\xrightarrow{\text{Amberlyst-15}}$$

6

7

As Lewis acid, titanium tetrachloride, boron trifluoride or ethylaluminum dichloride is often used. The stereochemical outcome of the reaction strongly depends on the Lewis acid used. The Sakurai reaction is a relatively new carbon–carbon forming reaction, that has been developed into a useful tool for organic synthesis.[2,3,6]

1. A. Hosomi, H. Sakurai, *J. Am. Chem. Soc.* **1977**, *99*, 1673–1675.
    A. Hosomi, *Acc. Chem. Res.* **1988**, *21*, 200–206;
    H. Sakurai, *Synlett* **1989**, 1.
2. I. Fleming, J. Dunogues, R. Smithers, *Org. React.* **1989**, *37*, 57–575; for competitive reaction pathways, and reactions with allyltriisopropylsilane *see*: H.-J. Knölker, J. Prakt. *Chem.* **1997**, *339*, 304–314.
3. D. Schinzer, *Synthesis*, **1988**, 263–273;
    E. Langkopf, D. Schinzer, *Chem. Rev.* **1995**, *95*, 1375–1408.
4. T. A. Blumenkopf, C. H. Heathcock, *J. Am. Chem. Soc.* **1983**, *105*, 2354–2358.
5. R. Pardo, J.-P. Zahra, M. Santelli, *Tetrahedron Lett.* **1979**, *20*, 4557–4560.
6. D. Schinzer, J. Kabbara, K. Ringe, *Tetrahedron Lett.* **1992**, *33*, 8017–8018.

Sandmeyer Reaction

Conversion of arenediazonium salts into aryl halides

The name *Sandmeyer reaction*[1,2] is used for the replacement of the diazonium group in an arenediazonium compound by halide or pseudohalide, taking place in the presence of a metal salt.[3] However this is not a strict definition, since the replacement of the diazonium group by iodide, which is possible without a metal catalyst, is also called a Sandmeyer reaction.

The reaction mechanism is not rigorously known, but is likely to involve the following steps.[4-6] First the arenediazonium ion species **1** is reduced by a reaction with copper-(I) salt **2** to give an aryl radical species **4**. In a second step the aryl radical abstracts a halogen atom from the CuX_2 compound **5**, which is thus reduced to the copper-I salt **2**. Since the copper-(I) species is regenerated in the second step, it serves as a catalyst in the overall process.

For the *in situ* preparation of the required arenediazonium salt from an aryl amine by application of the *diazotization reaction*, an acid HX is used, that corresponds to the halo substituent X to be introduced onto the aromatic ring. Otherwise—e.g. when using HCl/CuBr—a mixture of aryl chloride and aryl bromide will be obtained. The copper-(I) salt **2** (chloride or bromide) is usually prepared by dissolving the appropriate sodium halide in an aqueous solution of copper-(II) sulfate and then adding sodium hydrogensulfite to reduce copper-(II) to copper-(I). Copper-(I) cyanide CuCN can be obtained by treatment of copper-(I) chloride with sodium cyanide.

The Sandmeyer reaction generally permits the introduction of electron-withdrawing substituents onto an aromatic ring. Arenediazonium salts, as well as the Sandmeyer products derived thereof, are useful intermediates for the synthesis of substituted aromatic compounds. For example an aromatic nitrile, that is accessible by reaction of an arenediazonium salt with copper-(I) cyanide, can be further converted into a carboxylic acid through hydrolysis, or reduced to give a benzylic amine, or reacted with an organometallic reagent to yield a ketone on hydrolytic workup. The Sandmeyer reaction may also be used to synthesize regioisomerically pure halotoluenes **9**:

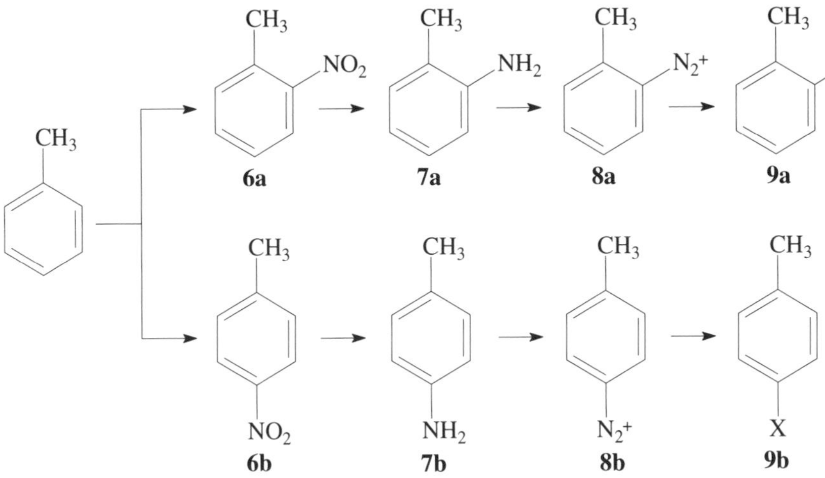

While the direct halogenation of toluene gives a mixture of isomers that is difficult to separate into the pure isomers, the isomeric *o*- and *p*-nitrotoluenes **6a** and **6b**, formed by nitration, are easy to separate from each other. Thus reduction of the single *o*- or *p*-nitrotoluene **6** to the *o*- or *p*-toluidine **7a** or **7b** respectively, followed by conversion into the corresponding diazonium salt **8** and a subsequent Sandmeyer reaction leads to the pure *o*- or *p*-halotoluene **9**.

1. T. Sandmeyer, *Ber. Dtsch. Chem. Ges.* **1884**, *17*, 1633–1635.
2. H. H. Hodgson, *Chem. Rev.* **1947**, *40*, 251–277.
3. E. Pfeil, *Angew. Chem.* **1953**, *65*, 155–158.
4. J. K. Kochi, *J. Am. Chem. Soc.* **1957**, *79*, 2942–2948.
5. C. Galli, *J. Chem. Soc., Perkin Trans. 2*, **1981**, 1461–1459.
6. C. Galli, *J. Chem. Soc., Perkin Trans. 2*, **1982**, 1139–1142.

Schiemann Reaction

Aryl fluorides from arenediazonium fluoroborates

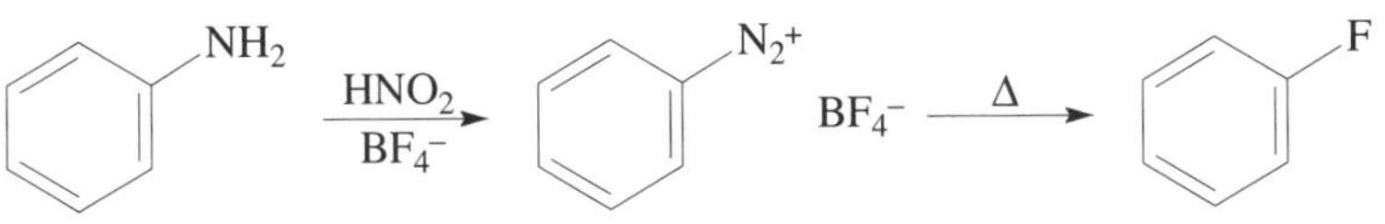

The preparation of an aryl fluoride—e.g. fluorobenzene **3**—starting from an aryl amine—e.g. aniline **1**—via an intermediate arenediazonium tetrafluoroborate **2**, is called the *Schiemann reaction* (also called the *Balz–Schiemann reaction*).[1,2] The *diazotization* of aniline **1** in the presence of tetrafluoroborate leads to formation of a benzenediazonium tetrafluoroborate **2** that can be converted into fluorobenzene **3** by thermolysis.

Treatment of aniline **1** with nitric acid in the presence of tetrafluoroboric acid leads to a relatively stable benzenediazonium tetrafluoroborate **2** by the usual diazotization mechanism. There are several variants for the experimental procedure.[3] Subsequent thermal decomposition generates an aryl cation species **4**, which reacts with fluoroborate anion to yield fluorobenzene **3**:[4]

In a general procedure the arenediazonium fluoroborate is isolated, and then heated without solvent. A modern variant permits the photochemical decomposition without initial isolation of the diazonium fluoroborate.[5]

The Schiemann reaction seems to be the best method for the selective introduction of a fluorine substituent onto an aromatic ring.[5] The reaction works with many aromatic amines, including condensed aromatic amines. It is however of limited synthetic importance, since the yield usually decreases with additional substituents present at the aromatic ring.

1. G. Balz, G. Schiemann, *Ber. Dtsch. Chem. Ges.* **1927**, *60*, 1186–1190.
2. A. Roe, *Org. React.* **1949**, *5*, 193–228.
3. M. P. Doyle, W. J. Bryker, *J. Org. Chem.* **1979**, *44*, 1572—1574.
4. C. G. Swain, R. J. Rogers, *J. Am. Chem. Soc.* **1975**, *97*, 799–800.
5. N. Yoneda, T. Fukuhara, T. Kikuchi, A. Suzuki, *Synth. Commun.* **1989**, *19*, 865–871.

Schmidt Reaction

Reaction of carboxylic acids, aldehydes or ketones with hydrazoic acid

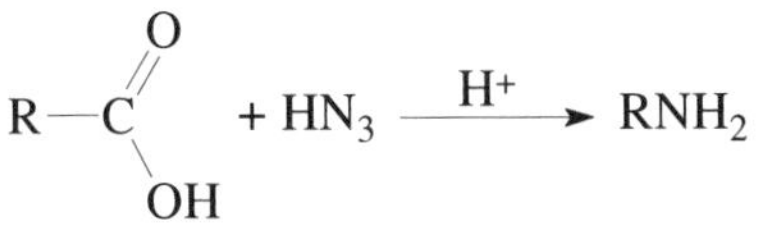

R—C(=O)OH + HN$_3$ $\xrightarrow{\text{H}^+}$ RNH$_2$

1   **2**

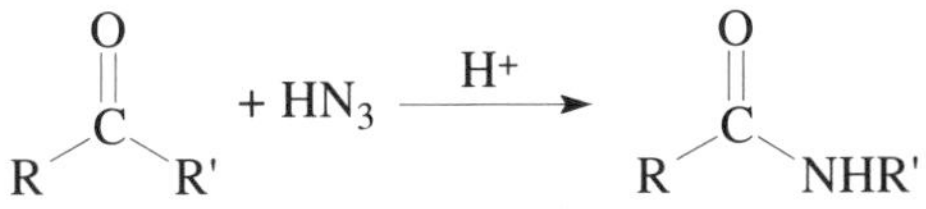

R—C(=O)—R' + HN$_3$ $\xrightarrow{\text{H}^+}$ R—C(=O)—NHR'

3   **4**

The reaction of carboxylic acids, aldehydes or ketones with hydrazoic acid in the presence of a strong acid is known as the *Schmidt reaction.*[1,2] A common application is the conversion of a carboxylic acid **1** into an amine **2** with concomitant chain degradation by one carbon atom. The reaction of hydrazoic acid with a ketone **3** does not lead to chain degradation, but rather to formation of an amide **4** by formal insertion of an NH-group.

For the different types of substrates, different reaction mechanisms are formulated.[3] For a carboxylic acid **1** as starting material the initial step is the protonation by a strong acid (most often sulfuric acid) and subsequent loss of water leading to formation of an acylium ion species **5**. Nucleophilic addition of hydrazoic acid to the acylium ion **5** gives an intermediate species **6** that further reacts by migration of the group R and concomitant loss of N$_2$ to the rearranged intermediate **7**. The latter reacts with water to give an unstable carbaminic acid that decomposes to the primary amine **2** and carbon dioxide:

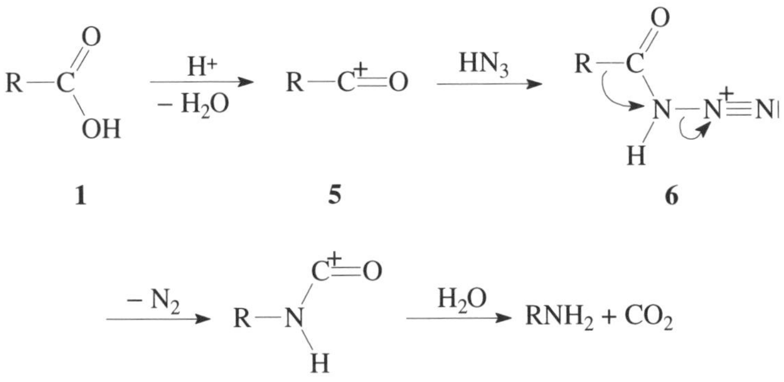

R—C(=O)OH $\xrightarrow[-\text{H}_2\text{O}]{\text{H}^+}$ R—C$^+$=O $\xrightarrow{\text{HN}_3}$ R—C(=O)—N(H)—N$^+$≡N|

1   **5**   **6**

$\xrightarrow{-\text{N}_2}$ R—N(H)—C$^+$=O $\xrightarrow{\text{H}_2\text{O}}$ RNH$_2$ + CO$_2$

7   **2**

The nucleophilic addition of hydrazoic acid to a ketone **3** to give **8** is promoted by the protonation of the carbonyl oxygen. Elimination of water from **8** leads to an intermediate **9**, from which a nitrilium ion intermediate **10** is formed by migration of group R and loss of N_2. Migration of R and cleavage of N_2 from **9** is likely to be a concerted process, since nitrenium ions have so far not been identified as intermediates with such a reaction. At this point of the reaction pathway there is a strong similarity with the *Beckmann rearrangement*, which proceeds by a similar rearrangement to give an intermediate nitrilium ion species **10**. Reaction of **10** with water, followed by a tautomerization, yields the stable amide **4**:

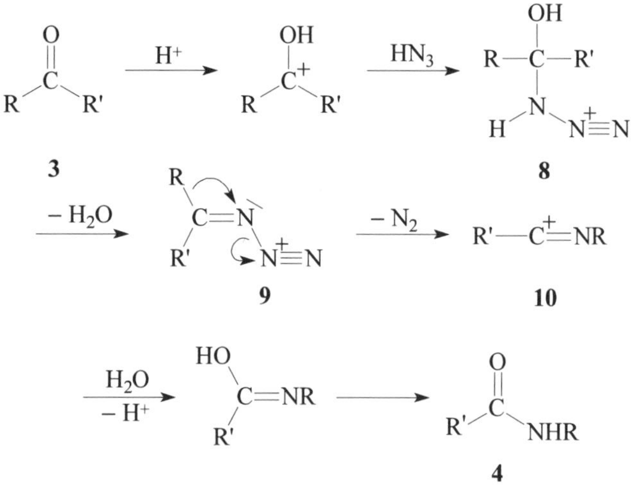

The cleavage of water from **8** usually leads to the isomer with the more voluminous substituent R *trans* to the diazonium group. It is that *trans*-substituent that will then migrate to the nitrogen-center in the rearrangement step; in case of an alkyl-aryl ketone as starting material, the aryl group usually will migrate.[4]

Aldehydes **11** react with hydrazoic acid to yield formamides **12**. The reaction pathway is similar to that formulated above for ketones. An important side-reaction however is the formation of nitriles **13**, resulting in a significant lower yield of amides, as compared to the reaction of ketones. Sometimes the nitrile **13** even is the major product formed:

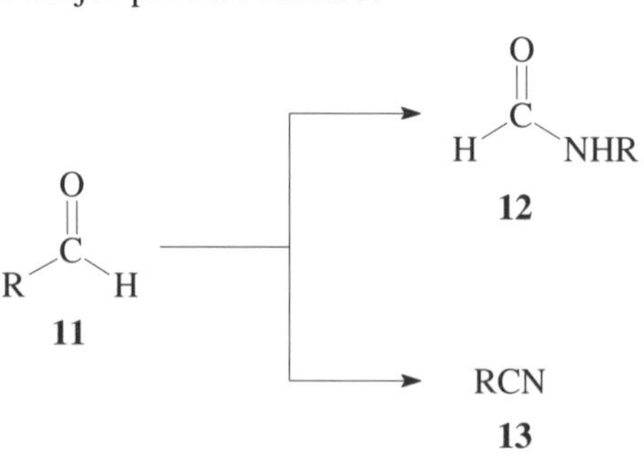

When applied to a cycloketone, the Schmidt reaction leads to formation of a ring-expanded lactam—e.g. **14** → **15**:[5]

14   **15**

In recent years the applicability of the Schmidt reaction for the synthesis of more complex molecules—especially the variant employing alkyl azides—has been further investigated. Cycloketones bearing an *azidoalkyl side-chain* at the α-carbon center have been shown to undergo, upon treatment with trifluoroacetic acid or titanium tetrachloride, an *intramolecular Schmidt reaction* to yield bicyclic lactams.[6] Intermolecular Schmidt reactions of *alkyl azides* and *hydroxyalkyl azides* with cycloketones in the presence of a Lewis acid, lead to formation of *N*-alkyl lactams and *N*-hydroxyalkyl lactams respectively in good yield.[7] The synthesis of chiral lactams by an *asymmetric Schmidt reaction* has also been reported.[8]

By application of the Schmidt reaction, the conversion of a carboxylic acid into an amine that has one carbon atom less than the carboxylic acid, can be achieved in one step. This may be of advantage when compared to the *Curtius reaction* or the *Hofmann rearrangement*; however the reaction conditions are more drastic. With long-chain, aliphatic carboxylic acids yields are generally good, while with aryl derivatives yields are often low.

The Schmidt reaction of ketones works best with aliphatic and alicyclic ketones; alkyl aryl ketones and diaryl ketones are considerably less reactive. The reaction is only seldom applied to aldehydes as starting materials. The hydrazoic acid used as reagent is usually prepared *in situ* by treatment of sodium azide with sulfuric acid. Hydrazoic acid is highly toxic, and can detonate upon contact with hot laboratory equipment.

1. K. F. Schmidt, *Angew. Chem.* **1923**, *36*, 511.
2. H. Wolff, *Org. React.* **1946**, *3*, 307–336.
3. G. I. Koldobskii, V. A. Ostrovskii, B. V. Gidaspov, *Russ. Chem. Rev.* **1978**, *47*, 1084–1094.
4. R. B. Bach, G. J. Wolker, *J. Org. Chem.* **1982**, *47*, 239–245.
5. G. R. Krow, *Tetrahedron* **1981**, *37*, 1283–1307.
6. J. Aubé, G. L. Milligan, *J. Am Chem. Soc.* **1991**, *113*, 8965–8966;
    G. L. Milligan, C. J. Mossman, J. Aubé, *J. Am. Chem. Soc.* **1995**, *117*, 10449–10459;
7. J. Aubé, G. L. Milligan, C. J. Mossman, *J. Org. Chem.* **1992**, *57*, 1635–1637;
8. V. Gracias, G. L. Milligan, J. Aubé, *J. Am. Chem. Soc.* **1995**, *117*, 8047–8048;
    V. Gracias, G. L. Milligan, J. Aubé, *J. Org. Chem.* **1996**, *61*, 10–11.

Sharpless Epoxidation

Asymmetric epoxidation of allylic alcohols

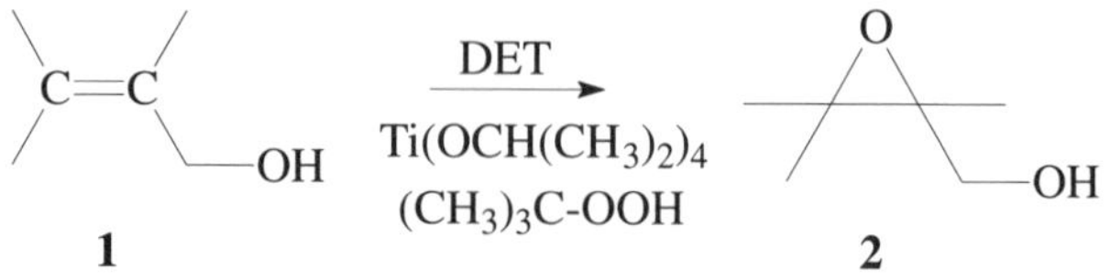

The asymmetric epoxidation of an allylic alcohol **1** to yield a 2,3-epoxy alcohol **2** with high enantiomeric excess, has been developed by *Sharpless* and *Katsuki*.[1] This enantioselective reaction is carried out in the presence of tetraisopropoxytitanium and an enantiomerically pure dialkyl tartrate—e.g. (+)- or (−)-diethyl tartrate (DET)—using *tert*-butyl hydroperoxide as the oxidizing agent.

With this epoxidation procedure it is possible to convert the achiral starting material—i.e. the allylic alcohol—with the aim of a chiral reagent, into a chiral, non-racemic product; in many cases an enantiomerically highly-enriched product is obtained. The desired enantiomer of the product epoxy alcohol can be obtained by using either the (+)- or (−)- enantiomer of diethyl tartrate as chiral auxiliary:[2–4]

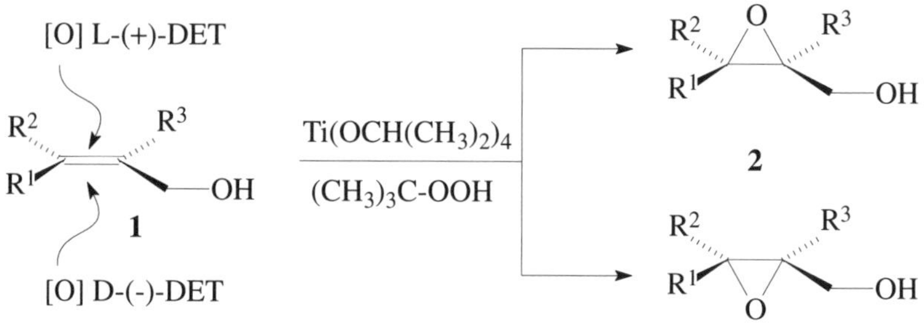

A model for the catalytically active species in the Sharpless epoxidation reaction is formulated as a dimer **3**, where two titanium centers are linked by two chiral tartrate bridges. At each titanium center two isopropoxide groups of the original tetraisopropoxytitanium-(IV) have been replaced by the chiral tartrate ligand:

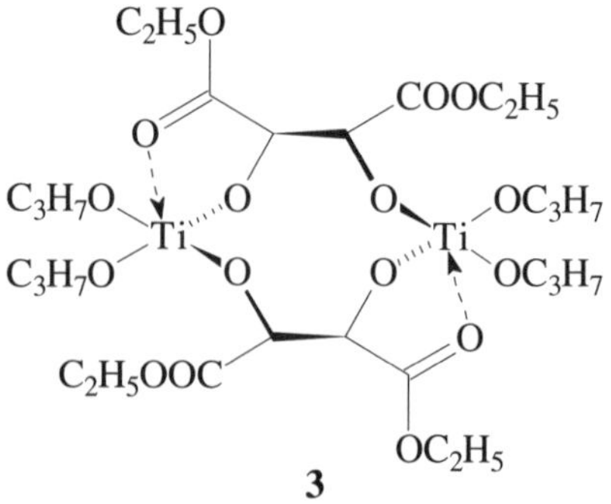

As the reaction proceeds, the two remaining isopropoxide groups at one titanium center are replaced by the allylic alcohol (the substrate) and *tert*-butyl hydroperoxide (the oxidizing agent) to give the complex **4**. Titanium-(IV) is suitable for such a reaction since it can form four covalent, but still reactive bonds, two bonds to bind the ends of two chiral bidentate tartrate ligands, one bond to the oxidizing agent, and one to the allylic alcohol substrate. The titanium thus serves as a template for the reactants; with the aim of the chiral ligands it has become a chiral template. The reactants are arranged geometrically in such a way to permit a facial selection, resulting in an enantioselective epoxidation step. Furthermore the tetraisopropoxytitanium-(IV) acts as a Lewis acid by coordinating to the other oxygen center—i.e. O-2 in the scheme below—of the *t*-butyl peroxy-ligand; as a result the oxygen center O-1 becomes more electrophilic. For the benefit of clarity the *bi*-centered titanium-tartrate moiety of the complex is shown simplified as Ti* in the following scheme:

The oxygen atom O-1 adds to the carbon–carbon double bond, while the oxygen O-2 forms a covalent bond to the titanium center. As a result complex **5** is formed, from which upon hydrolytic workup the epoxy alcohol and *t*-butanol are liberated.

   The reaction is limited to allylic alcohols; other types of alkenes do not or not efficiently enough bind to the titanium. The catalytically active chiral species can be regenerated by reaction with excess allylic alcohol and oxidant; however the titanium reagent is often employed in equimolar amount.

   In order to obtain good yields, it is important to use dry solvent and reagents. The commercially available *t*-butyl hydroperoxide contains about 30% water for stabilization. For the use in a Sharpless epoxidation reaction the water has to be removed first. The effect of water present in the reaction mixture has for example been investigated by *Sharpless et al.*[5] for the epoxidation of (*E*)-α-phenylcinnamyl alcohol, the addition of one equivalent of water led to a decrease in enantioselectivity from 99% e.e. to 48% e.e.

Titanium-IV compounds with their Lewis acid activity may catalyze an interfering rearrangement of the starting allylic alcohol or the epoxy alcohol formed. In order to avoid such side-reactions, the epoxidation is usually carried out at room temperature or below.

The Sharpless epoxidation is one of the most important of the newer organic reactions. Although limited to allylic alcohols, it has found wide application in natural product synthesis.

The 2,3-epoxy alcohols are often obtained in high optical purity (90% enantiomeric excess or higher), and are useful intermediates for further transformations. For example by nucleophilic ring opening the epoxide unit may be converted into an alcohol, a β-hydroxy ether or a vicinal diol.

1. T. Katsuki, K. B. Sharpless, *J. Am. Chem. Soc.* **1980**, *102*, 5974–5976;
    T. Katsuki, V. S. Martin, *Org. React.*, **1996**, *48*, 1–299.
2. A. Pfenninger, *Synthesis* **1986**, 89–116.
3. K. B. Sharpless, *Chemtech.* **1985**, *15*, 692–700.
4. D. Schinzer, *Nachr. Chem. Tech. Lab.* **1989**, *37*, 1294–1298.
5. J. G. Hill, B. E. Rossiter, K. B. Sharpless, *J. Org. Chem.* **1983**, *48*, 3607–3608.

Simmons–Smith Reaction

Cyclopropanes from alkenes

$$\text{1} \qquad \text{2} \qquad \text{3}$$

By application of the *Simmons–Smith reaction*[1-4] it is possible to synthesize a cyclopropane from an alkene by formal addition of carbene to the carbon–carbon double bond, without a free carbene being present in the reaction mixture; the usual side-reactions of free carbenes can thus be avoided. The cyclopropanation is carried out by treating the olefinic substrate **1** with diiodomethane **2** and zinc-copper couple.

By reaction of zinc–copper couple with diiodomethane **2** an organozinc species **4** is formed, similar to a Grignard reagent. Its structure cannot be fully described by a single structural formula. The actual structure depends on the reaction conditions—e.g. the solvent used; this corresponds to the *Schlenk equilibrium* as it is observed with the *Grignard reaction*:

$$2 \text{ CH}_2\text{I}_2 + 2 \text{ Zn} \longrightarrow 2 \text{ ICH}_2\text{ZnI} \rightleftharpoons (\text{ICH}_2)_2\text{Zn·ZnI}_2$$

$$\quad\quad \textbf{2} \quad\quad\quad\quad\quad\quad \textbf{4}$$

The addition reaction of the methylene to the carbon–carbon double bond is formulated as a one-step mechanism, where both new carbon-carbon bonds are formed simultaneously in a transition state of a structure like **5**:

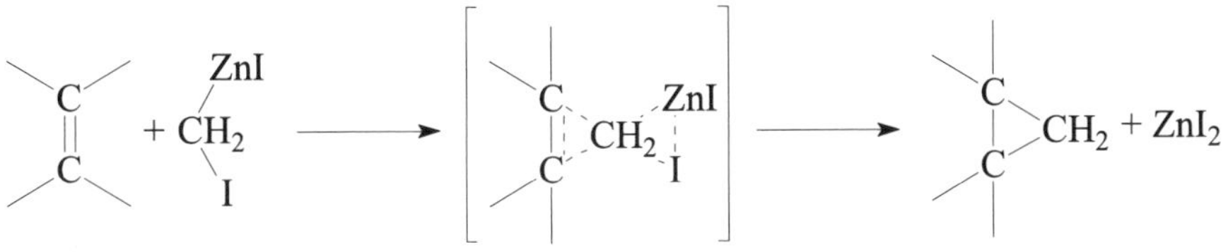

$$\textbf{1} \quad\quad \textbf{4} \quad\quad\quad\quad\quad \textbf{5} \quad\quad\quad\quad\quad \textbf{3}$$

The addition usually takes place from the sterically less hindered side of the alkene. The stereochemical course of the addition can be controlled by suitably positioned oxygen center that can coordinate to the organozinc reagent. For example the reaction with 4-hydroxycyclopentene **6** as substrate exclusively yields the *cis*-3-hydroxybicyclo [3.1.0] hexane **7**:

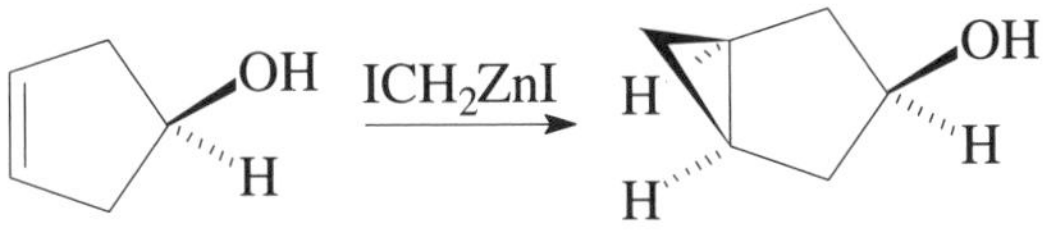

$$\textbf{6} \quad\quad\quad\quad\quad\quad\quad \textbf{7}$$

The Simmons–Smith reaction is well suited for the synthesis of spirocyclic compounds. It has for example been applied for the construction of the fifth cyclopropane ring in the last step of a synthesis of the rotane **8**:

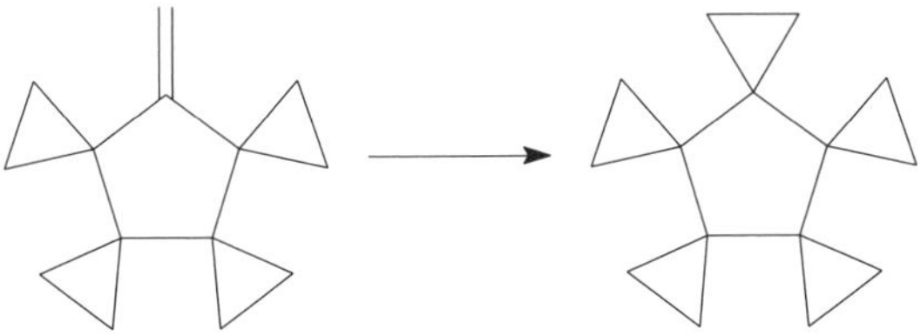

$$\textbf{8}$$

The Simmons–Smith cyclopropanation method has also found application for the α-methylation of ketones *via* an intermediate cyclopropane. The starting ketone—e.g. cyclohexanone **9**—is first converted into an enol ether **10**. Cyclopropanation of **10** leads to an alkoxynorcarane **11**, which on regioselective hydrolytic cleavage of the three-membered ring leads to the semiketal **12** as intermediate, and finally yields the α-methylated ketone **13**:

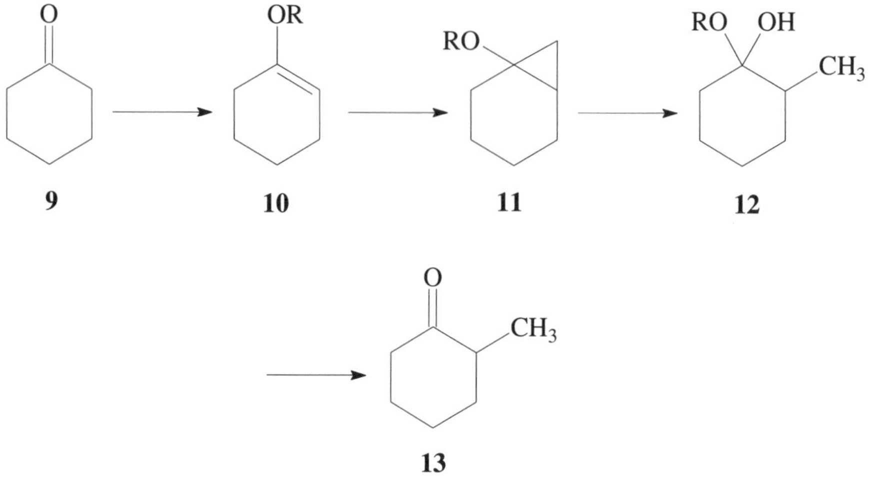

9   **10**   **11**   **12**

13

The zinc iodide formed in a Simmons–Smith reaction can act as Lewis acid, and thereby may catalyze rearrangement reactions; however interfering side-reactions are generally rare.

Yields are moderate to good. In addition to alkenes, the cyclopropanation also works with certain aromatic substrates.

1. H. E. Simmons, R. D. Smith, *J. Am. Chem. Soc.* **1959**, *81*, 4256–4264.
2. H. E. Simmons, T. L. Cairns, S. A. Vladuchick, C. M. Hoiness, *Org. React.* **1973**, *20*, 1–131.
3. J. Furukava, N. Kawabata, *Adv. Organomet. Chem.* **1974**, *12*, 83–134.
4. H. E. Simmons, E. P. Blanchard, R. D. Smith, *J. Am. Chem. Soc.* **1964**, *86*, 1347–1356.
5. J. L. Ripoll, J. M. Conia, *Tetrahedron Lett.* **1969**, 979–984.

Skraup Quinoline Synthesis

Quinolines by reaction of anilines with glycerol

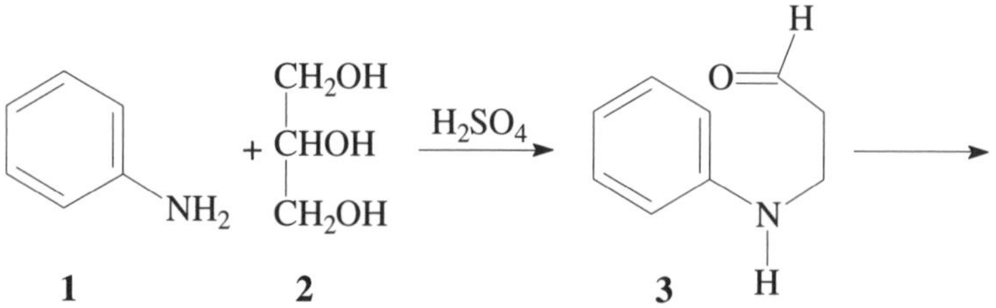

1   **2**   **3**   H

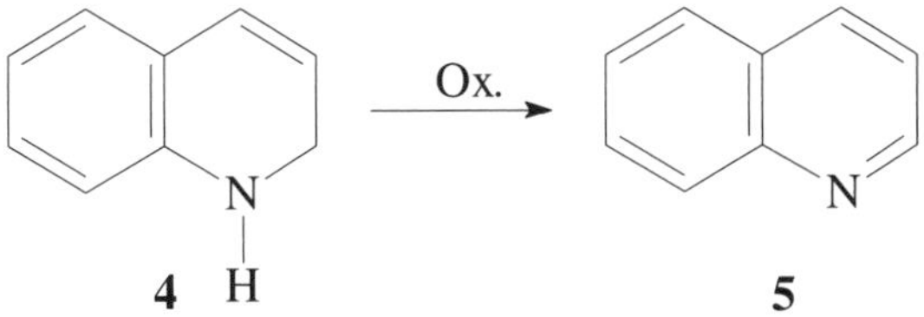

By reaction of a primary aromatic amine—e.g. aniline **1**—with glycerol **2**, and a subsequent oxidation of the intermediate product **4**, quinoline **5** or a quinoline derivative can be obtained.[1,2] As in the case of the related *Friedländer quinoline synthesis*, there are also some variants known for the *Skraup synthesis*, where the quinoline skeleton is constructed in similar ways using different starting materials.[3]

For the Skraup synthesis, glycerol **2** is used as starting material; in the presence of concentrated sulfuric acid (see scheme above) it is dehydrated to acrolein **6**. Although it is assumed that the reactive carbonyl component in the Skraup reaction actually is acrolein, attempts to use acrolein directly, instead of glycerol, proved to be unsuccessful.[4]

The formation of the quinoline is formulated to involve a conjugate addition of the primary aromatic amine to the acrolein **6**, to give a β-arylaminoaldehyde **3** as an intermediate:

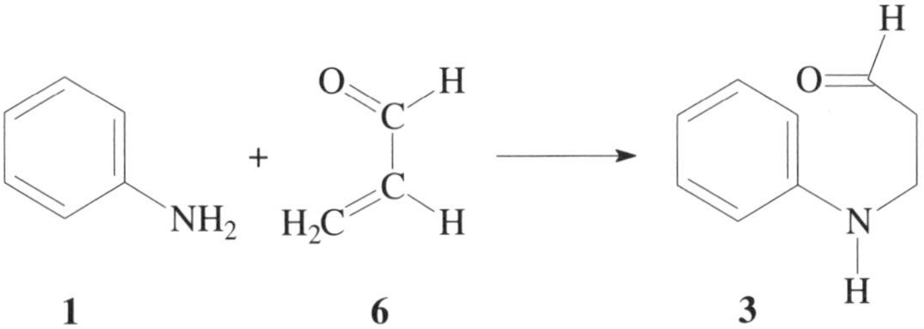

The β-arylaminoaldehyde **3** undergoes a ring-closure reaction with subsequent loss of water to give a dihydroquinoline **4**:

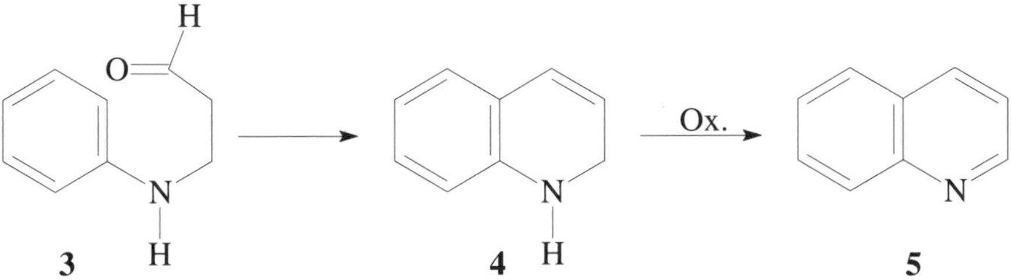

The final product—quinoline **5**—is formed by an oxidation of the dihydroquinoline **4**. As oxidant the aromatic nitro compound **7**, that corresponds to the aromatic amine **1**, can be employed. The aromatic nitro compound dehydrogenates the dihydroquinoline **4** to quinoline **5**, and in turn is reduced to the primary aromatic amine, which is then available as additional starting amine for reaction with excess glycerol:

Since the corresponding nitro derivative is not always available, other oxidants have also found application—e.g. arsenic pentoxide.

The Skraup reaction is of wide scope for the synthesis of substituted quinolines.[3] Certain primary amines, bearing a cyano, acetyl or methyl group, may however be subject to decomposition under the usual reaction conditions.

Quinolines substituted at the pyridine ring may be obtained by using a substituted α,β-unsaturated aldehyde or ketone instead of the glycerol as starting material. However often a large amount of the carbonyl component polymerizes under the reaction conditions.

1. Z. H. Skraup, *Ber. Dtsch. Chem. Ges.* **1880**, *13*, 2086–2087.
2. G. Jones, *Chem. Heterocycl. Compd.* **1977**, 32(1), 100–117.
3. R. H. F. Manske, M. Kulka, *Org. React.* **1953**, 7, 59–98.
4. B. C. Uff in *Comprehensive Heterocyclic Chemistry Vol. 2* (Eds. A. R. Katritzky, C. W. Rees), Pergamon, Oxford, **1984**, p. 465–470.

Stevens Rearrangement

Tertiary amines from quaternary ammonium salts by migration of an alkyl group

A quaternary ammonium species **1**, bearing an electron-withdrawing group Z α to the nitrogen center, can rearrange to a tertiary amine **3**, when treated with a strong base. This reaction is known as the *Stevens rearrangement*.[1,2]

Mechanistically the rearrangement is formulated to proceed *via* an intermediate radical-pair or ion-pair.[3] In either case the initial step is the formation of a nitrogen-ylide **2** by deprotonation of the ammonium species with a strong base. The abstraction of a proton from the α-carbon is facilitated by an electron-withdrawing group Z—e.g. an ester, keto or phenyl group:

$$\text{Z}-\overset{\overset{\text{H}}{|}}{\underset{|}{\text{C}}}-\overset{\overset{\text{R}}{|}}{\underset{|}{\text{N}^+}} \quad \xrightarrow{\text{base}} \quad \text{Z}-\overset{\overset{-}{\underset{|}{\text{C}}}}{}-\overset{\overset{\text{R}}{|}}{\underset{|}{\text{N}^+}}$$

1   **2**

Following the radical pathway[4,5] the next step is a homolytical cleavage of the N—R bond. The rearrangement to yield the tertiary amine **3** then proceeds *via* an intermediate radical-pair **4a**. The order of migration is propargyl > allyl > benzyl > alkyl:[2]

2   **4a**

3

Although the radical-pair is largely held together by the solvent cage, small amounts of the intermolecular coupling product R—R can be isolated sometimes.[5,6]

In certain cases,[7] e.g. with Z = *tert*-butyl, the experimental findings may better be rationalized by an ion-pair mechanism rather than a radical-pair mechanism. A heterolytic cleavage of the N—R bond will lead to the ion-pair **4b**, held together in a solvent cage:

2   **4b**   **3**

From a synthetic point of view the Stevens rearrangement is of minor importance. With Z being an ester or acyl group, an alkoxide will suffice as base. Usually a stronger base, such as sodium amide or an organolithium compound, is employed. In the latter case the use of a two-phase system may be necessary, since the quaternary ammonium salt may not be soluble in the solvent used for the organolithium reagent. The ammonium salts are for example soluble in liquid ammonia, dimethyl sulfoxide or hexamethylphosphoric triamide; however the use of those solvents may also give rise to side-reactions.

1. T. S. Stevens, E. M. Creighton, A. B. Gordon, M. Mac Nicol, *J. Chem. Soc.* **1928**, 3193–3197.
2. S. H. Pine, *Org. React.* **1970**, *18*, 403–464.
3. S. H. Pine, *J. Chem. Educ.* **1971**, *48*, 99–102.
4. U. Schöllkopf, U. Ludwig, *Chem. Ber.* **1968**, *101*, 2224–2230.
5. U. Schöllkopf, U. Ludwig, G. Ostermann, M. Patsch, *Tetrahedron Lett.* **1969**, 3415–3418.
6. G. F. Hennion, M. J. Shoemaker, *J. Am. Chem. Soc.* **1970**, *92*, 1769–1770.
7. S. H. Pine, B. A. Catto, F. G. Yamagishi, *J. Org. Chem.* **1970**, *35*, 3663.

Stork Enamine Reaction

Alkylation and acylation of enamines

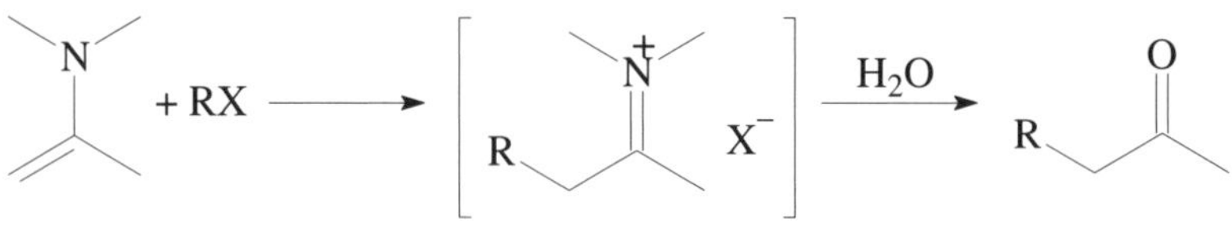

Enamines **1** are useful intermediates in organic synthesis. Their use for the synthesis of α-substituted aldehydes or ketones **3** by reaction with an electrophilic reactant—e.g. an alkyl halide **2**, an acyl halide or an acceptor-substituted alkene—is named after *Gilbert Stork*.[1-3]

The typical reactivity of an enamine **1** results from the nucleophilic character of the β-carbon center, as indicated by a resonance structure:

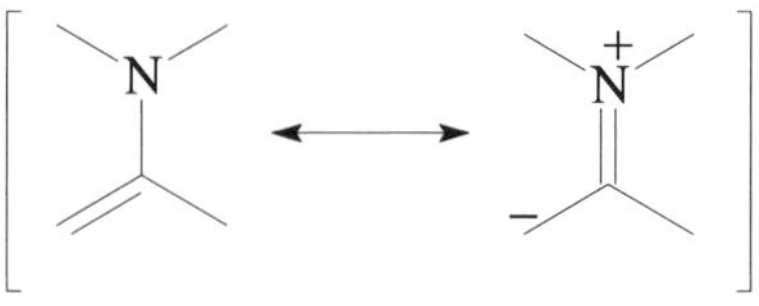

An enamine is easily prepared by reaction of the corresponding aldehyde or ketone **4** and a secondary amine **5**. A cyclic secondary amine like pyrrolidine, piperidine or morpholine is most often used. A general procedure has been reported by *Mannich* and *Davidsen*[4] in 1936:

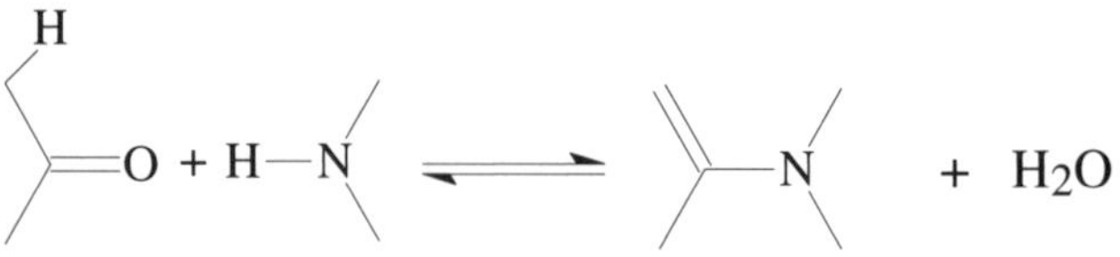

In order to shift the equilibrium, the water formed in that reaction is usually removed by azeotropic distillation with benzene or toluene.

In general the *Stork reaction* gives moderate yields with simple alkyl halides; better yields of alkylated product are obtained with more electrophilic reactants such like allylic, benzylic or propargylic halides or an α-halo ether, α-halo ester or α-halo ketone. An example is the reaction of 1-pyrrolidino-1-cyclohexene **6** with allyl bromide, followed by aqueous acidic workup, to yield 2-allylcyclohexanone:

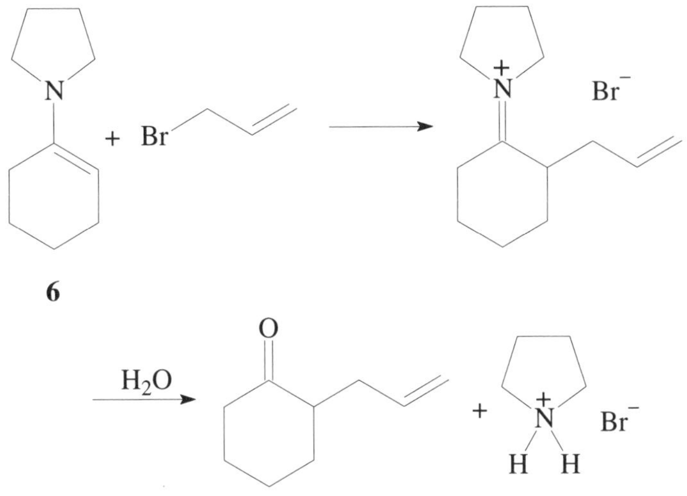

6

Generally the desired substituted carbonyl compound **3** is obtained after hydrolytic workup under acidic conditions. With simple alkyl halides an irreversible *N*-alkylation may take place as a side-reaction to give a quaternary ammonium salt **7**:

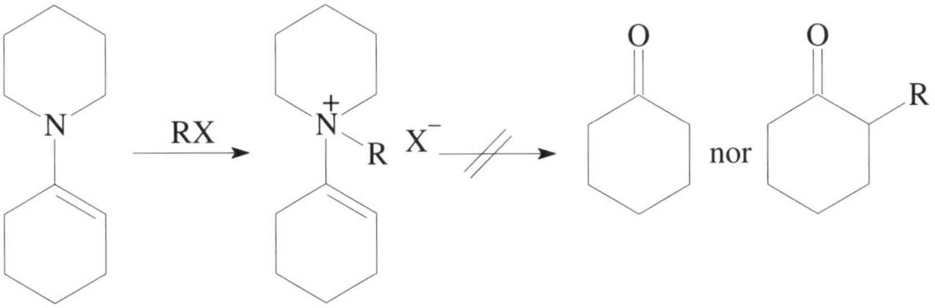

7

Enamines react with acceptor-substituted alkenes (Michael acceptors) in a conjugate addition reaction; for example with α,β-unsaturated carbonyl compounds or nitriles such as acrylonitrile **8**. With respect to the acceptor-substituted alkene the reaction is similar to a *Michael addition*:

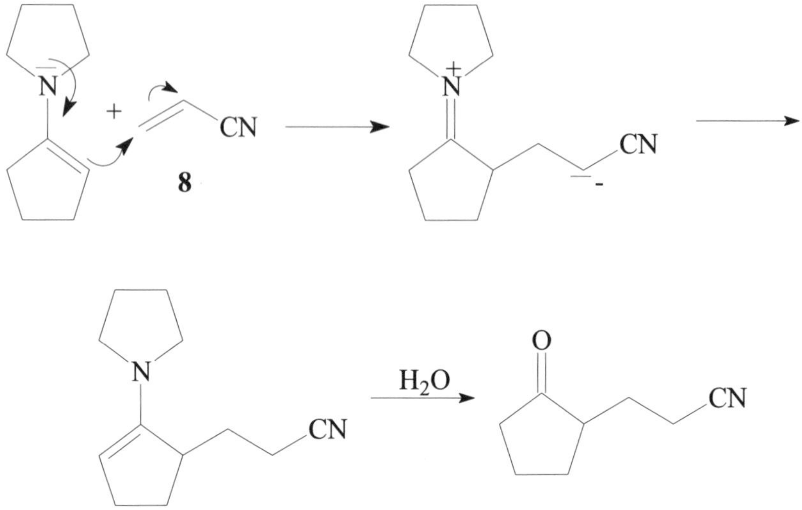

This type of reaction usually gives good yields; here the possible *N*-alkylation is reversible—through a *retro-Michael-type reaction*:

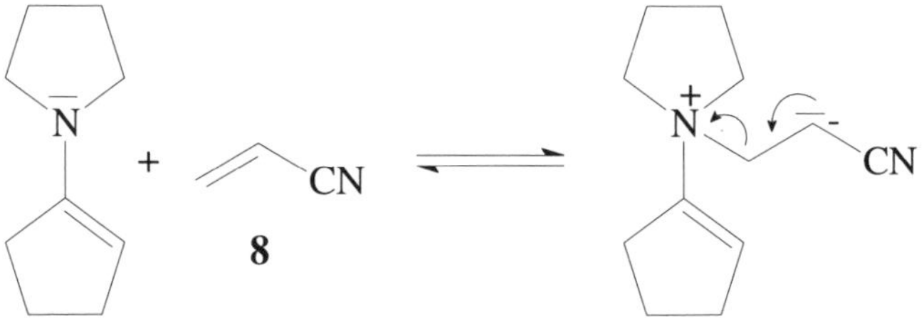

Another important application is the acylation of enamines **1** with an acyl chloride **9** to give a 1,3-dicarbonyl compound as final product:

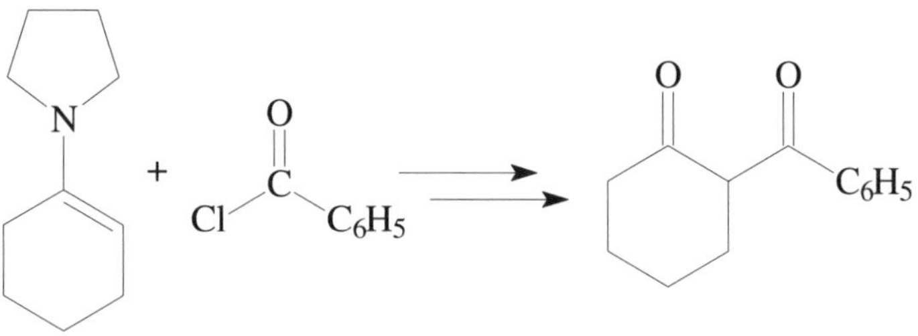

The Stork enamine reaction is an important and versatile method for the synthesis of α-substituted aldehydes and ketones. Such products should in principle also be available by reaction of the aldehyde or ketone with the electrophilic reactant in the presence of a base that is strong enough to convert the carbonyl substrate to the corresponding enolate. However this direct method often leads to formation of undesired products from various side-reactions. For example polysubstituted products are often obtained, as well as products from self-condensation of the starting aldehyde or ketone (see *Aldol reaction*). The latter is especially the case with cyclopentanone. Furthermore Michael-acceptor substrates often polymerize in the presence of base. In general the regiochemical outcome of the alkylation reaction is easier to control by using the enamine method.

1. G. Stork, R. Terrell, J. Szmuszkovicz, *J. Am. Chem. Soc.* **1954**, *76*, 2029–2030.
2. G. Stork, A. Brizzolara, H. Landesman, J. Szmuszkovicz, R. Trebell, *J. Am. Chem. Soc.* **1963**, *85*, 207–222.
3. J. K. Whitesell, M. A. Whitesell, *Synthesis* **1983**, 510–536.
4. C. Mannich, H. Davidsen, *Ber. Dtsch. Chem. Ges.* **1936**, *69*, 2106–2112.

Strecker Synthesis

α-Amino acids from aldehydes or ketones

An α-amino acid **3** can be prepared by treating aldehyde **1** with ammonia and hydrogen cyanide and a subsequent hydrolysis of the intermediate α-amino nitrile **2**. This so-called *Strecker synthesis*[1,2] is a special case of the *Mannich reaction*; it has found application for the synthesis of α-amino acids on an industrial scale. The reaction also works with ketones to yield α,α-disubstituted α-amino acids.

The formation of α-amino nitrile **2** is likely to proceed *via* a cyanohydrin **4** (an α-hydroxy nitrile) as intermediate, which is formed by the addition of hydrogen cyanide to the aldehyde **1**:

Reaction of cyanohydrin **4** with ammonia leads to formation of α-amino nitrile **2**, which can easily be hydrolyzed to give the corresponding α-amino acid **3**:

$$
\underset{\textbf{4}}{\underset{\underset{OH}{|}}{\overset{\overset{CN}{|}}{R-C-H}}} \xrightarrow{NH_3} \underset{\textbf{2}}{\underset{\underset{NH_2}{|}}{\overset{\overset{CN}{|}}{R-C-H}}} \xrightarrow{\hspace{1cm}} \underset{\textbf{3}}{\underset{\underset{NH_2}{|}}{\overset{\overset{COOH}{|}}{R-C-H}}}
$$

Alternatively a Mannich-like pathway may be followed (see *Mannich reaction*), where ammonia reacts with the aldehyde **1** to give an intermediate iminium species, that adds hydrogen cyanide to give the α-amino nitrile **2**. The actual mechanistic pathway followed depends on substrate structure and reaction conditions.

The scope of the reaction depends on the availability of the starting aldehyde (or ketone). A drawback is the toxicity of the hydrogen cyanide used as reactant.[2]

A variant of the Strecker synthesis is the *Bucherer–Bergs reaction*;[2] it gives better yields, and proceeds *via* formation of an intermediate hydantoin **5**:

$$
\underset{\textbf{2}}{\underset{\underset{NH_2}{|}}{\overset{}{R-CH-CN}}} + CO_2 \xrightarrow{\hspace{1cm}} \underset{\textbf{5}}{\underset{\underset{O}{\|}}{\overset{\overset{R}{\diagdown}\overset{}{\underset{CH}{}}\overset{O}{\diagup}}{HN\diagdown\diagup NH}}} \xrightarrow{\hspace{1cm}} \underset{\textbf{3}}{\underset{\underset{NH_2}{|}}{\overset{}{R-CH-COOH}}}
$$

$$
+ NH_3 + CO_2
$$

The importance of chemical syntheses of α-amino acids on industrial scale is limited by the fact that the standard procedure always yields the racemic mixture (except for the achiral glycine H_2N-CH_2-COOH and the corresponding amino acid from symmetrical ketones $R-CO-R$). A subsequent separation of the enantiomers then is a major cost factor. Various methods for the asymmetric synthesis of α-amino acids on laboratory scale have been developed, and among these are asymmetric Strecker syntheses as well.[3]

1. A. Strecker, *Justus Liebigs Ann. Chem.* **1850**, *75*, 27–45.
2. Th. Wieland, R. Müller, E. Niemann, L. Birkhofer, A. Schöberl, A. Wagner, H. Söll, *Methoden Org. Chem. (Houben-Weyl)*, **1959**, Vol. XI/2, p. 305–306.
3. H.-J-Altenbach In: J. Mulzer, H.-J. Altenbach, M. Braun, K. Krohn, H.-U. Reissig, *Organic Synthesis Highlights*, VCH, Weinheim, **1991**, p. 300–305.

$$\mathbf{T}$$

Tiffeneau–Demjanov Reaction

Ring enlargement of cyclic β-amino alcohols

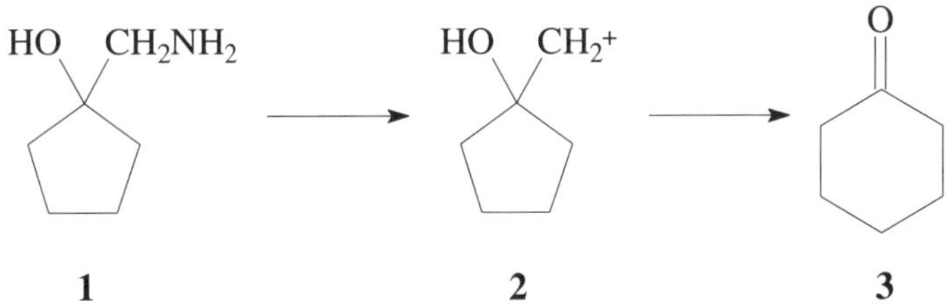

When a cyclic β-amino alcohol—e.g. **1**—is treated with nitrous acid, a deamination reaction can take place, to give a carbenium ion species **2**, which in turn can undergo a rearrangement and subsequent loss of a proton to yield a ring-enlarged cyclic ketone **3**. This reaction is called the *Tiffeneau–Demjanov reaction*;[1-3] it is of wider scope than the original *Demjanov reaction*.[2]

The reaction of nitrous acid with the amino group of the β-amino alcohol—e.g. 1-aminomethyl-cyclopentanol **1**—leads to formation of the nitrosamine **4**, from which, through protonation and subsequent loss of water, a diazonium ion species **5** is formed[2,4]—similar to a *diazotization reaction*:

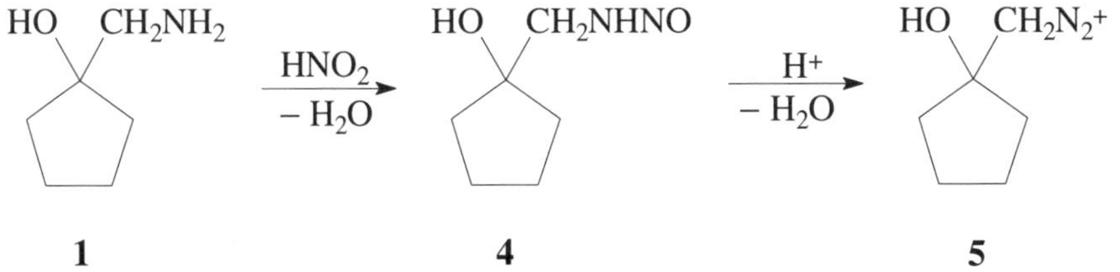

With diazonium species like **5**, dinitrogen (N_2) functions as a good leaving group. Loss of N_2 from **5** generates a carbenium ion species **2**, which rearranges by a 1,2-migration of a ring-CH_2 group to give the more stable hydroxy-substituted carbenium ion **6**. Loss of a proton from **6** yields a ring-enlarged ketone—e.g. cyclohexanone **3**—as the final product:

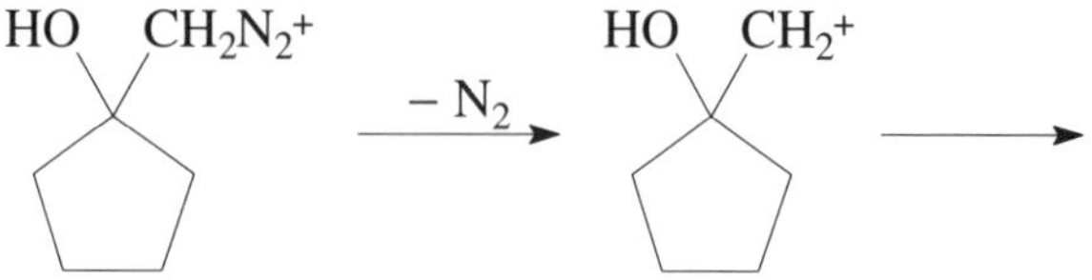

HO   CH$_2$N$_2$$^+$ HO   CH$_2$$^+$

$\xrightarrow{-\ N_2}$

5   **2**

OH   O

$\xrightarrow{-\ H^+}$

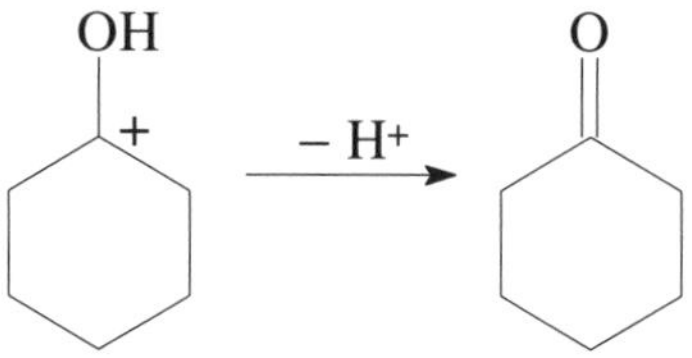

6   **3**

The starting material for the Tiffeneau–Demjanov reaction is available by various methods.[3] A common route is the addition of nitromethane to a cyclic ketone—e.g. cyclopentanone **7**—followed by a hydrogenation of the nitro group to give the β-amino alcohol, e.g. **1**:

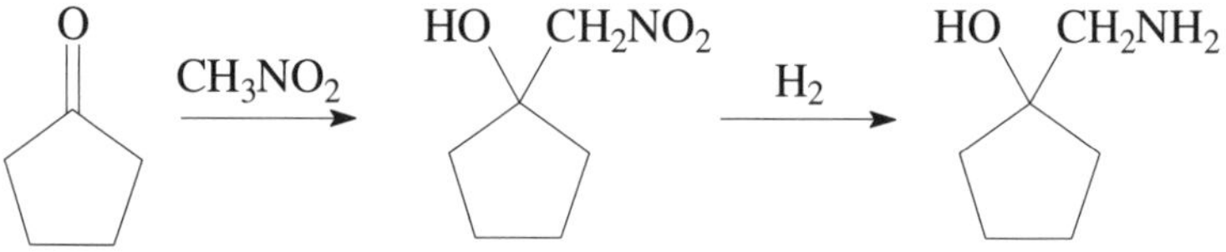

O   HO   CH$_2$NO$_2$ HO   CH$_2$NH$_2$

$\xrightarrow{CH_3NO_2}$ $\xrightarrow{H_2}$

7   **1**

The original *Demjanov reaction* is the conversion of an aminomethyl-cycloalkane into a cycloalkanol consisting of a carbocyclic ring that is expanded by one carbon center; e.g. the reaction of aminomethylcyclohexane **8** with nitrous acid leads to formation of cycloheptanol **9**:

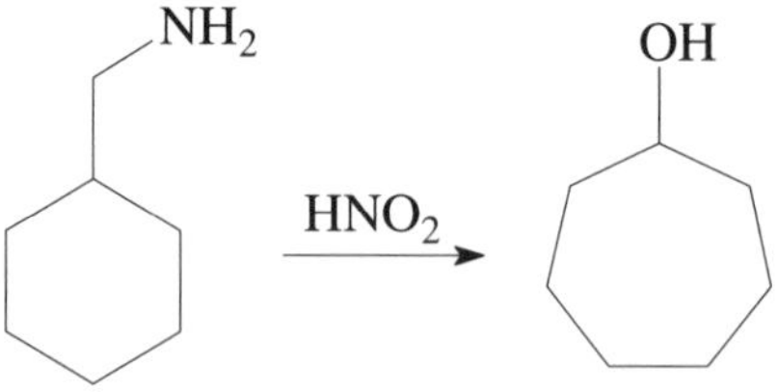

NH$_2$ OH

$\xrightarrow{HNO_2}$

8   **9**

Various side-reactions may be observed with the Demjanov reaction; the Tiffeneau–Demjanov reaction usually gives better yields of the ring-enlarged product.

The Tiffeneau–Demjanov reaction has found application for the construction of four- to nine-membered rings by a ring-enlargement route.[2,5] The presence of a heteroatom such as nitrogen or sulfur within the ring usually does not interfere with the reaction.[2] If the carbon center α to the amino group bears an additional substituent, yields may be low. A ring-enlargement may then even not take place at all, since the carbenium ion species is stabilized by that additional substituent. With starting materials bearing a substituent on the ring, mixtures of isomeric rearrangement products may be obtained; the synthetic importance is rather limited in those cases.

1. M. Tiffeneau, P. Weill, B. Tchoubar, *C. R. Acad. Sci.* **1937**, *205*, 144–146.
2. P. A. S. Smith, D. R. Baer, *Org. React.* **1960**, *11*, 157–188.
3. M. Hesse, *Ring Enlargement in Organic Chemistry*, VCH, Weinheim, **1991**, p. 9–10.
4. H. Stach, M. Hesse, *Tetrahedron* **1988**, *44*, 1573–1590.
5. H. N. C. Wong, M.-Y. Hon, C.-W. Tse, Y.-C. Yip, *Chem. Rev.* **1989**, *89*, 165–198.
6. M. A. McKinney, P. P. Patel, *J. Org. Chem.* **1973**, *38*, 4059–4067.

V

Vilsmeier Reaction

Formylation of aromatic compounds and of alkenes

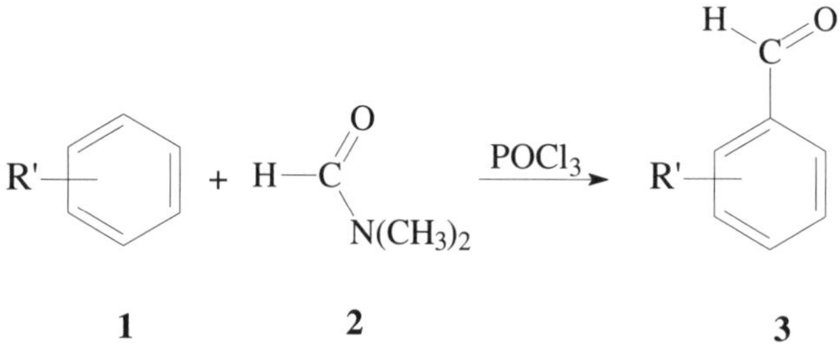

$$\mathbf{1} \qquad\qquad \mathbf{2} \qquad\qquad\qquad\qquad \mathbf{3}$$

The reaction of electron-rich aromatic compounds with N,N-dimethylformamide **2** and phosphorus oxychloride to yield an aromatic aldehyde—e.g. **3** from the substituted benzene **1**—is called the *Vilsmeier reaction*[1–3] or sometimes the *Vilsmeier–Haack reaction*. It belongs to a class of formylation reactions that are each of limited scope (see also *Gattermann reaction*).

In an initial step the reactive formylating agent is formed from N,N-dimethylformamide (DMF) **2** and phosphorus oxychloride. Other N,N-disubstituted formamides have also found application; for example N-methyl-N-phenylformamide is often used. The formylating agent is likely to be a chloromethyl iminium salt **4**—also called the *Vilsmeier complex*[4] (however its actual structure is not rigorously known)—that acts as the electrophile in an electrophilic substitution reaction with the aromatic substrate[5] **1** (see also *Friedel–Crafts acylation reaction*):

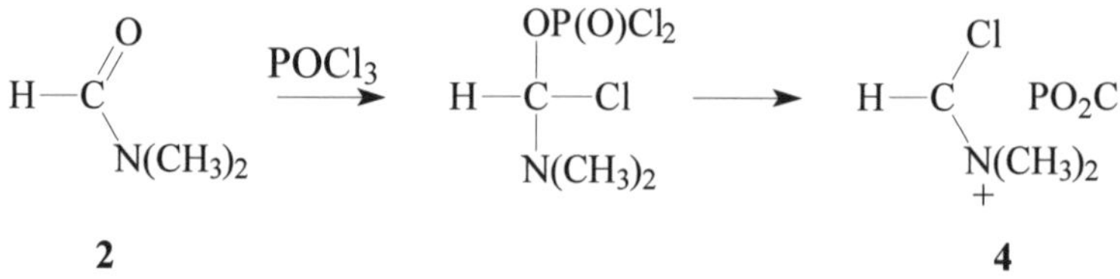

$$\mathbf{2} \qquad\qquad\qquad\qquad\qquad\qquad\qquad \mathbf{4}$$

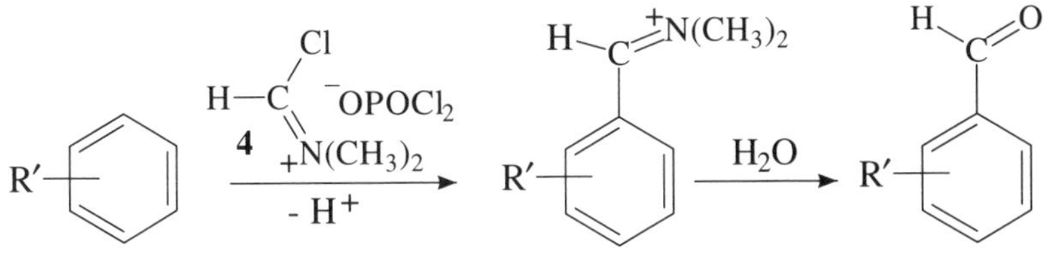

The initial product **5** of the electrophilic aromatic substitution step is unstable and easily hydrolyzes to yield the aromatic aldehyde **3** as the final reaction product. With *mono*-substituted aromatic substrates the *para*-substituted aldehyde is formed preferentially.

With respect to aromatic substrates, the Vilsmeier formylation reaction works well with electron-rich derivatives like phenols, aromatic amines and aromatic heterocycles like furans, pyrroles and indoles. However various alkenes are also formylated under Vilsmeier conditions. For example the substituted hexatriene **6** is converted to the terminal hexatrienyl aldehyde **7** in 70% yield:[6]

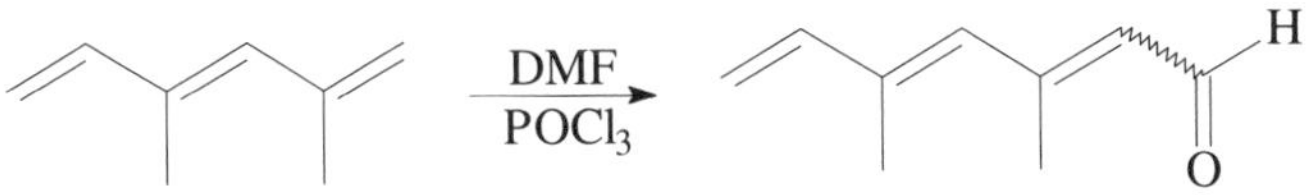

An elegant application of the Vilsmeier reaction is the synthesis of substituted biphenyls as reported by *Rao* and *Rao*.[7] Starting with homoallylic alcohol **8**, the biphenyl derivative **9** was obtained from a one-pot reaction in 80% yield:

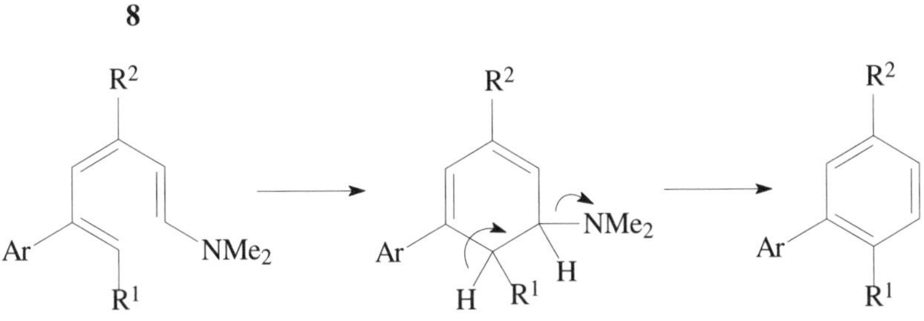

Although limited to electron-rich aromatic compounds and alkenes, the Vilsmeier reaction is an important formylation method. When N,N-dimethylformamide is used in excess, the use of an additional solvent is not necessary. In other cases toluene, dichlorobenzene or a chlorinated aliphatic hydrocarbon is used as solvent.[8]

1. A. Vilsmeier, A. Haack, *Ber. Dtsch. Chem. Ges.* **1927**, *60*, 119–122.
2. C. Jutz, *Adv. Org. Chem.* **1976**, *9*, Vol. 1, 225–342.
3. S. S. Pizey, *Synthetic Reagents* **1974**, Vol. 1, p. 54–71.
4. J. C. Tebby, S. E. Willetts, *Phosphorus Sulfur* **1987**, *30*, 293–296.
5. G. Jugie, J. A. S. Smith, G. J. Martin, *J. Chem. Soc., Perkin Trans. 2*, **1975**, 925–927.
6. P. C. Traas, H. J. Takken, H. Boelens, *Tetrahedron Lett.* **1977**, 2129–2132.
7. M. S. C. Rao, G. S. K. Rao, *Synthesis* **1987**, 231–233.
8. G. Simchen, *Methoden Org. Chem. (Houben-Weyl)*, **1983**, Vol. E3, p. 36–85.

Vinylcyclopropane Rearrangement

Cyclopentenes by rearrangement of vinylcyclopropanes

1   **2**

The thermal rearrangement of vinylcyclopropanes **1** to yield cyclopentenes **2** is called the *vinylcyclopropane rearrangement*.[1–3]

For the mechanistic course of that reaction two pathways are discussed:[2,4] a concerted [1,3]-sigmatropic rearrangement, and a pathway *via* an intermediate diradical species.[5] Experimental findings suggest that both pathways are possible. The actual pathway followed strongly depends on substrate structure; the diradical pathway appears to be the more important.

The direction of ring opening by homolytic cleavage of a cyclopropane bond is controlled by the stability of the diradical species formed. Upon heating of the mono-deuterated vinylcyclopropane **3**, a mixture of the two isomeric mono-deuterated cyclopentenes **4** and **5** is formed:

In addition to cyclopentenes, other types of compounds may be formed upon heating of vinylcyclopropanes. For example pentadienes **6a/b** may be formed by a competitive route from a diradical intermediate.

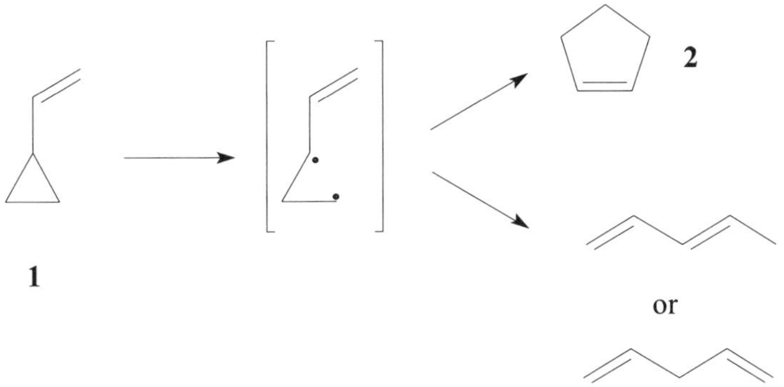

With a substitution pattern as found in 1-vinyl-2-methylcyclopropane **7**, a retro-ene reaction (see *Ene reaction*) may take place to yield hexa-1,4-diene **8**:

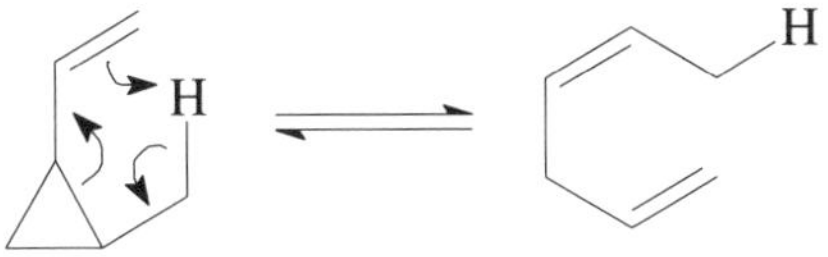

An illustrative example for the generation of cyclopentenes from vinyl-cyclopropanes is the formation of bicyclo[3.3.0]oct-1-ene **10** from 1,1-dicyclopropylethene **9** by two consecutive vinylcyclopropane → cyclopentene rearrangements.[6]

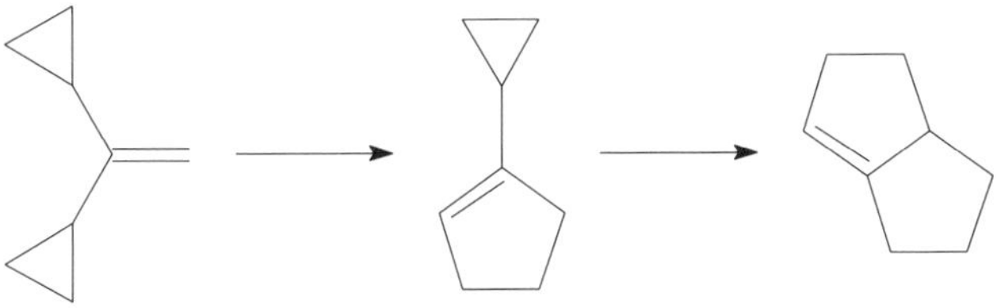

The vinylcyclopropane rearrangement is an important method for the construction of cyclopentenes. The direct 1,4-addition of a carbene to a 1,3-diene to give a cyclopentene works only in a few special cases and with poor yield.[7] The desired product may instead be obtained by a sequence involving the 1,2-addition of a carbene to one carbon–carbon double bond of a 1,3-diene to give a vinylcyclo-propane, and a subsequent rearrangement to yield a cyclopentene:

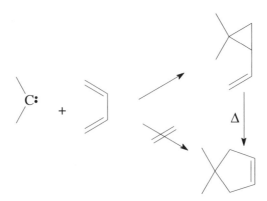

Apart from the carbene-1,2-addition route starting from 1,3-dienes, vinylcyclo-propanes may be obtained from 1,4-dienes through a *di-π-methane rearrangement.*

The vinylcyclopropane rearrangement is of synthetic importance, as well as of mechanistic interest—i.e. the concerted vs. the radical mechanism. A reaction temperature of 200 to 400 °C is usually required for the rearrangement; however, depending on substrate structure, the required reaction temperature may range from 50 to 600 °C. Photochemical[8] and transition metal catalyzed[2] variants are known that do not require high temperatures.

1. N. P. Neureiter, *J. Org. Chem.* **1959**, *24*, 2044–2046.
2. T. Hudlicky, T. M. Kutchan, S. Naqvi, *Org. React.* **1985**, *33*, 247–335.
3. H. M. Frey, R. Walsh, *Chem. Rev.* **1969**, *69*, 103–124.
4. E. M. Mil'vitskaya, A. V. Tarakanova, A. F. Plate, *Russ. Chem. Rev.* **1976**, *45*, 469–478.
5. G. McGaffin, A. de Meijere, R. Walsh, *Chem. Ber.* **1991**, *124*, 939–945.
6. G. R. Branton, H. M. Frey, *J. Chem. Soc. A* **1966**, *31*, 1342–1343.
7. C. J. Moody, G. H. Whitham, *Reactive Intermediates*, Oxford Science Publications, Oxford, **1992**, p. 38–39.
8. H. E. Zimmerman, S. A. Fleming, *J. Am. Chem. Soc.* **1983**, *105*, 622–625.

Wagner–Meerwein Rearrangement

Rearrangement of the carbon skeleton *via* carbenium ions

$$R^2-\underset{\underset{R^3}{|}}{\overset{\overset{R^1}{|}}{C}}-\underset{\underset{H}{|}}{\overset{\overset{R^4}{|}}{C}}-OH \longrightarrow R^2-\underset{\underset{R^3}{|}}{\overset{\overset{R^1}{|}}{C}}-\underset{\underset{H}{|}}{\overset{\overset{R^4}{|}}{C}^+} \longrightarrow \underset{R^3}{\overset{R^2}{}}C=C\underset{R^4}{\overset{R^1}{}}$$

$$\textbf{1} \qquad\qquad\qquad \textbf{2} \qquad\qquad\qquad \textbf{3}$$

Skeletal rearrangements of carbenium ion species **2**, that involve nucleophilic 1,2-migrations of alkyl groups, are called *Wagner–Meerwein rearrangements.*[1–3]

In an initial step the carbenium ion species **2** has to be generated, for example by protonation of an alcohol **1** at the hydroxyl oxygen under acidic conditions and subsequent loss of water. The carbenium ion **2** can further react in various ways to give a more stable product—e.g. by addition of a nucleophile, or by loss of a proton from an adjacent carbon center; the latter pathway results in the formation of an alkene **3**.

In the case of an appropriate substrate structure, the carbenium ion species can undergo a 1,2-alkyl shift, thus generating a different carbenium ion—e.g. **4**. The driving force for such an alkyl migration is the formation of a more stable carbenium ion, which in turn may undergo further rearrangement or react to a final product by one of the pathways mentioned above—e.g. by loss of a proton to yield an alkene **3**:

$$R^2-\underset{\underset{R^3}{|}}{\overset{\overset{R^1}{|}}{C}}-\underset{\underset{H}{|}}{\overset{\overset{R^4}{|}}{C}}-OH \underset{\overset{- H_2O}{}}{\overset{+ H^+ /}{\rightleftharpoons}} R^2-\underset{\underset{R^3}{|}}{\overset{\overset{R^1}{|}}{C}}-\underset{\underset{H}{|}}{\overset{\overset{R^4}{|}}{C}^+} \longrightarrow R^2-\underset{\underset{R^3}{|}}{\overset{\overset{R^1}{|}}{\overset{+}{C}}}-\underset{\underset{H}{|}}{\overset{\overset{R^1}{|}}{C}}-R^4$$

$$\textbf{1} \qquad\qquad\qquad\qquad \textbf{2} \qquad\qquad\qquad \textbf{4}$$

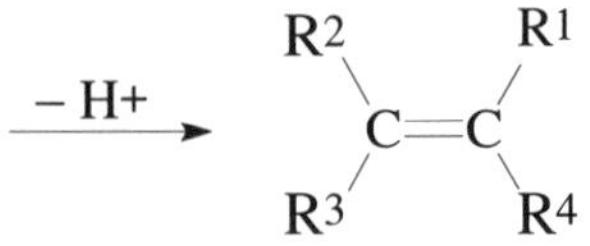

$$\xrightarrow{-\,H+} \quad \underset{R^3}{\overset{R^2}{\diagdown}} C = C \underset{R^4}{\overset{R^1}{\diagup}}$$

3

A carbenium ion center can be stabilized by an alkyl substituent through hyper-conjugation, or by an aryl substituent through resonance. A tertiary carbenium ion is more stable than a secondary or a primary one. The order of migration for a few selected groups R is: phenyl > *tert*-butyl > ethyl > methyl. The migrating group R generally does not fully dissociate from the rest of the molecule, but is rather bound in a π-complex or a S_N2-like transition state or a tight ion pair.

Of synthetic importance is the Wagner–Meerwein rearrangement especially in the chemistry of terpenes and related compounds.[4,5] For example isoborneol **5** can be dehydrated and rearranged under acidic conditions to yield camphene **6**:

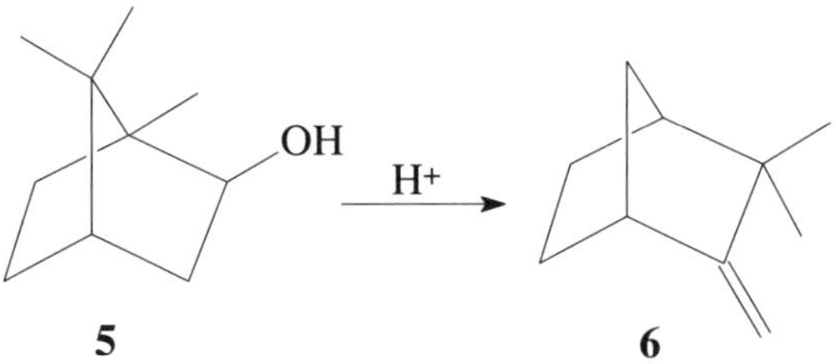

5   **6**

With appropriate substrates, two or more consecutive rearrangements may take place.[6,7] The carbon skeleton of the starting material may then suffer a major structural reorganization:

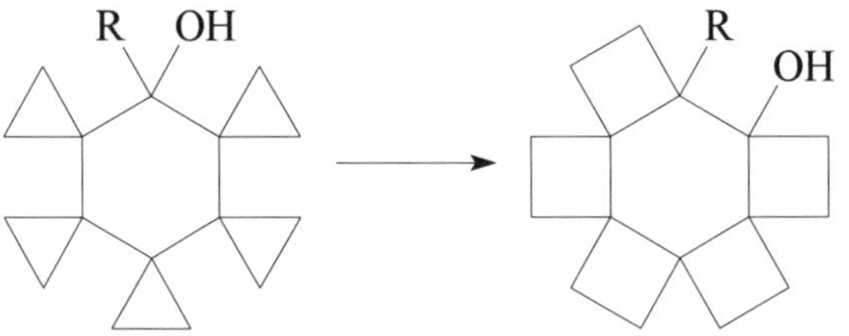

R = H or CH$_3$

The leaving group doesn't have to be a water molecule; any group or substituent which upon cleavage from the carbon skeleton under appropriate reaction conditions leaves behind a carbenium ion—e.g. a halogen substituent—will suffice. The other substituents can be hydrogen, alkyl or aryl.[3]

Except for terpene chemistry, the Wagner–Meerwein rearrangement is of limited synthetic importance. It is rather found as an undesired side-reaction with other reactions, for example in the synthesis of alkenes by elimination reactions.

1. H. Meerwein, W. Unkel, *Justus Liebigs Ann. Chem.* **1910**, *376*, 152–163.
2. A. Streitwieser, Jr., *Chem. Rev.* **1956**, *56*, 698–713.
3. H. Hogeveen, E. M. G. A. v. Kruchten, *Top. Curr. Chem.* **1979**, *80*, 89–124.
4. T. S. Sorensen, *Acc. Chem. Res.* **1976**, *9*, 257–265.
5. L. A. Paquette, L. Waykole, H. Jendralla, C. E. Cottrell, *J. Am. Chem. Soc.* **1986**, *108*, 3739–3744.
6. L. Fitjer, D. Wehle, M. Noltemeyer, E. Egert, G. M. Sheldrick, *Chem. Ber.* **1984**, *117*, 203–221.
7. M. Hesse, *Ring Enlargement in Organic Chemistry*, VCH, Weinheim, **1991**, p. 8–9.

Weiss Reaction

A synthesis of the bicyclo[3.3.0]octane skeleton

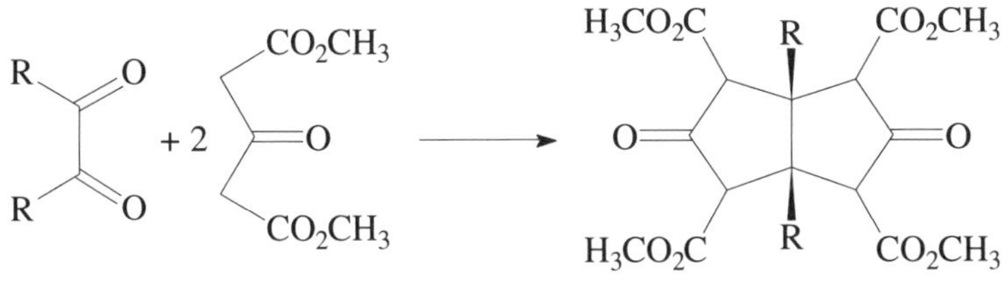

The formation of bicyclo[3.3.0]octane-3,7-diones **3** by reaction of an α-diketone **1** with a 3-oxoglutaric diester **2** is called the *Weiss reaction*.[1–3]

Four carbon–carbon bonds are formed in a one-pot reaction that involves two *aldol reactions* and two *Michael addition reactions*.

The initial step is an intermolecular aldol addition of 3-oxoglutaric diester **2** to the α-diketone **1**. A second—now intramolecular—aldol reaction leads to formation of the five-membered ring intermediate **4**. Elimination of water from **4** leads to a cyclopentenone derivative **5**, which then reacts with a second 3-oxoglutaric diester molecule in an intermolecular Michael addition. A second dehydration step then again generates a cyclopentenone derivative, which undergoes a second—now intramolecular—Michael addition reaction to yield the *cis*-bicyclo[3.3.0]octan-3,7-dione **3** as the final reaction product:[1,2]

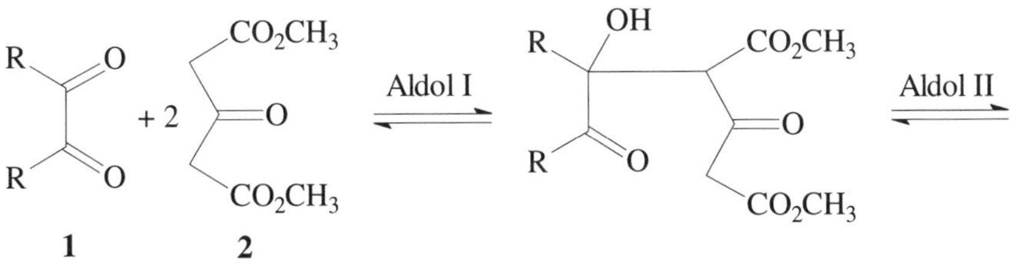

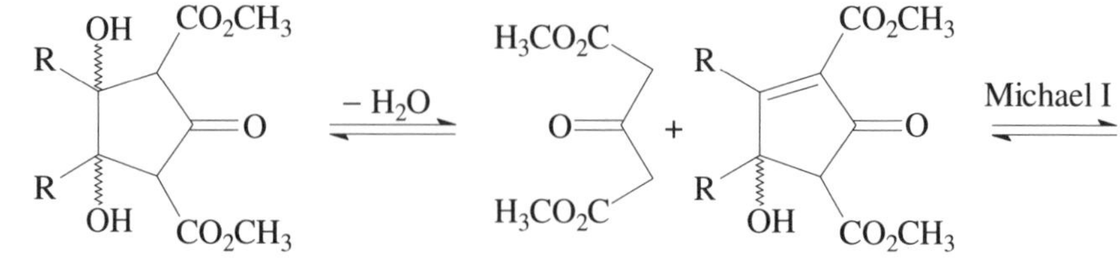

4   **1**   **5**

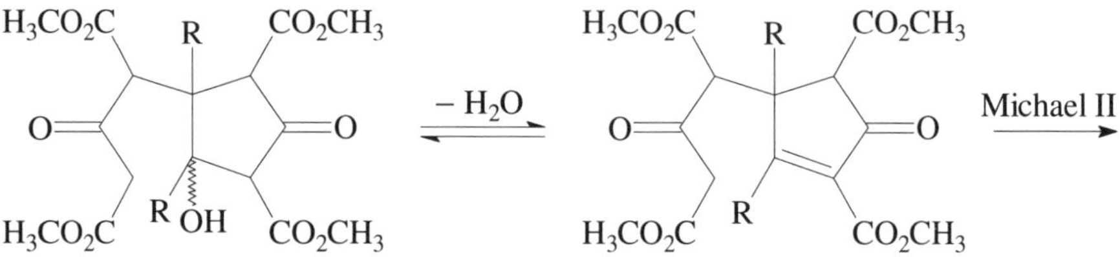

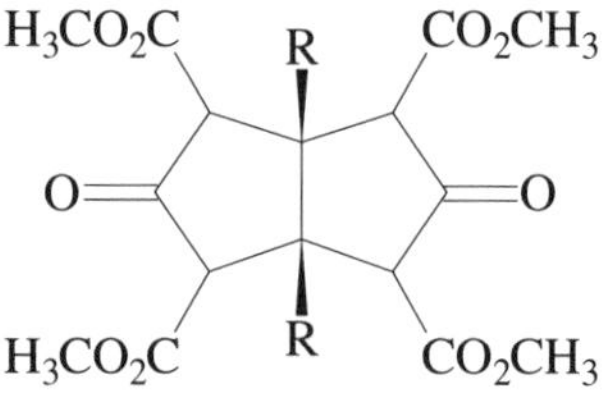

3

The reasonable mechanism outlined above has not yet been rigorously proven in every detail, but is supported by the fact that a 1 : 1-intermediate **5** has been isolated.[4] The ester groups are essential for the Weiss reaction; because of the β-keto ester functionalities however, the ester groups can be easily removed from the final product by ester hydrolysis and subsequent decarboxylation.

An illustrative example for the usefulness of the Weiss reaction for the construction of complex cyclopentanoid carbon skeletons is the synthesis of the all-*cis* [5.5.5.5]fenestrane **7** after *Cook et al.*,[5] starting from the α-diketone[6]:

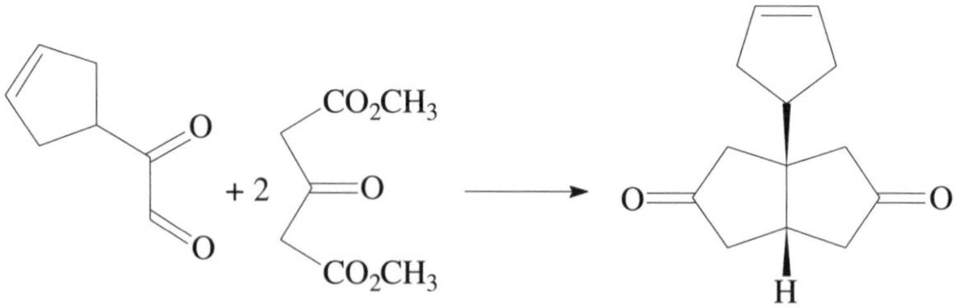

6

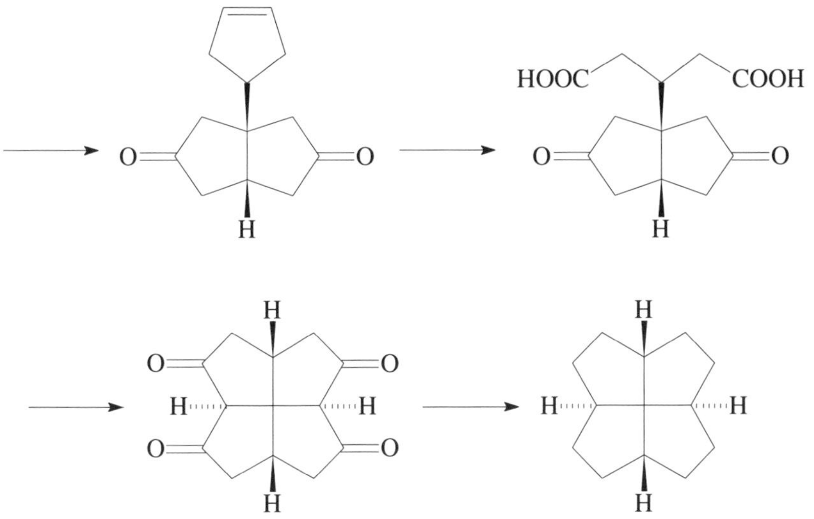

7

In order to obtain good yields from a Weiss reaction sequence, the H^+-concentration has to be adjusted properly in the reaction mixture. The reaction is usually carried out in a buffered, weakly acidic or weakly basic solution. By the Weiss reaction simple starting materials are converted into a complex product of defined stereochemistry. There is no simpler procedure for the synthesis of the 1,5-*cis*-disubstituted bicyclo[3.3.0]octane skeleton; it has for example found application in the synthesis of polyquinanes.[6]

1. U. Weiss, J. M. Edwards, *Tetrahedron Lett.* **1968**, 4885–4887.
2. J. Mulzer, H.-J. Altenbach, M. Braun, K. Krohn, H.-U. Reissig, *Organic Synthesis Highlights*, VCH, Weinheim, **1991**, p. 121–125.
3. X. Fu, J. M. Cook, *Aldrichimica Acta* **1992**, *25*, 43–54.
4. G. Kubiak, J. M. Cook, *Tetrahedron Lett.* **1985**, *26*, 2163–2166.
5. G. Kubiak, X. Fu, A. Gupta, J. M. Cook. *Tetrahedron Lett.* **1990**, *31*, 4285–4288.
6. A. Gupta, X. Fu, J. P. Snyder, J. M. Cook, *Tetrahedron* **1991**, *47*, 3665–3710.

Willgerodt Reaction

ω-Arylalkane carboxylic amides from aryl alkyl ketones

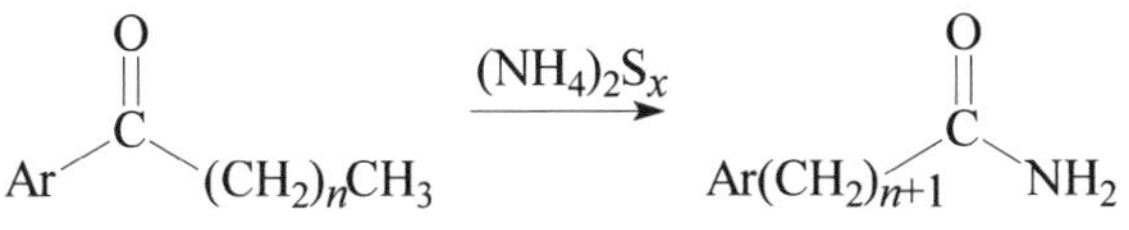

1   **2**

An aryl alkyl ketone **1** can be converted into an ω-arylalkane carboxylic amide **2** by employing the *Willgerodt reaction*.[1-3] The number of carbon centers is retained. The reaction is carried out by treating the ketone with an aqueous solution of ammonium polysulfide. A variant that has been developed by *Kindler*,[4] and which is called the *Willgerodt–Kindler reaction*, uses a mixture of sulfur and a secondary amine instead of the ammonium polysulfide.

The Willgerodt reaction starts with the formation of an enamine **4** from the ketone, e.g. from acetophenone **3**. The further course of the reaction cannot be described by a single mechanism that would apply to all examples known.[2,3,5] For aryl methyl ketones **3** the mechanism for the *Kindler variant* is formulated as follows:

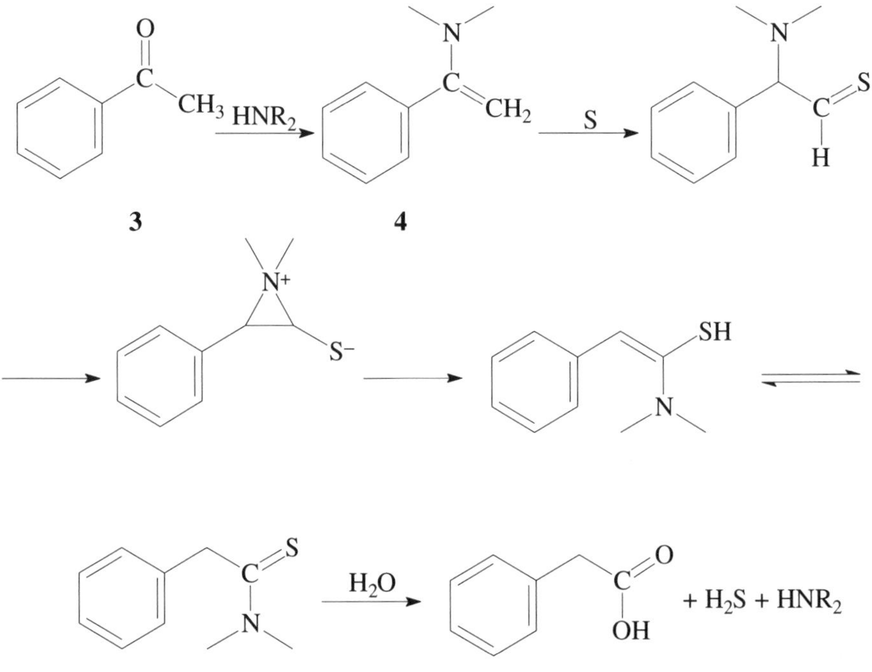

The Willgerodt reaction yields amides **2** as products, while the Willgerodt–Kindler reaction yields N,N-disubstituted thioamides **5**. Both types of products can be converted to the corresponding carboxylic acid **6** by alkaline hydrolysis.

The Willgerodt reaction is usually carried out under high pressure, thus requiring special laboratory equipment, while with the Kindler variant this is not necessary. The Kindler variant is of wider scope, and yields are generally better. In addition aromatic compounds with vinyl substituents may be employed as substrates instead of the ketone, e.g. styrene **7**:[2]

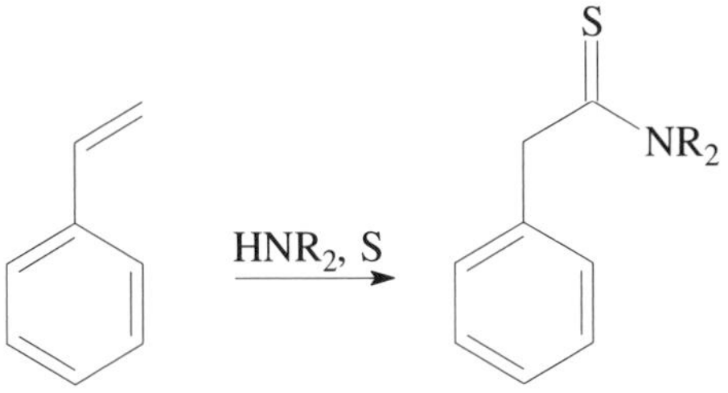

$$ 7 $$

The Willgerodt reaction also works with hetaryl alkyl ketones, but often gives unsatisfactory yields. Yields generally decrease with increasing chain length of the alkyl group.

1. C. Willgerodt, *Ber. Dtsch. Chem. Ges.* **1888**, *21*, 534–536.
2. E. V. Brown, *Synthesis* **1975**, 358–375.
3. M. Carmack, M. A. Spielman, *Org. React.* **1946**, *3*, 83–107.
4. K. Kindler, *Justus Liebigs Ann. Chem.* **1923**, *431*, 187–207.
5. F. Asinger, W. Schäfer, K. Halcour, A. Saus, H. Triem, *Angew. Chem.* **1963**, *75*, 1050–1059; *Angew. Chem. Int. Ed. Engl.* **1964**, *3*, 19.

Williamson Ether Synthesis

Ethers by reaction of alkyl halides with alkoxides

This reaction, which is named after *W. Williamson*,[1,2] is the most important method for the synthesis of unsymmetrical ethers **3**. For this purpose an alkoxide or phenoxide **1** is reacted with an alkyl halide **2** (with $R' =$ alkyl, allyl or benzyl). Symmetrical ethers can of course also be prepared by this route, but are accessible by other routes as well.

For the classical Williamson synthesis an alcohol is initially reacted with sodium or potassium to give an alkoxide, e.g. **1**. Alternatively an alkali hydroxide or amide may be used to deprotonate the alcohol. Phenols are more acidic, and can be converted to phenoxides by treatment with an alkali hydroxide or with potassium carbonate in acetone.[2]

In most cases the alkoxide or phenoxide **1** reacts with the alkyl halide **2** by a bimolecular nucleophilic substitution mechanism:

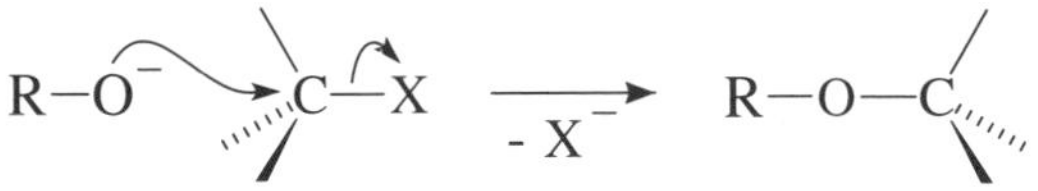

With secondary and tertiary alkyl halides an E_2-elimination is often observed as a side-reaction. As the alkyl halide reactant an iodide is most often employed, since alkyl iodides are more reactive than the corresponding bromides or chlorides. With phenoxides as nucleophiles a C-alkylation can take place as a competing reaction. The ratio of O-alkylation versus C-alkylation strongly depends on the solvent used. For example reaction of benzylbromide **4** with β-naphth-oxide **5** in N,N-dimethylformamide (DMF) as solvent yields almost exclusively the β-naphthyl benzylether **6**, while the reaction in water as solvent leads *via* intermediate **7** to formation of the C-benzylated product—1-benzyl-2-naphthol **8**—as the major product:[3]

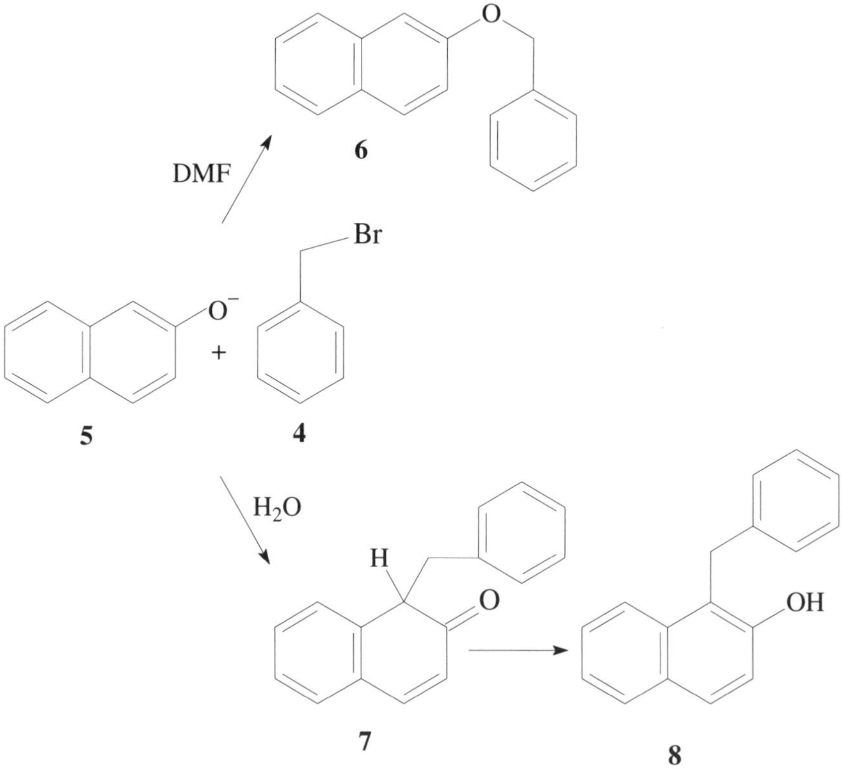

An inert solvent such as benzene, toluene or xylene, or an excess of the alcohol corresponding to the alkoxide is often used as solvent. When a dipolar aprotic solvent such as N,N-dimethylformamide (DMF) or dimethylsulfoxide (DMSO) is used, the reaction often proceeds at higher rate.

As alkylating agent an alkyl halide, alkyl tosylate or dialkyl sulfate is used in most cases; the latter type of reagent is often used in the preparation of methyl and ethyl ethers by employing dimethyl sulfate and diethyl sulfate respectively. Dimethyl sulfate is an excellent methylating agent, but is acutely toxic as well as carcinogenic.[4]

A variant of the Williamson ether synthesis uses thallium alkoxides.[5] The higher reactivity of these can be of advantage in the synthesis of ethers from diols, triols and hydroxy carboxylic acids, as well as from secondary and tertiary alcohols; on the other hand however thallium compounds are highly toxic.

$$\text{Ph-X} + \text{KO-Ph} \xrightarrow[-\text{KX}]{\text{Cu}} \text{Ph-O-Ph}$$

9

A variant for the synthesis of diaryl ethers—e.g. diphenyl ether **9**, where an aryl halide and a phenoxide are reacted in the presence of copper or a copper-(I) salt, is called the *Ullmann ether synthesis*.[6,7]

1.   W. Williamson, *Justus Liebigs Ann. Chem.* **1851**, *77*, 37–49.
2.   H. Feuer, J. Hooz in *The Chemistry of the Ether Linkage* (Ed.: S. Patai), Wiley, New York, **1967**, p. 445–498.
3.   N. Kornblum, R. Seltzer, P. Haberfield, *J. Am. Chem. Soc.* **1963**, *85*, 1148–1154.
4.   L. Roth, *Krebserzeugende Stoffe*, Wissenschaftliche Verlagsgesellschaft, Stuttgart, **1983**, pp. 49,54.
5.   H.-O. Kalinowski, G. Grass, D. Seebach, *Chem. Ber.* **1981**, *114*, 477–487.
6.   F. Ullmann, P. Sponagel, *Ber. Dtsch. Chem. Ges.* **1905**, *38*, 2211–2212.
7.   A. A. Moroz, M. S. Shvartsberg, *Russ. Chem. Rev.* **1974**, *43*, 679–689.

Wittig Reaction

Alkenes (olefins) from reaction of phosphonium ylides with aldehydes or ketones

$$\underset{1}{\overset{R^1}{\underset{R^2}{>}}C=PR_3} + \underset{2}{O=\overset{R^3}{\underset{R^4}{<}}C} \longrightarrow \underset{3}{\overset{R^1}{\underset{R^2}{>}}C=\overset{R^3}{\underset{R^4}{<}}C} + \underset{4}{R_3P=O}$$

The reaction of an alkylidene phosphorane **1** (i. e. a phosphorus ylide) with an aldehyde or ketone **2** to yield an alkene **3** (i.e. an olefin) and a phosphine oxide **4**, is called the *Wittig reaction* or *Wittig olefination reaction*.[1–5]

   Phosphorus ylides like **1** can be prepared by various routes. The most common route is the reaction of triphenylphosphine **5** with an alkyl halide **6** to give a triphenylphosphonium salt **7**, and treatment of that salt with a base to give the corresponding ylide **1**:

$$(C_6H_5)_3P \ + \ X - \underset{|}{\overset{|}{\underset{\displaystyle \mathbf{6}}{C}}} - \overset{\displaystyle H}{} \quad\longrightarrow\quad (C_6H_5)_3\overset{+}{P} - \underset{|}{\overset{|}{\underset{\displaystyle \mathbf{7}}{C}}} - \overset{\displaystyle H}{} \ + X^-$$

$$\mathbf{5} \qquad\qquad \mathbf{6} \qquad\qquad\qquad\qquad\qquad \mathbf{7}$$

$$\xrightarrow{\text{base}} \quad \left[(C_6H_5)_3\overset{+}{P} - \overset{-}{\underset{|}{C}} - \quad\longleftrightarrow\quad (C_6H_5)_3P = C \overset{\diagup}{\diagdown} \right]$$

$$\mathbf{1}$$

The phosphonium salt **7** is usually isolated, and in most cases is a crystalline material, while the ylide **1** is usually prepared in solution and used directly for reaction with the carbonyl substrate.

The initial step of olefin formation is a nucleophilic addition of the negatively polarized ylide carbon center (see the resonance structure **1** above) to the carbonyl carbon center of an aldehyde or ketone. A betain **8** is thus formed, which can cyclize to give the oxaphosphetane **9** as an intermediate. The latter decomposes to yield a trisubstituted phosphine oxide **4**—e.g. triphenylphosphine oxide (with R = Ph) and an alkene **3**. The driving force for that reaction is the formation of the strong double bond between phosphorus and oxygen:

$$R_3P = C\overset{\diagup}{\diagdown} \ + \ \overset{\diagup}{\diagdown}C = O \quad\longrightarrow\quad \begin{array}{c} R_3\overset{+}{P} - \underset{|}{\overset{|}{C}} - \\ {}^-O - \underset{|}{\overset{|}{C}} - \end{array} \quad\longrightarrow\quad \begin{array}{c} R_3P - \underset{|}{\overset{|}{C}} - \\ | \\ O - \underset{|}{\overset{|}{C}} - \end{array}$$

$$\mathbf{1} \qquad\qquad \mathbf{2} \qquad\qquad\qquad \mathbf{8} \qquad\qquad\qquad \mathbf{9}$$

$$\longrightarrow\quad R_3P = O \ + \ \overset{\diagdown}{\diagup}C = C\overset{\diagup}{\diagdown}$$

$$\mathbf{4} \qquad\qquad \mathbf{3}$$

Evidence for the four-membered ring intermediate—the oxaphosphetane **9**—comes from ^{31}P-NMR experiments;[6] betaines of type **8** have in some cases been isolated.

The reactivity of the phosphorus ylide **1** strongly depends on substituents R^1, R^2. For preparative use R often is a phenyl group. When R^1 or R^2 is an electron-withdrawing group, the negative charge can be delocalized over several centers, and the reactivity at the ylide carbon is reduced. The reactivity of the carbonyl compound towards addition of the ylide increases with the electrophilic character of the C=O group. R^1, R^2 are often both alkyl, or alkyl and aryl.

Simple ylides are sensitive towards water as well as oxygen. By reaction with water, the ylide is hydrolyzed to give the trisubstituted phosphine oxide **4** and the hydrocarbon **10**:

$$R_3P = C \overset{\diagup}{\diagdown} \quad \xrightarrow{H_2O} \quad R_3P = O \; + \; \overset{H}{\underset{H}{\diagdown}} C \overset{\diagup}{\diagdown}$$

$$\textbf{1} \qquad\qquad\qquad \textbf{4} \qquad \textbf{10}$$

By reaction with oxygen the ylide is cleaved to the trisubstituted phosphine oxide **4** and a carbonyl compound. The latter can undergo a Wittig reaction with excess ylide to give an alkene. This route can be used to prepare symmetrical alkenes by passing oxygen through a solution of excess phosphorus ylide; oxidants other than molecular oxygen may be used instead.[7] In the following scheme this route is outlined for the oxidative cleavage of benzylidene triphenylphosphorane to give triphenylphosphine oxide **4** and benzaldehyde **11**, and subsequent Wittig reaction of the latter with excess benzylidene phosphorane to yield stilbene **12**:

$$(C_6H_5)_3P = C \overset{\overset{H}{\diagup}}{\underset{C_6H_5}{\diagdown}} \quad \xrightarrow{O_2} \quad (C_6H_5)_3P = O \; + \; O = C \overset{\overset{H}{\diagup}}{\underset{C_6H_5}{\diagdown}}$$

$$\textbf{4} \qquad\qquad \textbf{11}$$

$$(C_6H_5)_3P = C \overset{\overset{H}{\diagup}}{\underset{C_6H_5}{\diagdown}} \; + \; O = C \overset{\overset{H}{\diagup}}{\underset{C_6H_5}{\diagdown}} \quad \longrightarrow \quad C_6H_5CH = CHC_6H_5$$

$$\textbf{12}$$

Important and widely used variants of the Wittig reaction are based on carbanionic organophosphorus reagents, and are known as the *Wadsworth–Emmons reaction*, *Wittig–Horner reaction* or *Horner–Wadsworth–Emmons reaction*.[8,9] As first reported by Horner,[10] carbanionic phosphine oxides can be used; today carbanions from alkyl phosphonates **13** are most often used. The latter are easily prepared by application of the *Arbuzov reaction*. The reactive carbanionic species—e.g. **14**—is generated by treatment of the appropriate phosphonate with base, e.g. with sodium hydride:

$$(EtO)_2\overset{\overset{O}{\|}}{P}CH_2CO_2C_2H_5 \quad \xrightarrow{NaH} \quad (EtO)_2\overset{\overset{O}{\|}}{P}\bar{C}HCO_2C_2H_5 \; Na^+$$

$$\textbf{13} \qquad\qquad\qquad \textbf{14}$$

$$(EtO)_2\overset{\overset{O}{\|}}{P}\bar{C}HCO_2C_2H_5 \; Na^+ \; + \; \overset{\diagup}{\underset{\diagdown}{C}} \overset{\diagup}{\diagdown} \quad \longrightarrow \quad \overset{\diagup}{\diagdown}C = C \overset{\overset{H}{\diagup}}{\underset{CO_2C_2H_5}{\diagdown}} \; + \; (EtO)_2\overset{\overset{O}{\|}}{P}O^- \; Na^+$$

$$\textbf{14} \qquad\qquad\qquad \textbf{2} \qquad\qquad\qquad \textbf{15} \qquad\qquad \textbf{16}$$

The carbanionic deprotonated phosphonate thus obtained—e.g. **14**—can be reacted with a carbonyl substrate **2** just like a phosphorus ylide. However such carbanions are stronger nucleophiles then the corresponding ylides. As reaction products an alkene and a dialkyl phosphate salt are obtained—e.g. α,β-unsaturated ester **15** and sodium diethyl phosphate **16**. The dialkyl phosphate salt is soluble in water, and therefore can in most cases be easily removed from the desired olefination product, which is usually much less soluble in water.

The (Horner–)Wadsworth–Emmons reaction generally is superior to the Wittig reaction, and has found application in many cases for the synthesis of α,β-unsaturated esters, α,β-unsaturated ketones and other conjugated systems. Yields are often better then with the original Wittig procedure. However the Wadsworth–Emmons method is not suitable for the preparation of alkenes with simple, non-stabilizing alkyl substituents.

The Wittig reaction is one of the most important reactions in organic synthesis. The synthetic importance of the Wittig reaction and its variants and related reactions is based on the fact that the new carbon–carbon double bond in the product molecule is generated at a fixed position. Other methods for the formation of carbon–carbon double bonds, e.g. elimination of water or HX, or pyrolytic procedures often lead to mixtures of isomers. The Wittig and related reactions have found application in the synthesis of many organic target molecules, for example in natural product synthesis. As an illustrating example a step from a synthesis of β-carotene **17** is outlined:[11]

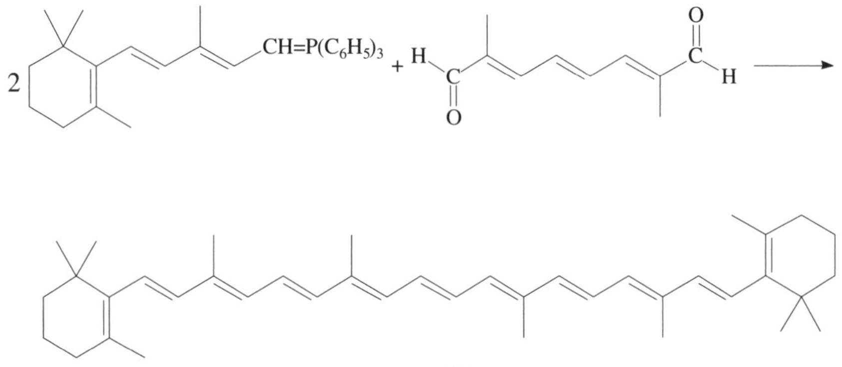

17

With respect to the carbonyl substrate, a variety of additional functional groups is tolerated, e.g. ester, ether, halogen. With compounds that contain an ester as well as a keto or aldehyde function, the latter usually reacts preferentially. Due to its mild reaction conditions the Wittig reaction is an important method for the synthesis of sensitive alkenes, as for example highly unsaturated compounds like the carotinoid **17** shown above.

1. G. Wittig, G. Geissler, *Justus Liebigs Ann. Chem.* **1953**, *580*, 44–57.
2. A. W. Johnson, *Ylid Chemistry*, Academic Press, New York, **1979**.
3. A. Maercker, *Org. React.* **1965**, *14*, 270–490.

4. H. Pommer, *Angew. Chem.* **1977**, *89*, 437–443; *Angew. Chem. Int. Ed. Engl.* **1977**, *16*, 423.
5. B. E. Maryanoff, A. B. Reitz, *Chem. Rev.* **1989**, *89*, 863–927.
6. B. E. Maryanoff, A. B. Reitz, M. S. Mutter, R. R. Inners, H. R. Almond, R. R. Whittle, R. A. Olofson, *J. Am. Chem. Soc.* **1986**, *108*, 7664–7678.
7. H. J. Bestmann, R. Armsen, H. Wagner, *Chem. Ber.* **1969**, *102*, 2259–2269.
8. W. S. Wadsworth, Jr., W. D. Emmons, *J. Am. Chem. Soc.* **1961**, *83*, 1733–1738.
9. W. S. Wadsworth, Jr., *Org. React.* **1977**, *25*, 73–253.
    J. Clayden, S. Warren, *Angew. Chem.* **1996**, *108*, 261–291; *Angew. Chem. Int. Ed. Engl.* **1996**, *35*, 241–270.
10. L. Horner, H. Hoffmann, H. G. Wippel, G. Klahre, *Chem. Ber.* **1959**, *92*, 2499–2505.
11. G. Wittig, H. Pommer, DBP 954 247, **1956**; *Chem. Abstr.* **1959**, *53*, 2279.

Wittig Rearrangement

Rearrangement of ethers to yield alcohols

The rearrangement of an ether **1** when treated with a strong base, e.g. an organo-lithium compound RLi, to give an alcohol **3** *via* the intermediate α-metallated ether **2**, is called the *Wittig rearrangement*.[1,2] The product obtained is a secondary or tertiary alcohol. R^1, R^2 can be alkyl, aryl and vinyl. Especially suitable substrates are ethers where the intermediate carbanion can be stabilized by one of the substituents R^1, R^2; e.g. benzyl or allyl ethers.

In contrast to the related *Stevens rearrangement*, experimental findings suggest a radical or a concerted reaction pathway. The mechanism of the radical [1,2]-Wittig rearrangement is formulated as follows. In the initial step the ether **1** is deprotonated α to the oxygen by a strong base, e.g. an organolithium compound or sodium amide, to give a carbanion **2**. Homolytic cleavage of the α-carbon–oxygen bond leads to formation of a radical-pair **4**,[3] which then recombines to the rearranged alkoxide **5**. Aqueous workup finally yields the alcohol **3**.

$$R^1-\overset{\overset{\displaystyle R^2}{|}}{\underset{\underset{\displaystyle H}{|}}{C}}-O^-$$

5

Driving force for the Wittig rearrangement is the transfer of the negative charge from carbon to the more electronegative oxygen.

In certain cases the reaction may proceed by a concerted mechanism. With allyl ethers a concerted [2,3]-sigmatropic rearrangement *via* a five-membered six-electron transition state is possible:[4,5]

Recently this [2,3]-Wittig rearrangement has received much attention and has been developed into a useful method for the stereoselective synthesis of homoallylic alcohols.[4-7]

1. G. Wittig, L. Löhmann, *Justus Liebigs Ann. Chem.* **1942**, *550*, 260–268.
2. D. L. Dalrymple, T. L. Kruger, W. N. White, in *The Chemistry of the Etherlinkage* (Ed.: S. Patai), Wiley, New York, **1967**, p. 617–628.
3. U. Schöllkopf, *Angew. Chem.* **1970**, *82*, 795–805; *Angew. Chem. Int. Ed. Engl.* **1970**, *21*, 763.
4. T. Nakai, K. Mikami, *Chem. Rev.* **1986**, *86*, 885–902. T. Nakai, K. Mikami, in *Organic Reactions*, Vol. 46, (Ed.: L. Paquette) Wiley: New York, **1994**, p. 105–209.
5. R. Brückner, *Nachr. Chem. Tech. Lab.* **1990**, *38*, 1506–1510.; R. Brückner, In *Comprehensive Organic Synthesis*, Vol. 4, (Eds.: B.M. Trost, I. Fleming) Pergamon: Oxford, **1991**, ch. 4.6, p. 873–892.
6. J. A. Marshall, in *Comprehensive Organic Synthesis*, Vol. 3, (Eds.: B. M. Trost, I. Fleming) Pergamon: Oxford, **1991**, ch. 3.11, p. 975–1014. J. Kallmerten, in *Houben-Weyl*, 4th Ed., Vol. E21d, (Eds.: G. Helmchen, R. W. Hoffmann, J. Mulzer, E. Schaumann) Thieme: Stuttgart, **1995**; pp 3758 and 3821.
7. for recent examples see: D. Enders, D. Backhaus, J. Runsink, *Tetrahedron* **1996**, *52*, 1503–1528; and references therein.

Wohl–Ziegler Bromination

Allylic bromination with *N*-bromosuccinimide

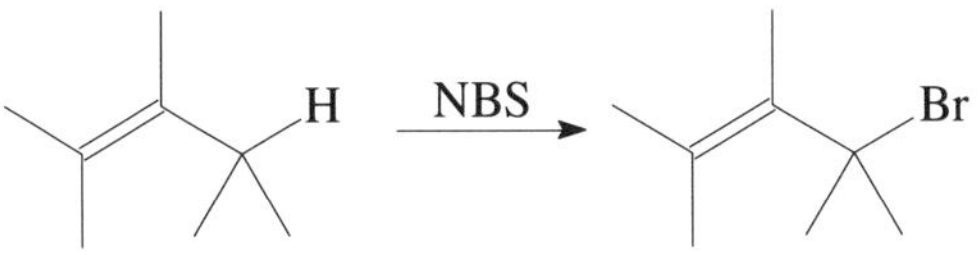

Olefins react with bromine by addition of the latter to the carbon–carbon double bond. In contrast the *Wohl–Ziegler bromination reaction*[1–4] using *N*-bromosuccinimide (NBS) permits the selective substitution of an allylic hydrogen of an olefinic substrate **1** by a bromine atom to yield an allylic bromide **2**.

The allylic bromination of an olefin with NBS proceeds by a free-radical chain mechanism.[5,6] The chain reaction initiated by thermal decomposition of a free-radical initiator substance that is added to the reaction mixture in small amounts. The decomposing free-radical initiator generates reactive bromine radicals by reaction with the *N*-bromosuccinimide. A bromine radical abstracts an allylic hydrogen atom from the olefinic substrate to give hydrogen bromide and an allylic radical **3**:

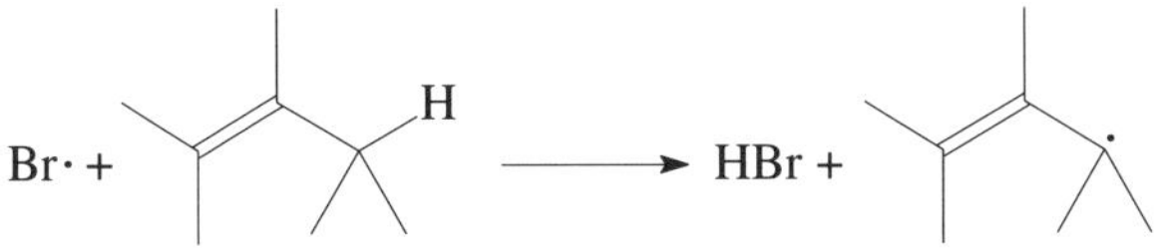

The chain propagation step consists of a reaction of allylic radical **3** with a bromine molecule to give the allylic bromide **2** and a bromine radical. The intermediate allylic radical **3** is stabilized by delocalization of the unpaired electron due to resonance (see below). A similar stabilizing effect due to resonance is also possible for benzylic radicals; a benzylic bromination of appropriatly substituted aromatic substrates is therefore possible, and proceeds in good yields.

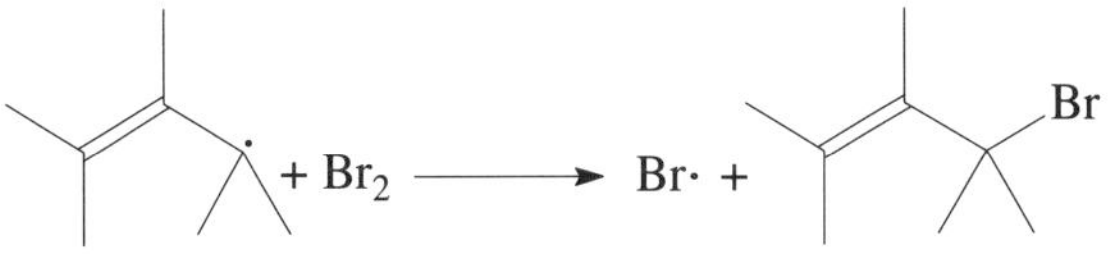

The low concentration of elemental bromine required for the chain propagation step is generated from NBS **4** by reaction with the hydrogen bromide that has been formed in the first step:

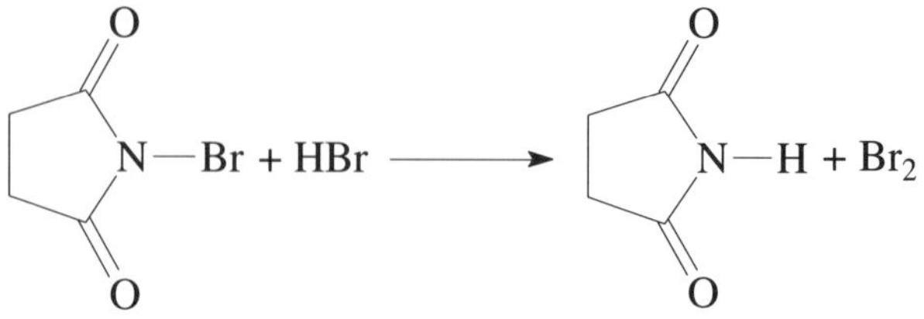

4

By this reaction a constantly low concentration of elemental bromine is supplied. With higher concentrations of free bromine, an addition to the carbon–carbon double bond is to be expected.

The allylic resonance may give rise to formation of a mixture of isomeric allylic bromides, e.g. **6** and **8** from but-1-ene. The product ratio depends on the relative stability of the two possible allylic radical species **5** and **7**:

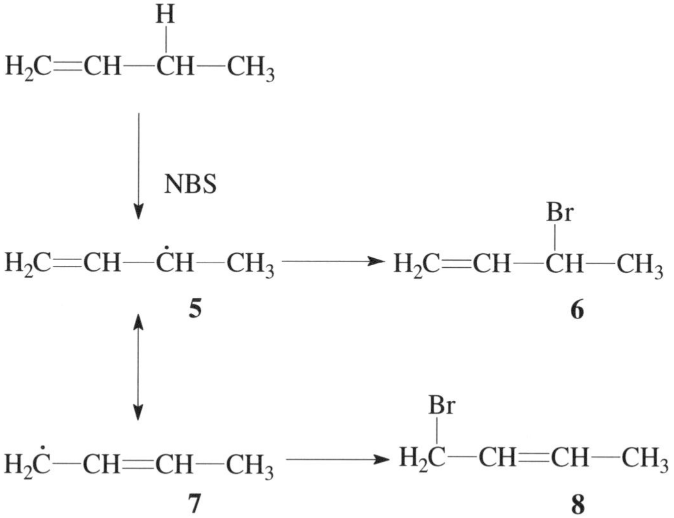

With two competing allylic species, a secondary center $-CH_2-$ is brominated preferentially over a primary center $-CH_3$.

The free-radical chain reaction may also be terminated by coupling of two carbon-radical species. As solvent carbon tetrachloride is commonly used, where the N-bromosuccinimide is badly soluble. Progress of reaction is then indicated by the decrease of the amount of precipitated NBS and the formation of the succinimide that floats on the surface of the organic liquid layer.

In order to induce the free-radical chain reaction, a starter compound such as dibenzoyl diperoxide, azo-*bis*-(isobutyronitrile) or *tert*-butyl hydroperoxide or UV-light is used. The commercially available, technical grade N-bromosuccinimide contains traces of bromine, and therefore is of slight red-brown color. Since a small amount of elemental bromine is necessary for the radical chain-propagation step, the usual slightly colored NBS needs not to be purified by recrystallization.

1. A. Wohl, *Ber. Dtsch. Chem. Ges.* **1919**, *52*, 51–63.
2. K. Ziegler, A. Späth, E. Schaaf, W. Schumann, E. Winkelmann, *Justus Liebigs Ann. Chem.* **1942**, *551*, 80–119.
3. H. J. Dauben, Jr., L. L. McCoy, *J. Am. Chem. Soc.* **1959**, *81*, 4863–4873.
4. L. Horner, E. H. Winkelmann, *Angew. Chem.* **1959**, *71*, 349–365.
5. C. Walling, A. L. Rieger, D. D. Tanner, *J. Am. Chem. Soc.* **1963**, *85*, 3129–3134.
6. J. C. Day, M. J. Lindstrom, P. S. Skell, *J. Am. Chem. Soc.* **1974**, *96*, 5616–5617.

Wolff Rearrangement

Ketenes from α-diazo ketones

An α-diazo ketone **1** can decompose to give a ketocarbene, which further reacts by migration of a group R to yield a ketene **2**. Reaction of ketene **2** with water results in formation of a carboxylic acid **3**. The *Wolff rearrangement*[1,2,6] is one step of the *Arndt–Eistert reaction*. Decomposition of diazo ketone **1** can be accomplished thermally, photochemically or catalytically; as catalyst amorphous silver oxide is commonly used:

The ketocarbene **4** that is generated by loss of N_2 from the α-diazo ketone, and that has an electron-sextet, rearranges to the more stable ketene **2** by a nucleophilic 1,2-shift of substituent R. The ketene thus formed corresponds to the isocyanate product of the related *Curtius reaction*. The ketene can further react with nucleophilic agents, that add to the C=O-double bond. For example by reaction with water a carboxylic acid **3** is formed, while from reaction with an alcohol R'−OH an ester **5** is obtained directly. The reaction with ammonia or an amine R'−NH$_2$ leads to formation of a carboxylic amide **6** or **7**:

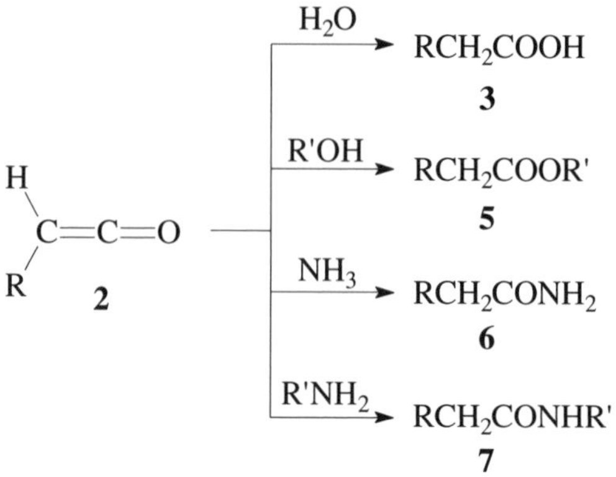

The intermediacy of a ketocarbene species **4** is generally accepted for the thermal or photochemical Wolff rearrangement;[3] oxirenes **8** that are in equilibrium with ketocarbenes, have been identified as intermediates:

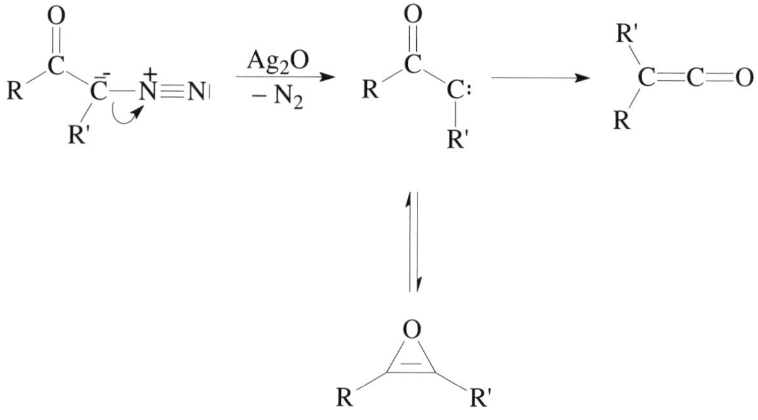

With cyclic α-diazo ketones, e.g. α-diazo cyclohexanone **9**, the rearrangement results in a ring contraction by one carbon:[4,5]

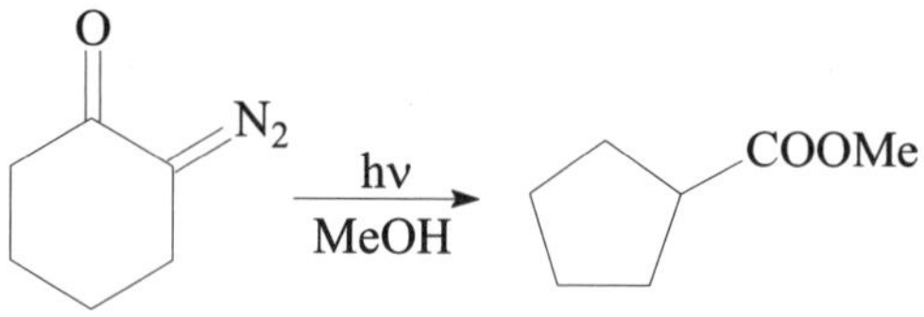

The Wolff rearrangement is a versatile reaction: R can be alkyl as well as aryl; most functional groups do not interfere. The generally mild reaction conditions permit an application to sensitive substrates.

1. L. Wolff, *Justus Liebigs Ann. Chem.* **1912**, *394*, 23–59.
2. W. E. Bachmann, W. S. Struve, *Org. React.* **1942**, *1*, 38–62.
3. M. Torres, J. Ribo, A. Clement, O. P. Strausz, *Can. J. Chem.* **1983**, *61*, 996–998.
4. M. Jones, Jr., W. Ando, *J. Am. Chem. Soc.* **1968**, *90*, 2200–2201.
5. W. D. Fessner, G. Sedelmeier, P. R. Spurr, G. Rihs, H. Prinzbach, *J. Am. Chem. Soc.* **1987**, *109*, 4626–4642.
6. S. Motallebi, P. Müller, *Chimia* **1992**, *46*, 119–122.

Wolff–Kishner Reduction

Hydrocarbons by reduction of aldehydes or ketones

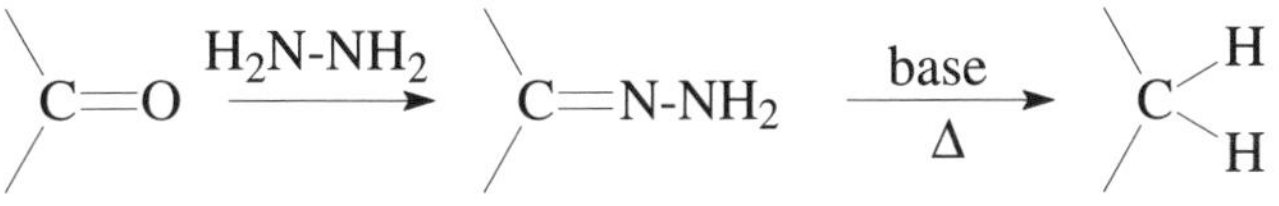

An aldehyde or ketone **1** can react with hydrazine to give a hydrazone **2**. The latter can be converted to a hydrocarbon—the methylene derivative **3**—by loss of N_2 upon heating in the presence of base. This deoxygenation method is called the *Wolff–Kishner reduction*.[1–3]

The initial step is the formation of hydrazone **2** by reaction of hydrazine with aldehyde or ketone **1**:

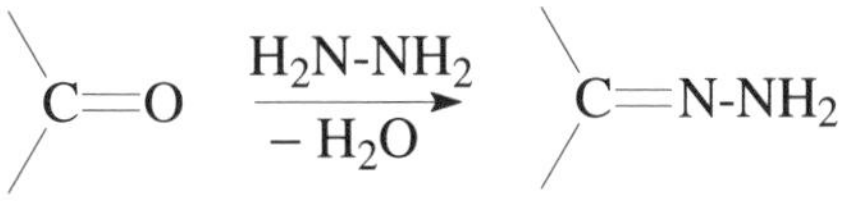

The subsequent steps are a sequence of base-induced H-shifts to give the anionic species **5**, from which loss of nitrogen (N_2) leads to a carbanionic species **6**. The latter is then protonated by the solvent to yield hydrocarbon **3** as the final product:

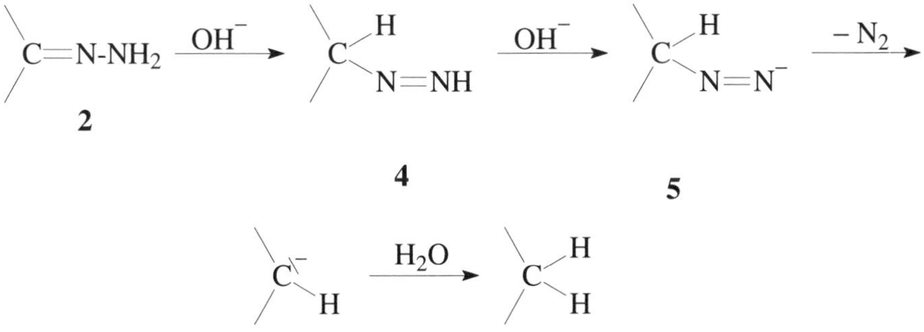

The classical procedure for the Wolff–Kishner reduction—i.e. the decomposition of the hydrazone in an autoclave at 200 °C—has been replaced almost completely by the modified procedure after *Huang-Minlon*.[4] The isolation of the intermediate is not necessary with this variant; instead the aldehyde or ketone is heated with excess hydrazine hydrate in diethyleneglycol as solvent and in the presence of alkali hydroxide for several hours under reflux. A further improvement of the reaction conditions is the use of potassium *tert*-butoxide as base and dimethyl sulfoxide (DMSO) as solvent; the reaction can then proceed already at room temperature.[5]

The Wolff–Kishner reduction is an important alternative method to the *Clemmensen reduction*, and is especially useful for the reduction of acid-labile or high-molecular substrates.[3] Yields are often below 70%, due to various side-reactions such as elimination or isomerization reactions.[2]

1. L. Wolff, *Justus Liebigs Ann. Chem.* **1912**, *394*, 86–108.
2. H. H. Szmant, *Angew. Chem.* **1968**, *80*, 141–149; *Angew. Chem. Int. Ed. Engl.* **1968**, *7*, 120.
3. D. Todd, *Org. React.* **1948**, *4*, 378–422.
4. Huang-Minlon, *J. Am. Chem. Soc.* **1946**, *68*, 2487–2488.
5. D. J. Cram, M. R. V. Sahyun, G. R. Knox, *J. Am. Chem. Soc.* **1962**, *84*, 1734–1735.

Wurtz Reaction

Hydrocarbons by coupling of alkyl halides

$$2\ RX + 2\ Na \longrightarrow R-R + 2\ NaX$$

$$\mathbf{1} \qquad\qquad\qquad \mathbf{2}$$

The coupling of alkyl halides **1** upon treatment with a metal, e.g. elemental sodium, to yield symmetrical alkanes **2**, is called the *Wurtz reaction*.[1–4] Aryl alkanes can be prepared by the *Wurtz–Fittig reaction*, i.e. the coupling of aryl halides with alkyl halides.

Mechanistically the reaction can be divided into two steps.[5] Initially the alkyl halide **1** reacts with sodium to give an organometallic species **3**, that can be isolated in many cases. In a second step the carbanionic R^- of the organometallic compound **3** acts as nucleophile in a substitution reaction with alkyl halide **1** to replace the halide:

$$RX + 2\ Na \longrightarrow R^-Na^+ + NaX$$

$$\mathbf{1} \qquad\qquad\qquad \mathbf{3}$$

$$R^-Na^+ + X-R \longrightarrow R-R + Na$$

$$\mathbf{3} \qquad \mathbf{1} \qquad\qquad\qquad \mathbf{2}$$

Alternatively a radical mechanism is discussed. There is no uniform mechanism that would apply to all kinds of substrates and the various reaction conditions. The synthetic applicability is rather limited, due to the various side-reactions observed, such as eliminations and rearrangement reactions. The attempted coupling of two different alkyl halides in order to obtain an unsymmetrical hydrocarbon, usually gives the desired product in only low yield. However the coupling reaction of an aryl halide with an alkyl halide upon treatment with a metal (the *Wurtz–Fittig reaction*) often proceeds with high yield. The coupling of two aryl halides usually does not occur under those conditions (see however below!) since the aryl halides are less reactive.

In the case of an intramolecular Wurtz reaction less side-reactions are observed; this variant is especially useful for the construction of strained carbon skeletons.[6] For example bicyclobutane **5** has been prepared from 1-bromo-3-chlorocyclobutane **4** in a yield of > 90%:[7]

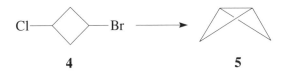

$$\qquad \textbf{4} \qquad\qquad\qquad \textbf{5}$$

In addition to sodium, other metals have found application for the Wurtz coupling reaction, e.g. zinc, iron, copper, lithium, magnesium. The use of ultrasound can have positive effect on reactivity as well as rate and yield of this two-phase reaction;[8] aryl halides can then even undergo an aryl–aryl coupling reaction to yield biaryls.[9]

1. A. Wurtz, *Justus Liebigs Ann. Chem.* **1855**, *96*, 364–375.
2. B. Tollens, R. Fittig, *Justus Liebigs Ann. Chem.* **1864**, *131*, 303–323.
3. H. F. Ebel, A. Lüttringhaus, *Methoden Org. Chem. (Houben-Weyl)* **1970**, Vol. 13/1, p. 486–502.
4. H. Fricke, *Methoden Org. Chem. (Houben-Weyl)* **1972**, Vol. 5/1b, p. 451–465.
5. T. L. Kwa, C. Boelhouwer, *Tetrahedron* **1969**, *25*, 5771–5776.
6. R. K. Freidlina, A. A. Kamyshova, E. T. Chukovskaya, *Russ. Chem. Rev.* **1982**, *51*, 368–376.
7. K. B. Wiberg, G. M. Lampman, *Tetrahedron Lett.* **1963**, 2173–2175.
8. C. Einhorn, J. Einhorn, J.-L. Luche, *Synthesis* **1989**, 787–813.
9. B. H. Han, P. Boudjouk, *Tetrahedron Lett.* **1981**, *22*, 2757–2758.

Index

acetates, pyrolysis of, 97
acetoacetic ester condensation, 45
acetoacetic ester synthesis, 178
acyloin condensation, 1–3, 47, 60
aldol reaction, 3–10, 40, 164, 176, 190, 228, 253, 265
   crossed, 5
   directed, 7
alkene metathesis, 10–12, 66
amino acids, α-, 122, 148, 253–254
amino-Claisen rearrangement, 50
aminoxides, 54
anionic *oxy*-Cope rearrangement, 58
annulenes, 126
Arbuzov reaction, 12–13, 273
Arndt–Eistert synthesis, 13–15, 172, 279
aryl hydrazones, 161
asteranes, 69
asymmetric induction, 8
automerization, 57
aziridines, 72
azo coupling *see* diazo coupling

Baeyer–Villiger oxidation, 16–19, 220
Balz–Schiemann reaction, 237
Bamford–Stevens reaction, 19–22
Beckmann rearrangement, 22–24, 240
benzidine rearrangement, 24–26
benzilic acid rearrangement, 26–27
benzilic ester rearrangement, 27
benzoin condensation, 27–29
Bergman cyclization, 30–33
bicyclobutane, 21
Birch reduction, 33–35
Blanc reaction, 36–37
boranes, 157–160
   9-borabicyclo[3.3.1]nonane (9-BBN), 159
Bucherer–Bergs reaction, 254
Bucherer reaction, 37–39
butterfly mechanism, 219

Cadiot–Chodkiewicz reaction, 127
calcium antagonists, 143
calicheamicine, 31
Cannizzaro reaction, 40–42
cascade reactions, 166
catenanes, 3
Chugaev reaction, 42–44
cinnamic acids, 213
Claisen ester condensation, 45–48, 166
   crossed, 46
Claisen–Schmidt reaction, 5
Claisen rearrangement, 48–52, 56, 104
Clemmensen reduction, 52–53, 282
collidine, 141
conjugate addition, 189
Cope elimination, 54–56, 149
Cope rearrangement, 48, 56–59
Corey lactone, 17
Corey–Winter fragmentation, 59–61
cracking distillation, 79
crossed aldol reaction, 5
crossed Claisen condensation, 46
cubane synthesis, 101
Curtius reaction, 61–63, 153, 176, 241, 279
cyanoacetic ester synthesis, 178
cycloadditions, 206
[2 + 2] cycloaddition, 67–70, 80, 203, 209
[4 + 2] cycloaddition, 78–85
cyclohexadienone synthesis, 228
cyclooctene, *E*-, 152
[2.2.2]cyclophane, 108
cyclopropenes, 21, 186
cycloreversion, 65, 206

Dakin reaction, 18
Darzens glycidic ester condensation, 71–72
Delépine reaction, 73
Demjanov reaction, 256
di-π-methane rearrangement, 86–88, 262

diazo coupling, 74–76, 78, 161
diazotization, 77–78, 236, 238, 258
dibenzofurans, 131
diborane, 157
Dieckmann condensation, 2, 45, 47
Diels–Alder reaction, 49, 78–85, 166
dienophile, 78
1,4-dihydropyridines, 142
diisopinocampheylborane, 159
dimethyldioxirane, 220
1,3-dipolar cycloaddition, 64–67, 206
dipolarophile, 65
1,3-dipole metathesis, 66
directed aldol reaction, 6
disiamylborane, 158
Doebner modification, 164
domino reactions, 84, 166, 167
double asymmetric synthesis, 8
Dötz reaction, 88–91

Eglinton reaction, 125
Elbs reaction, 92–93
ene reaction, 93–97, 261
endo rule, 81
enophile, 94
epoxidation, 219, 244
Erlenmeyer–Plöchl azlactone synthesis,
214
esperamicine, 30
ester pyrolysis, 42, 54, 97–99

Favorskii rearrangement, 100–102
fenestranes, 212, 260
Finkelstein reaction, 102–103
Fischer indole synthesis, 103–106, 162
flavaniline, 115
formylation, 123
Friedel–Crafts acylation, 106–109, 123,
258
Friedel–Crafts alkylation, 110–114
Friedländer quinoline synthesis,
114–116, 247
Fries rearrangement, 116–119
photo, 118

Gabriel synthesis, 73, 120–122
Gattermann–Koch reaction, 124
Gattermann synthesis, 123–125, 226,
258
german cockroach, sex pheromone of the,
171

Glaser coupling, 125–127
glycol cleavage, 127–129
Gomberg–Bachmann reaction, 129–131
Grignard reaction, 132–138, 225, 244
reduction, 135

haloform reaction, 139–141
Hantzsch pyridine synthesis, 141–144
Heck reaction, 144–147
Hell–Volhard–Zelinskii reaction,
147–149
hetero-Diels–Alder reaction, 84, 166,
167
hexamethylenetetramine, 73
Hey reaction, 131
Hoesch reaction, 124
Hofmann elimination reaction, 54,
149–153
1,6-Hofmann elimination, 153
Hofmann rearrangement, 153–154
Houben–Hoesch reaction, 124
Huang–Minlon variant, 282
Hunsdiecker reaction, 155–157
hydroboration, 157–160
hydroxamic acids, 176

indoles, 103
Ing–Manske procedure, 120
iodoform test, 140
Ireland–Claisen rearrangement, 51

Japp–Klingemann reaction, 76, 105,
161–163

Knoevenagel/hetero-Diels–Alder sequence,
166, 167
Knoevenagel reaction, 142, 164–167,
213
Knorr pyrrole synthesis, 168–170
Kochi reaction, 156
Kolbe elektrolytic synthesis, 155,
170–172
Kolbe synthesis of nitriles, 172–173
Kolbe–Schmitt reaction, 173–174
Kornblum's rule, 173

Leuckart–Wallach reaction, 175–176
Lossen reaction (Lossen degradation),
62, 153, 176–177

m-chloroperbenzoic acid,   220
magnesium ene reaction,   95
malonic ester synthesis,   178–181
Mannich base,   182
Mannich reaction,   182–184
McMurry reaction,   184–187
Meerwein–Ponndorf–Verley reduction,
     187–188
Michael reaction,   166, 189–192, 199,
     228–232, 252
Michaelis–Arbuzov reaction,   12
Mitsunobu reaction,   192–194
Myers rearrangement,   32

Nazarov cyclization,   195–196
Neber rearrangement,   169, 197–198
Nef reaction,   170, 198–200
neocarzinostatine,   31, 32
nitrendipine,   143
norcarane, alkoxy-,   245
Norrish type I reaction,   200–203, 204,
     210
Norrish type II reaction,   203–205, 210

olefin metathesis *see* alkene metathesis
Oppenauer oxidation,   187
oxa-di-π-methane rearrangement,   87
oxy-Cope rearrangement,   58
ozonides,   206
ozonolysis,   65, 129, 206–208

Paal–Knorr reaction,   169
pagodane,   84
para-Claisen rearrangement,   49
[2.2]paracyclophanes,   83, 103, 153
[n]paracyclophanes,   2
Paterno–Büchi reaction,   203, 209–210
Pauson–Khand reaction,   210–213
Perkin reaction,   166, 213–215
Peterson olefination,   215–217
phosphonium ylides,   271–274
photo Curtius rearrangement,   63
photo-Fries rearrangement,   118
photosensitizer,   67, 203
pinacol rearrangement,   217–218
pressure dependent reactions,   66, 83
Prilezhaev reaction,   218–220
Prins reaction,   220–222
[1.1.1]propellane,   156
Pschorr reaction,   129
pyridines,   141
pyrroles,   168

quinolines,   114, 246

Ramberg–Bücklund reaction,   223, 224
Reformatsky reaction,   166, 224–226
Reimer–Tiemann reaction,   125, 226–227
retro Claisen condensation,   46
retro Diels–Alder reaction,   79
retro ene reaction,   96, 261
Rieche formylation,   125
Rieke procedure,   226
ring-closing metathesis,   10, 11
ring-opening metathesis polymerization,
     10, 11
Robinson annulation,   228–232
Rosenmund reduction,   232–233
Rosenmund–von Braun reaction,   172
rotanes,   245
Rühlmann variant,   2

Sakurai reaction,   234–235
Sandmeyer reaction,   78, 236–237
Schiemann reaction,   237–238
Schlenk equilibrium,   133, 244
Schmidt reaction,   239–242
Shapiro reaction,   21
Sharpless epoxidation,   219, 242–244
Simmons–Smith reaction,   244–246
single electron transfer mechanism (SET
     mechanism),   133
Skraup quinoline synthesis,   114,
     246–248
Stevens rearrangement,   248–250
Stork enamine reaction,   230, 250–253
Strecker synthesis,   253–254

tandem reactions,   166
tetraasterane,   69
thexylborane,   159
Tiffeneau–Demjanov reaction,   255–257
1,2-transposition of a carbonyl group,
     199
tropinone,   184
twistene,   60

Ullmann ether synthesis,   271
umpolung,   132, 199

Vilsmeier–Haack reaction,   258
Vilsmeier reaction,   125, 258–260
vinylcarbene,   21

vinylcyclopropane rearrangement, 86, 260–262

Wadsworth–Emmons reaction, 273
Wagner–Meerwein rearrangement, 20, 217, 263–265
Weiss reaction, 265–267
Willgerodt–Kindler reaction, 268
Willgerodt reaction, 268–269
Williamson ether synthesis, 269–271

Wittig–Horner reaction, 13, 273
Wittig reaction, 271–275
Wittig rearrangement, 275–276
Wohl–Ziegler bromination, 277–279
Wolff–Kishner reduction, 53
Wolff rearrangement, 15, 281–282
Wurtz–Fittig reaction, 282
Wurtz reaction, 137, 282–283

xanthates (xanthogenates), 42